Nature, Economy and Society

Nilanjan Ghosh • Pranab Mukhopadhyay
Amita Shah • Manoj Panda
Editors

Nature, Economy and Society

Understanding the Linkages

Springer

Editors
Nilanjan Ghosh
Observer Research Foundation
Kolkata
West Bengal
India

Amita Shah
Gujarat Institute of Development Research
Ahmedabad
Gujarat
India

Pranab Mukhopadhyay
Department of Economics
Goa University
Panaji
Goa
India

Manoj Panda
Institute of Economic Growth
University of Delhi
New Delhi
Delhi
India

ISBN 978-81-322-2403-7 ISBN 978-81-322-2404-4 (eBook)
DOI 10.1007/978-81-322-2404-4

Library of Congress Control Number: 2015940754

Springer New Delhi Heidelberg New York Dordrecht London

Printed on acid-free paper

Springer (India) Pvt. Ltd. is part of Springer Science+Business Media (www.springer.com)

Contents

Contributors

Ravindra Adusumilli Watershed Support Services and Activity Network (WASSAN), Secunderabad, India

S. Chandrasekaran Department of Plant Sciences, School of Biological Sciences, Madurai Kamaraj University, Madurai, India

Kanchan Chopra Formerly Institute of Economic Growth, Delhi, India

Indrani Roy Chowdhury Department of Economics, Jamia Millia Islamia (Central University), New Delhi, India

Shyamasree Dasgupta Department of Economics, Jadavpur University, Kolkata, India

Federico Demaria Institute of Environmental Sciences and Technology (ICTA), Autonomous University of Barcelona (UAB), Bellaterra (Cerdanyola del Vallès), Barcelona, Spain

Centre for Studies in Science Policy (CSSP), Jawaharlal Nehru University (JNU), New Delhi, India

Nilanjan Ghosh Observer Research Foundation, Kolkata Chapter, Kolkata, India

Environmental Economics, World Wide Fund for Nature, New Delhi, India

U. K. Jayasinghe-Mudalige Department of Agribusiness Management, Faculty of Agriculture and Plantation Management, Wayamba University of Sri Lanka, Gonawila (NWP), Sri Lanka

Ravirajan Karunanidhi Department of Agricultural Economics, Pandit Jawaharlal Nehru College of Agriculture and Research Institute, Karaikal, Puducherry, India

Omar Hattab Kasim Department of Soil Science and Agricultural Chemistry, Pandit Jawaharlal Nehru College of Agriculture and Research Institute, Karaikal, Puducherry, India

P. C. Kotwal Ex-Faculty, Indian Institute of Forest Management, Bhopal, Madhya Pradesh, India

Umamaheswari Leelakrishnan Department of Agricultural Economics, Pandit Jawaharlal Nehru College of Agriculture and Research Institute, Karaikal, Puducherry, India

Kiran P. Mali Indian Institute of Forest Management, Bhopal, India

Joan Martinez-Alier Institut de Ciència i Tecnologia Ambientals ICTA, Universitat Autònoma de Barcelona, Barcelona, Spain

Trupti Mishra Shailesh J. Mehta School of Management (SJMSOM), Indian Institute of Technology (IIT) Bombay, Mumbai, India

Pranab Mukhopadhyay Department of Economics, Goa University, Goa, India

South Asian Network for Development and Environmental Economics, Kathmandu, Nepal

M. N. Murty Institute of Economic Growth, New Delhi, India

Nandan Nawn Department of Policy Studies, TERI University, New Delhi, India

M. D. Omprakash Indian Institute of Forest Management, Bhopal, India

Manoj Panda Institute of Economic Growth, New Delhi, India

Nasurudeen Pathusha Department of Agricultural Economics, Pandit Jawaharlal Nehru College of Agriculture and Research Institute, Karaikal, Puducherry, India

Krishna Raj Center for Economic Studies and Policy (CESP), Institute for Social and Economic Change, Bangalore, India

A. Vijay Mohan Rao Department of Animal Husbandry, Hyderabad, India

B. Suresh Reddy Division for Sustainable Development Studies, Centre for Economic and Social Studies, Hyderabad, India

Joyashree Roy Jadavpur University, Kolkata, India

Bhagya Laxmi S Watershed Support Services and Activity Network (WASSAN), Secunderabad, India

Frank van der Salm Utrecht University School of Economics, Utrecht, The Netherlands

Tapas Kumar Sarangi Institute of Economic Growth (IEG), New Delhi, India

Kailas Sarap National Institute of Rural Development (NIRD), Hyderabad, India

Amita Shah Gujarat Institute of Development Research, Ahmedabad, India

Katar Singh Ex Director, Institute of Rural Management Anand, Gujarat, India

P. S. Swamy Department of Plant Sciences, School of Biological Sciences, Madurai Kamaraj University, Madurai, India

Leah Temper Institut de Ciència i Tecnologia Ambientals ICTA, Universitat Autònoma de Barcelona, Barcelona, Spain

Menuka Udugama Department of Agribusiness Management, Faculty of Agriculture and Plantation Management, Wayamba University of Sri Lanka, Gonawila (NWP), Sri Lanka

P. K. Viswanathan Gujarat Institute of Development Research, Ahmedabad, India

About the Editors

Nilanjan Ghosh is a Senior Fellow (Professor), Observer Research Foundation, Kolkata Chapter, India. He is also a Senior Advisor–Environmental Economics, World Wide Fund (WWF) for Nature, New Delhi, India. Prior to joining ORF, he served as the Chief Economist at the Multi Commodity Exchange of India Limited and a senior research professional at The Energy and Resources Institute (TERI), New Delhi. He is also a visiting fellow at Department of Political Science, School of Social Sciences, Växjö University (now Linnaeus University), Växjö, Sweden. A natural resource economist and econometrician by training, he obtained his doctorate degree from the Indian Institute of Management Calcutta, Kolkata. He has been working in the domains of ecological economics, water resources, and commodity derivative markets, conducting his research by combining frameworks of neoclassical economics and institutions. His publications include 6 books, 30-odd peer-reviewed research papers in top international journals (like *Water Policy, Conservation and Society, Journal of Industrial Ecology, Journal of Index Investing, Economic & Political Weekly*). He was also the Vice President of the Indian Society for Ecological Economics (INSEE) during 2012–2014, and delivered the presidential address at the seventh INSEE biennial conference in 2013.

Pranab Mukhopadhyay is a professor and former head of the department of economics at the Goa University, India. He worked as an environmental economist with the South Asian Network for Development & Environmental Economics (SANDEE), based at the International Union for Conservation of Nature, (IUCN, Nepal), Kathmandu in 2006–2008. He is a SANDEE Fellow and continues to be actively involved with the Network's activities. He has published in journals such as *Ecology and Society, Economic & Political Weekly, Environment & Development Economics*, and some edited volumes. His work in environmental economics focuses on common property resources and institutional economics. He was the Secretary of the Indian Society for Ecological Economics (INSEE) during 2012–2014.

Amita Shah is a professor and the former Director at the Gujarat Institute of Development Research, Ahmedabad, India. Her major areas of research interest include natural resource development with special focus on dry-land agriculture and forestry, environmental impact assessment, gender and environment, the agriculture-industry interface, small scale and rural industries, diffusion of technologies, and employment-livelihood issues. More recently, she has been involved in a number of studies pertaining to participatory watershed development, protected area management, economic valuation of bio-diversity, and chronic poverty in remote rural areas, migration and status of women in agriculture. She has published nearly 100 research papers in professional journals and books and participated in a number of academic seminars. She has co-authored a book and jointly edited three books pertaining to agriculture, sustainable development and gender. She has been invited as a visiting fellow/scholar to academic institutions in the United Kingdom, China, France, the Netherlands and Canada. She was the President of the Indian Society for Ecological Economics in 2012–2014.

Manoj Panda holds a PhD in economics from the Indian Statistical Institute. He is currently Director of the Institute of Economic Growth, Delhi. He has served as the Director of the Centre for Economic and Social Studies (CESS), Hyderabad (2008–2012), as a Professor and an Associate Professor at the Indira Gandhi Institute of Development Research (IGIDR), Mumbai (1992–2012) and as a senior economist at the National Council of Applied Economic Research, New Delhi (1982–1992). He spent a year at Yale University, USA on a post-doctoral fellowship (1996–1997). His research areas span monitoring and analysis of macroeconomic trends and prospects, evaluation of alternative trade and fiscal policy options from the point of view of growth and distribution, development and application of computable general equilibrium (CGE) models, construction of Social Accounting Matrices (SAM), measurement and policy issues in poverty and human development and interaction of carbon emission with economic growth and its pattern. He has undertaken short-term assignments for several international organizations including the World Bank, ADB, UNDP, FAO, UN-ESCAP, IFPRI, MacArthur Foundation and Carnegie Endowment. He has published extensively in journals and edited books.

Chapter 1
Ecological Economics: At the Interface of Nature, Economy, and Society

Nilanjan Ghosh, Pranab Mukhopadhyay, Amita Shah and Manoj Panda

1.1 Introduction

The evolution of ecological economics as a discipline has a chequered history, more aptly classified by cognitive dissonance, bitter debates, and scholastic antagonism. The conflict essentially emerges from the very scope of the discipline, which, according to detractors of the discipline, is still amorphous and, according to practitioners and believers of the discipline, is evolving and growing. It is a fact that till now although broad contours of the discipline have been drawn, any significant research in the disciplines of economics, ecology, sociology, geography, social anthropology to describe any ecological phenomenon affecting human, social, and economic endeavours pushes the frontiers of the discipline or brings about a separate dimension to its scope.

According to certain definitions, ecological economics is referred to as both a transdisciplinary and interdisciplinary field of academic research addressing the critical problems at the interface of nature, economy, and society. In the process of doing so, ecological economics emerges as a discipline which acknowledges the interdependence and co-evolution of human societies and the natural ecosystems over time and space. However, there is another body of thought that defines ecological economics as an attempt to expand economic theory to integrate the earth's

---

P. Mukhopadhyay (✉)
Department of Economics, Goa University, Goa, India
e-mail: pm@unigoa.ac.in

South Asian Network for Development and Environmental Economics, Kathmandu, Nepal

N. Ghosh
Observer Research Foundation, Kolkata Chapter, Kolkata, India

Environmental Economics, World Wide Fund for Nature, New Delhi, India

M. Panda
Institute of Economic Growth, New Delhi, India

A. Shah
Gujarat Institute of Development Research, Ahmedabad, India

N. Ghosh et al. (eds.), *Nature, Economy and Society*,
DOI 10.1007/978-81-322-2404-4_1

natural systems, human values, health and well-being. This definition brings about a sense that ecological economics is either a subdiscipline of economics or a discipline that makes use of economic techniques to analyse problems (and offer solutions) at the socioecological interface of human endeavour.

There exists another body of thought (that is sometimes used synonymously with ecological economics) labelled as environmental economics, with which it overlaps in certain ways. Environmental economics is recognised in the mainstream economic analysis as being firmly placed within the neoclassical marginalist framework and differs in that sense from the scope and methodological foundations from ecological economics.

While ecological economics saw the economy as embedded in, and supported by, natural systems, environmental economics treats nature as a pool of resources that act as inputs in economic activity. Ecological economics goes beyond this not only by integrating models from ecology and economics but also by looking at the interplay of various other socioecological and institutional factors that governed an economic system.

In a recently published article in the *Breakthrough Journal,* Sagoff (2012) postulates, "... Ecological economics aimed to be revolutionary, but it is now ignored by the sciences it had hoped to transform. Both ecology and economics have changed, but not because of the rise of ecological economics." This position, however, is not true as a large component of Sagoff's arguments are based on the contention that ecological economics has attempted to place a price tag on the ecosystem services and functions. This position would not be true if we refer to the larger literature in environmental economics and even less so if we refer to ecological economics, which is further removed from issues of values and market prices.

Sagoff (2012) was right in pointing out the lack of advances made in this domain, as also the inherent reductionism that dominated the policy and academic spheres, because of the adherence to these measures. What Sagoff missed out in his argument is the advancement of ecological economics in the other direction that is independent of cost-benefit analysis of ecosystem services and functions.

The Indian Society for Ecological Economics (INSEE), so far, has not typecast any specific definition of ecological economics. Neither has it discarded the neoclassical framework for the sake of defining the domain of ecological economics. Rather, if one goes through the various conference papers, and the publications of the Society, the reader would recognise the broad framework INSEE has adopted. Ecological economics has been acknowledged by the Society to subsume the neoclassical framework of environmental economics, apart from considering the broader body of the literature emerging at the interface of economics, ecological sciences, hydrology, geology, geography, sociology, political science, anthropology, etc.

1.2 Ecological Economics: The Pathway to Interdisciplinarity

All the resources used by humans are embedded in complex social-ecological systems (SESs; Ostrom 2009). It is the SESs that define the critical interface of nature, economy, and society. The complexity of this interface arises from the limited (often non-existent) human knowledge about the ecosystems and their role in the social welfare function, thereby leading to inappropriate societal response to the problems of natural resource management. Human society, as such, therefore remains embedded in the SESs, gets affected by the stimuli created by them, and releases forces that again affect the SESs. This brings to the fore the critical linkage between ecosystems and livelihoods on the one hand and the ecosystem and sustainability on the other (Chopra et al. 2005).

Just as there have been attempts to analyse these relations from a systems approach, there have also been attempts to understand this dynamics from an institutional perspective. Just as social anthropological perspectives on natural resource management have gained prominence so has neoclassical economics through its offshoot of environmental economics, which tries to provide solutions to the natural resource allocation problem (Dasgupta 2001).

Yet, there is widespread recognition that such problems are too complex to be understood through the lens of a single discipline. The problems have multidimensional characteristics. These are not only intrinsic to the sub-systems of society, ecology and economy but also affected by external sources. Some of these problems also emerge from the interactive dynamics of the three. Therefore, there is a clear need to create the pathways towards interdisciplinarity in analysing the problems at the interface.

It is with this idea that INSEE conceived of its Sixth Biennial Conference's theme as "Nature, Economy, and Society: Understanding the Linkages", which was held in Centre for Economic and Social Studies (CESS), Hyderabad, 20–22 October, 2011. The objective of the conference was to deliberate on the various disciplinary approaches to analyse the interface of the three pillars of human civilisation (nature, economy, and society), and how best to understand this interaction. This volume is an outcome of selected papers presented in the conference.

1.3 About the Volume

The volume brings together a wide variety of methodological approaches to expand the frontiers of ecological economics and to find solutions to natural resource management. It begins with a theoretical overview of the dilemmas that researchers in ecological economics face—conceiving of values and fair exchange not just between human agents but also with nature.

Chopra in Chap. 2 poses the dilemma that philosophers have often raised about social exchange—how do we create equal exchange in a world of unequal relations?

Exchange in economics would translate into a simple problem of prices, but in the real world not all commodity relations lie in the domain of prices. Some commodities are non-traded, not just because there is no market for them but also because there are no prices that bring all transactors onto a common platform.

When the market is unable to mediate an exchange, then agents resort to alternative ways of acquisition: They either possess the resource by means outside the legal framework and in violation of it or acquire the resource by using state power from within the legal system—but both ways are probably violative of a system of individual or community rights not adequately or explicitly recognised by our legal framework. This then results in conflict between those who have traditionally regarded these as their natural inheritance and those who wish to acquire it to cater to a larger market mechanism for profits.

Martinez-Alier, Temper and Demaria (Chap. 3) document the transition in social metabolism in India and the conflicts over resources. Their aim is to critically evaluate the material flows in the economic system. The normal mechanism for looking at material flows is by way of trade valued in monetary terms. However, from the perspective of "social metabolism" (which examines the exchange of energy and materials with nature) it is the physical measure that needs to be examined since prices need not indicate a true value of exchange.

The authors find a growing correspondence between higher social metabolism (extraction of resources) and growing conflict in many parts of India—especially the resource-rich states which display a high interstate export of resources but also are burdened with high levels of poverty. The question that the authors leave us with is whether once can find a way to a steady state that is not ecologically damaging?

Chopra approaches these questions in the context of governance and the use of science in negotiating policy dilemmas while being cognisant that scientific knowledge itself is a disputed domain. In such a situation, she suggests a three-pronged movement forward—creation of an equal-rights-based framework for the global commons, transition to green energy sources, and a more comprehensive measure of wellbeing than GDP that better incorporates notions of sustainability and the nature–human relation.

Adjusting the current measures of aggregate income by incorporating incomes from resources (which are currently not included in the national accounting framework) requires a careful estimation and valuation of such flows. Murty and Panda (Chap. 4) examine the status of green accounting in India. They provide an overview of experiences in other countries that have generated green accounts and give a pathway for India to establish the United Nations' (UN)integrated System of Environmental-Economic Accounting (SEEA) as part of its national income accounting system.

In an effort in this direction Mali, Singh, Kotwal, and Omprakash (Chap. 5) examine the contribution of forests to local incomes in three forest divisions of Kerala. By critically examining the existing forest resource accounting (FRA) system in Kerala, they find that there is a 100 % increment in accounted income from forestry a comprehensive accounting system is adopted. This has important implications for policy-making as well as resource management practices. They further emphasise the need for green accounting, which would give perspective to the contribution of nature to the economy and society, especially regarding the provision of livelihood choices to the poor.

The process of accounting is just another way of recognising the importance of resources that we use for our well-being. It is a mechanism that helps us in more efficient management of natural resources. Sometimes, even the physical accounting of resources helps us in understanding the extent of losses we incur due to poor management. Krishna Raj (Chap. 6) presents the case of urban water supply in a rapidly growing city in India—Bengaluru. He finds that both the quantity and the quality of water supply are a matter of concern. The utility provider runs at a large economic loss, and by the author's estimates there is as much as 48 % difference between water supply and metered consumption. With the rapidly declining water table due to over-exploitation of ground water sources in the city and its neighbouring areas, there is need for urgent intervention. The question Raj poses is what would be a good mechanism to sustainably manage urban water supply in Bengaluru.

Management of water is an equally large concern in rural areas especially since large proportions of our agrarian systems are rain fed. Nandan Nawn (Chap. 7) attempts to find the use value of rainwater in rain-fed zones of agriculture in a framework where irrigation water is valuable, but the infrastructure to store rainwater is limited. Nawn argues that one could use a cost-benefit analysis to see the economic feasibility of creating rainwater storage systems. He proposes that the rural employment guarantee schemes could be used to financially support the labour costs of creating rainwater-harvesting systems. However, planning and execution of such works remain low in his study area of West Bengal. The issue that remains is how to create this infrastructure: Should this be a community initiative or should it by the state?

1.3.1 Market Incentives Versus Command and Control

There has been a long-standing debate over the institutional efficiency of the market and the state in managing natural resources. This question was put to the test in understanding the solid waste management practices by firms in Sri Lanka. Udugama and Jayasinghe-Mudalige (Chap. 8) find that firms were less influenced by market-based incentives, while regulatory instruments were more effective. They, however, suggest that regulation need not necessarily be undertaken by a third party. The evolution of co-regulation framework could achieve similar outcomes.

However, command and control policies have known problems. One of these is their inability to adapt policies to local situations. Bhagya Laxmi, Adusumilli, and Rao (Chap. 9) find that in the case of livestock policies, dependence on imported breeds to the exclusion of local breeds is ill-advised, especially in ecologically fragile semi-arid regions. They recommend that livestock policies also need to take cognisance of improving feed and fodder for increased buffalo production.

The command and control in agriculture is best displayed in the large-scale roll out of the green revolution technology in India in the 1970s. The policy push for adoption of this technology helped India increase its food production tremendously. However, the adoption of this technology was possible only in irrigated areas, which left out a major part of India's agricultural sector—the rain-fed zones. The soil condition in green revolution areas is now deteriorating to an extent where yields are falling. Therefore, there is an urgent need to understand soil management practices. Suresh Reddy (Chap. 10) emphasises that financial support to farmers wanting to use traditional inputs and technology would help farmers practice sustainable agriculture best suited for these regions. Further, Reddy argues that government policies should not undermine practices that have evolved over generations.

Policy interventions in the forestry sector also have had adverse impacts. The management of forest resources in India have de jure been vested with the state for a long time since the first pre-independence Forest Act of 1865. However, the forest dwelling and dependent populations have compelled the state to re-examine its strategy of exclusion of such communities in the management strategies. A series of amendments in the forest policy have emerged – a radical law giving land rights through the Forest Rights Act (2006), and the extension of the local government institutions (*panchayats*) to scheduled areas. This would enable them to function as self-governing institutions under the Panchayat Extensions to Scheduled Areas Act (PESA 1996). Sarap and Sarangi (Chap. 11) undertake a critical evaluation of the evolution of forest governance institutions and their shortfalls in Odisha, which has large populations and areas that fall under forest governance.

Forests yield multiple benefits from provisioning and regulating services to recreational and aesthetic services. One variety of forests—mangroves—is of critical importance in coastal zones but has been under threat, leading to policy interventions to restore it. P. K. Viswanathan (Chap. 12) examines one such experiment in Gujarat with a detailed household survey on tangible and non-tangible benefits derived by the beneficiary population. He finds that state intervention has to be more creative in removing impediments from full-benefit accrual from mangrove restoration.

1.3.2 Trade and Environmental Management

In an era of globalisation, economies are open not only to flow of goods and services but also to "bads"—pollution for example. It has long been argued that trade is an engine of growth, but it is increasingly being scrutinised for its impacts on the environment. Mishra (Chap. 13) addresses this issue in the context of developing countries. Using a panel dataset for 15 developing countries for the period 1990–2005, Mishra finds that CO_2 emissions have risen as developing countries opened their trade in this time period.

The trade and environment linkage has also been examined by Demaria (Chap. 14) in the context of pollution dumping in developing countries and its impact on local populations. His study focuses on the ship-breaking industry in Alang. 80 % of global goods are carried by sealines and therefore a critical component of world trade. Unlike CO_2 that has much wider global impacts, the ship-breaking industry has much more localised impact—both are consequences of increased world trade.

Liberal trade regimes increase imports and this probably brings pressure on developing countries to increase their exports, and very often this occurs in primary products. Such efforts may impose adverse consequences in developing countries in unexpected ways. Fisheries have been an important export sector in India, and shrimp farming along the coast has been a popular economic activity. However, intensive shrimp farming techniques are known to have adverse effects on agriculture. Umamaheswari, Nasurudeen, Hattab, and Ravirajan (Chap. 15) examine the trade-off between paddy and prawn farming in Tamil Nadu. They find that paddy farms closer to the shrimp farms exhibited significantly lower returns due to salinisation of the soil.

Can the market mechanism play a role in pollution management? Dasgputa, van der Salm, and Roy (Chap. 16) examine the European Union's (EU) emission trading mechanism and its applicability in India with the Perform-Achieve-Trade (PAT) mechanism. Europe has been a forerunner in experimenting with emission trading schemes that caps pollution and yet awards cleaner performers by giving them tradable permits. The authors identify critical issues in the PAT mechanism design that Indian policymakers need to bear in mind, learning from the experiences of its implementation in the EU. These policies allow market mechanism to reach optimal solutions.

1.3.3 Social Capital

The nature–economy–society linkage cannot be complete without discussing the role of social capital. It is often argued that social capital is a catalyst in economic development. However, its presence could on occasions be a hindrance as suggested by Chowdhury (Chap. 17). She attempts to explain the persistence of Jhum cultivation in the North East in a game-theoretic framework.

The presence of a strong social network, which ties in local families to Jhum cultivation, restrains allocation of labour in any other activity and therefore leaves it immune to other economic forces. In a situation where there is no pressure on forests, this would not be a reason for concern. But given the anthropogenic pressure on forests, and the shortening cycles of Jhum cultivation, the continuance of Jhum poses serious forest management issues. Chowdhury argues that a reduced level of social capital would help wean families away from Jhum cultivation and thereby reduce pressure on forests.

Chandrasekaran and Swamy (Chap. 18) introduce a new angle to natural resource management (NRM) in a rural economy that is impacted by an invasive species.

They examine both the ecological and the economic impact of *Prosopis juliflora* (an invasive species) on biodiversity and its adoption by farmers as an economical profitable alternative to paddy. While this creeper has deleterious effects on grazing and medicinal herbs, it compensates farmers by being a profitable input for charcoal. This once again highlights the tension of balancing between biodiversity and livelihood objectives.

1.4 Conclusion

This volume is an attempt to understand the linkages between ecosystem, economy, and the social forces. It attempts to establish that causality between them is not unidirectional or linear. While economic forces affect the ecosystem, the ecosystem also creates stimuli to affect economic variables and social institutions. Sometimes, institutions emerge endogenously to manage anthropogenic pressures. History provides sufficient evidence (and climate change is an imminent reminder) of the impacts of indiscriminate anthropogenic interventions in the bio-geo-chemical cycle. Often, impacts on nature are not perceptible in the short run and, therefore economic agents get a false sense of security about their actions. Policymakers in developed countries have used this knowledge to take corrective action but this policy sensitivity is yet to emerge significantly in the developing world. It is hoped that this volume would be a step in that direction.

References

Chopra K, Leemans R, Kumar P, Simons H (2005) Ecosystem and human well-being: policy responses (Findings of the responses working group of millennium ecosystem assessment). Island Press, Washington D.C

Dasgupta P (2001) Human well-being and the natural environment. Oxford University Press, Oxford

Ostrom E (2009) A general framework for analyzing sustainability of social-ecological systems. Science 325:419–422.

Sagoff M (2012) The rise and decline of ecological economics Breakthr J 1(2). http://thebreakthrough.org/index.php/journal/past-issues/issue-2/the-rise-and-fall-of-ecological-economics

Chapter 2
Nature, Economy and Society: Of Values, Valuation and Policy-Making in an Unequal World

Kanchan Chopra

In this chapter, some aspects of the linkages between nature, economy and society (the theme of the conference) are examined at different levels. The first is a conceptual one, which begins from and goes beyond stressing the urgent need for dealing with the complexity of nature and society interactions from diverse discipliary perspectives: I intend to postulate that whichever discipline we treat as the starting point of the analysis the ethical undertones and assumptions drive the analysis in directions which acquire meaning in terms of the quality and legitimacy of decision-making. In other words, methodologies acquire meaning only when interfaced with or interpreted in the context of value systems. Continuing in the same strain, I intend to examine briefly the emerging literature on valuation of ecosystems and ecosystem services, both as a methodology and as a tool for providing policy direction.

The last issue I propose to touch on deals with the major environmental challenges that face us humans today and for alleviating which specific policy directions at international and national levels are needed. The choices which face India and South Asia, as development and environment both need to be addressed aptly, span a large number of these challenges. The question is: What directions does the current level or state of knowledge give to help us to emerge with meaningful policy directions?

Revised and updated version of presidential address delivered at the Sixth Biennial Conference of the Indian Society for Ecological Economics (INSEE), October 20–22, 2011 held at Centre for Economic and Social Studies, Hyderabad.

K. Chopra (✉)
Formerly Institute of Economic Growth, Delhi, India
e-mail: choprakanchan14@gmail.com

N. Ghosh et al. (eds.), *Nature, Economy and Society*,
DOI 10.1007/978-81-322-2404-4_2

2.1 Of Methodology and Values: Towards Deliberative Processes as Links

Methodologies in all disciplines are driven by underlying values and assumptions. Take for instance economics and ecology. Both share at least one common methodological approach, that of modelling. And, underlying all models are assumptions; in economics with respect to human behaviour and in ecology with respect to community evolution and/or stability. Economics proceeds to create macro perspectives, and ecology moves towards case-by-case recommendations. In both disciplines, the models used can be classified in several ways. A division of models into descriptive, normative and decision theoretic is useful for our purposes.

Descriptive models attempt to define (and perhaps explain) the behaviour of some aspects of the environment or the economy. "Normative models" go further. They prescribe how things should be. The prime examples of normative mathematical models in ecology are formal decision models used in conservation management. In economics, the standard optimizing model that counsels an agent to maximize utility/expected utility is the representative example of a normative mathematical model.

Both descriptive and normative models by themselves can only provide images of reality and its driving forces. Most of the time, they select certain aspects of nature or economy and formalize the relationships. Normative or optimizing models can succeed in providing guidelines for environmental decision-making, but only by using ethical principles to provide an objective function. These principles could be: maximizing welfare, protecting the vulnerable, reducing inequality or adopting a right-based approach to livelihoods not to mention several others. The guidance provided for policy depends on which of these principles are taken on board by the decision-makers and stakeholders who they give recognition to. And since constituent stakeholders may have different preferences on the issue at hand, multiple solutions, non-commensurability and consensus become significant in reaching solutions. It is then correct to conclude that methodologies encompassing decision-theoretic tools and deliberative approaches are of the essence in most practical policy-making situations.

In other words, as soon as we concede that multiple criteria for arriving at decision rules exist and are to be taken into account, the emphasis in methodology has to shift to models of consensus building[1]. Such models which examine real-life processes of decision-making need to augment those rooted in a Platonic ideal in which technical elites take decisions through the so-called dispassionate use of scientific knowledge. For example, in any conservation related decision-making, regarding biodiversity as an end in itself raises a variety of ethical questions about the ultimate source of the value it has. Alternatively, a purely technical approach may limit itself

[1] For a discussion of these and related issues, see Colyvan et al. (2009).

to balancing the value of biodiversity as an embodiment of one stream of goods and services against that of other streams.

The point that rational action may better be captured through models of procedural rather than substantive rationality has of course been made earlier by, among others, Simon (1972). Developing this line of argument further, deliberative monetary valuation suggests that revealed or stated preference methods be followed up with group deliberation to arrive at values which a society holds, cherishes and intends to promote. Such approaches are also based on the presumption that a deliberative process does not only elicit preferences but forms them too. While case studies using deliberative processes have been carried out at different ecosystem levels, the methods of deliberative analysis have been received with apprehension. Some claim that they are time consuming and vague. However, even if we are never able to fully model the change in preferences and the subsequent effect on decision-making which a deliberative process brings about, we will enable decisions which take into account most stakeholders' interest. In other words, to provide institutions for preference changing behaviour to emerge is an important contribution of the process. And as Norgaard (2008) has maintained, "the lines between *scientific* ways of knowing and *democratic* ways of learning and choosing continue to blur as scientists acknowledge the role of judgement in science".

Deliberative processes also, at times, illustrate the rich nature of decision-making outcomes when institutional arrangements at different levels interact and learn from each other. Witness the positive outcomes when formal legal arrangements relating to natural resources learn from and amalgamate critical lessons from community management regimes. On the contrary, when schemes such as those for payments for ecosystem services (PES) are introduced without understanding that their success involves a reconfiguration of the roles of the community and the state, they are bound to be unsuccessful. In other words, deliberstion between the state's hierarchical structures, markets and community management is of critical significance in setting up systems of rights and payments for ecosystem services.[2]

2.2 Of Valuation Within and Outside the Market: Will Valuation of Resources or of Ecosystem Services Help?

Economists have maintained, and rightly so, that a large number of environmental problems often arise because of the absence of market value for some kinds of resources and services. An outcome of this line of argument has been a whole literature on valuation, including a focus on methodologies of nonmarket valuation. This literature has strengthened our understanding of how natural resources contribute to livelihoods of the poor and to the wellbeing of all sections of society. It has also provided refinements in methodology by giving rise to contingent valuation, hedonic

[2] See Vatn (2010) for details.

pricing and several other techniques. In the particular case of extensions of valuation methods to linked ecological and economic systems, in particular regulating services of ecosystems, details of ecological interactions have also sometimes been taken into account. Most of these valuation exercises fall back on an interpretation or extension of the notion of prices in the market, directly or indirectly as the anchor of the methodology.

In interpreting and applying these exercises however, it is of critical importance to remember that markets as institutions are defined within a set of assumptions relating to the knowledge, information and economic power that participants in them command. Information and power asymmetry corrode the efficient functioning of markets. More importantly, where these are asymmetrically distributed, valuation cannot take us far in terms of guidance for policy. Further, overriding considerations of societal norms and values play an important role in decisions on conservation. I wish to draw attention here to Geoffrey Heal's statement *"Valuation is neither necessary nor sufficient for conservation. We conserve much that we do not value and do not conserve much that we value" (in monetary/economic terms).*

Biological and ecological findings have often supported the conservation of areas also without any resort to valuation. Arthur Cooper argued that there were numerous examples of the way that ecology has directed environmental ethics and policy. The best illustration, he said, has been the role that findings about estuarine ecosystems have played in stimulating government programs for coastal zone management. Ecological findings were directly responsible for environmental decisions to limit the use of dichlorodiphenyltrichloroethane (DDT), to promote multispecies forests, and to publicize the problem of acid rain (Cooper 1982). In other words, ecological "facts" provide, at least, part of the basis for inferring what ethical, political and practical "values" ought to characterize environmental decision-making.

Ecological drivers by themselves are perhaps likely to be more compelling in developed countries where the drive for growth leading to perceived poverty eradication is not as paramount as in regions such as South Asia. Further, there exist other kinds of conflicts in developing countries too. The unequal distribution of income and power together with low levels of living has resulted in a focus on conflicts over resources and their appropriation by privileged sections. This is in particular emphasized by a large part of the literature on political and social ecology and for resources such as land and water. It is claimed that aggregative valuation often ignore distributional impacts and does not give due significance to the underlying relationship between ecosystems and multiple stakeholders.[3] Consequently, in many cases, monetization aimed at resolving a conflict in the use of ecosystem services may, in fact, lead to the perpetuation of the conflict.[4]

---

[3] For attempts to extend the literature on valuation to take into account stakeholder perspectives see Lele (2009 and 2013).

[4] See, for example Martinez-Alier (2002).

Take another example, that of land; with increasing urbanization, the tensions of the interface between value of land for different uses and by different stakeholders are rapidly increasing. In the face of a huge demand from urban use, driven by purchasing power, retaining land for agricultural or ecological use is not going to be easy. The underlying asymmetries referred to above, result in very high capacity to pay resulting in high demand driven prices.[5] *While valuation of land for different ecosystem services may provide additional inputs, an understanding that there exist "inviolate areas", whether for ecological or distributional justice reasons will have to be a critical component of policy.* Once again, we are led to conclude that although ecologists and economists in the past have frequently employed a notion of "scientific or economic rationality", current environmental problem solving requires them also to use "ethical rationality".

2.3 Of Emergent Issues Facing the Region and the World: And the Way Ahead

During the last century, an array of natural resource scarcity related issues have often led to emergent situations in parts of the world. At the same time, it is true that not all resources are equally threatened or scarce, whether from the perspective of nation states or of the world at large. There are large areas where knowledge and ingenuity are likely to alleviate resource shortages. In particular, we know that in some cases the more nuanced the technology for exploration, the more the reserves that become known. Also, substitutes are used when resources become scarce and there often occurs substantial reduction in resources used per unit of output. Such developments have led to more production per unit of a resource and alleviated scarcity. On the other hand, certain other kinds of resources and the ecosystems within which they are found may be nearing critical thresholds of change and sometimes moving towards extinction due to overuse.

Keeping in view all such possibilities, scientists have identified nine areas which are in need of a limit on human resource use, what they term as areas in which human use is straining boundaries at the planetary level (Rockstorm et al. 2009). Climate change and biodiversity loss are high on the list. Global consumption of

[5] Driven by India's high rates of urbanization, demand for land for urban construction and infrastructure continues to push the price of land to very high levels. Undoubtedly, these are higher than values yielded by "traditional" valuation of ecosystem services. Tensions from differing land values associated with different uses and different stakeholders are rapidly increasing. For instance, supply side ecosystem services-based approach to valuation yielded estimates of forest and deemed forest land between ₹ 7 to 9 lakhs per hectare for dense natural forests (See Chopra et al. (2006) Supreme Court Expert Committee 2007, Chopra and Dasgupta 2008). This was much higher than the compensatory afforestation payment of ₹ 50,000 per hectare paid for conversion. But demand-driven urban land use could garner a price of up to ₹ 90 lakhs per hectare or more. A similar situation exists for land diverted to mining. The drivers in this case may be high export prices.

fresh water, air pollution and changes in land use also figure. Closer home, at the national and regional levels too we witness many conflicts related to these areas to deal with.

Simultaneously, pressures from increased demands for natural resources continue. Countries such as India and China are undergoing major economic and social transformations. As a consequence, many boundaries between sectors (e.g. between the rural and the urban) are getting blurred. Changes in use of natural resources are driven by market determined drivers operating across sectors and ecosystemic scales. This change takes place so fast that we do not have the time to sit back to examine or rectify the dynamics of the processes which are simultaneously changing ecosystems, depriving us of important services and thereby increasing the costs to the economic and social systems of providing them. We destroy floodplains, (the natural sponges for excess water) through indiscriminate urban construction and then set up elaborate systems for flood relief. We create pollution and then clean-up expenditure needs to be incurred. Such quick fix solutions are often harmful. In the longer run, we also need to understand how to increase and sustain the capability of people, economies and nature to deal with fast changes in economic systems.

Are governments showing any long run leadership in this direction? A few policy pronouncements of the Indian government such as the Forest Rights Act (2006) and the recognition of places of ecological value as "no-go" areas or more recently "inviolate areas" incorporate some of the movement towards such leadership roles. Newspaper headlines sometimes (refreshingly) contain allusions to principles of environmental economics, for example in the case of the oil spill off the west coast, *"polluter pays principle to be used"*. Or to lack of enforcement of Acts as in the case of the Vedanta mining project, the Saxena committee report pointed out *"violations of the Forest Rights Act, the Forest Conservation Act and the Environment Protection Act (in the case of the associated aluminium refinery)"*. Further, *"the government ordered the closure of the Loharinag Palla hydro electric project" and the Minister for Power and for the Environment and Forests jointly decided this.* Also, lately, an incentive for conservation approach seems to have found favour as when we hear that *"green states to get a bonus"*.

More often than not, however, the emergence of trade-offs between environmental and development concerns has witnessed short-run expediency gaining the upper hand in policy-making. The moves to enable mining in certain forest areas and to proceed with hydel power development in pristine ecologically sensitive areas are examples of such policy directions.

In other words, while a certain amount of dynamism in the acceptance of these developments by policy-makers has been observed, an understanding of the underlying long-run linkages between human wellbeing and wise use of nature and its reflection in the design of policy seems a far cry at the present moment of time.

There is in fact, a focus on "the two cultures" when referring, in particular to the environment development debate.[6]

This preponderance of "two cultures" is true at all levels of governance; local, national and global. At the same time, learning in social science from policy implementation is itself the moving force behind the progress of some aspects of social science. All this happens in a somewhat piecemeal fashion with very little of an analytical framework to define it. The discourse is not a part of regular policy-making.

What is the way ahead in such situations? The term "green growth" has been used extensively of late, in particular in the context of the Rio + 20 conference. The United Nations Environment Programme (UNEP) defines a green economy as *"one that results in improved human well-being and social equity, while significantly reducing environmental risks and ecological scarcities. In its simplest expression, a green economy can be thought of as one which is low carbon, resource efficient and socially inclusive"* (UNEP 2011). The "green economy" approach is nothing but a reiteration of the viewpoint that since developing countries are at the point where massive investments are being undertaken, they can choose to invest in ecofriendly technologies. Such a choice will reduce the human footprint on areas in which human use is straining at planetary boundaries. In other words, these countries have the opportunity of "tunnelling through" the Environmental Kuznets curve, which postulates that the initial stages of development have seen a deterioration in environmental quality, with the relationship being reversed later.

However, investments in pathways to a green economy may not be easy to come by. They will require the framing of a *compelling and committed global interest*, in investing at least 1–2 % of global gross domestic product (GDP) in greening the economy[7] in order to shift development and unleash public and private capital flows onto *a low-carbon resource-efficient path.* They may also require specifics in policies such as: reducing or eliminating environmentally harmful or perverse subsidies (e.g. at the global level, around US$ 235 billion per year way back in 1992); creating markets for ecosystem goods and services; providing market-based incentives, opportunities and enabling institutions through appropriate regulatory framework. There is every possibility that potential investors under the banner of corporate social responsibility may find the best options to undertake green investments only in such countries,

a. Which are very high in carbon and suffering from "brown economic growth".
b. Which have all the essential infrastructures such as transport, communication and markets for investment.
c. Where the returns or turnover on investments are higher. One is not sure if developing countries like India, Nepal or Brazil will become their first candidates.[8]

[6] See Jairam Ramesh (2010), for a succinct exposition of the state of the debate and the policymakers' consequent dilemma.

[7] The Stern review (2006) places the figure at 1 %.

[8] See Kadekodi (2012) for details.

In other words, while technologies that use renewable resources efficiently and achieve distributive justice exist at least in some sectors, the challenges presented by their widespread dissemination seem formidable. They may require large one time investments with low returns in the short run. Alternatively, if indeed constraints to unabashed maximization of short run growth rates are presented by natural capital, they may need to be tackled using the "reduce, reuse and recycle" route. In other words, developed and developing countries may be confronted with the question of whether some sections of the population are indeed consuming too much.

This significant underlying question leads us to the second component of the way ahead. We need to monitor macroeconomic parameters in all countries to inform us on the nature of production and consumption in economies. Are the present levels of consumption and production "sustainable"? This question can be answered only if a few macroeconomic parameters relating to the environment and natural resources are monitored in conjunction with standard macroeconomic indicators covered in the System of National Accounts (SNA). The SNA focuses policy attention on parameters such as Gross Domestic Product (GDP), Gross Fixed Capital Formation (GFCF) and others. Yearly and indeed quarterly assessments of the state of the economy are based on these parameters. A similar statistical accounting of critical components of natural capital needs to be undertaken to ensure that year to year economic growth is not eating into the natural capital of the country. In recent times, developments in environmental economic accounting have taken place both at the international and national levels (See, for example, Government of India CSO (2013)) which attempt at providing a common basic framework for such an exercise. Most of this work is an attempt at using the theoretical developments in "sustainable income" to develop the outlines of a System of Environmental Economic Accounting (SEEA). The United Nations presented SEEA (2003) as a possible starting point. This has been revised subsequently and now provides the following two tier framework:

a. SEEA central framework which starts from the perspective of the economy and its economic units.
b. A SEEA experimental ecosystem accounting which links ecosystems to economic and other activity. This approach understands and states upfront that placing ecosystems in an accounting context requires the disciplines of ecology, ecological economics and statistics to come together and think of measurement and policy issues in new ways. It does not give precise instructions on how to compile ecosystem accounts but it represents a strong and clear movement towards a convergence across the disciplines on many core aspects.

"Genuine savings" is one of the most documented macro parameters, both internationally and in India. Estimates for different countries from the World Bank of genuine savings indicate that for India, the number is 24.64 % in 2008 and 24.56 %

in 2009 as against net savings of 29.68 and 26.60 % (of the GDP).[9] These estimates indicate that green GDP is lower than conventional GDP by 8.22 % in 2008 and by 5.1 % in 2009.[10] Though these estimates are partial, they provide the way forward with respect to the Central SEEA framework mentioned above.

Two approaches to assist in policy-making which accounts for nature–economy relations have been outlined above: Taking "the green economy" route to resource conservation and monitoring macro parameters for "sustainable development". But this is not enough. These policy changes need to be nested in a larger blueprint for a longer term development future, both for the planet and for the country. In such a blueprint, the following components could provide a good starting point:

1. The first is a framework of equal rights and entitlements for all humans to share the global commons resources, in particular atmospheric space. Either legally binding (for developed nations) or voluntary commitments (for e.g. the BRIC nations) to limit the use of that space by limiting greenhouse gas (GHG) emissions is then the next step. This will lead us on to consumption and production patterns that are feasible. We may be able to respond to the question of whether we are consuming too much. Or who is?
2. The second component of the blueprint is the transition to a green source of energy. In other words, the new technology has to be based on renewable sources and away from fossil fuels. A global as well as national agenda for moving in that direction can then be drawn up.
3. To understand the significance of the first two components of the blueprint, a new economic indicator of wellbeing which complements macro measures such as GDP is urgently needed. We need to measure relative prosperity of nations in a more inclusive and carbon liable manner. Then, alone can the implication of a relentless pursuit of individual material prosperity be made transparent. Though steps are being taken to estimate "green" GDP, real savings and related measures, in some countries it has to be globally mandated.

To enable nations to move towards the vision inherent in the blueprint mentioned above, new and reformed institutions are needed for facilitating a change in human behaviour, to increase local appreciation of shared global concerns and to correct collective action failures that cause global-scale problems. However, this change in behaviour assumes acceptance of a common international norm. Such norms are more likely to emerge with decreases in inequalities in distribution of income and power and more interactions across the globe which facilitate emergence of international institutions. (Walker et al. in Science 2009). This brings us back, full circle, to the issue of distributional justice and shared norms. In conclusion, it is true that shared values with regard to the linkages between economy, society and nature need to inform policy-making at both macro and project levels. It is these that we need to move towards, as we simultaneously strengthen the information and data bases that enable us to do so.

---

[9] Corresponding gross national savings are 38.17 and 35.20 % for 2008 and 2009, respectively. See World Bank, World Development Indicators (2010 and 2011).

[10] See Murty and Panda (2012).

References

Chopra K, Dasgupta P (2008) Assessing economic and ecosystem services contribution of forests: issues in modelling and an illustration. Int For Rev 10(2):376–386 (Special Edition)
Chopra K, Dasgupta P, Eswaran VB, Kadekodi GK (2006) Report of the expert committee on net present value of forest land, submitted to the Supreme Court of India
Colyvan M, Linquist S, Grey W, Griffiths, PE, Odenbaugh J, Possingham HP (2009) Philosophical issues in ecology: recent trends and future directions. Ecol Soc 14(2):22
Cooper A (1982) Why doesn't anyone listen to ecologists—and what can ESA do about it? Bull Ecol Soc Am 63:348
Government of India, Central Statistical Organisation (2013) Green national accounts in India: A Framework, Report of the Expert Group
Jairam R (16 October 2010) The two cultures revisited: the environment-development debate in India. Econ Polit Wkly XLV(42):13–16
Kadekodi GK (2012) Possibilities of going on the green economic path Kamath Helekar Memorial lecture at Goa University, 21st February, 2012
Lele S (2009) Hydrological services of western ghats: lessons learned for integrated hydro-economic analysis. In: valuation of regulating services: from theory to practice, SCAPES International Workshop. Organised by University of Liverpool and ATREE, under the Strengthening Capacity to Alleviate Poverty through Ecosystem Services (SCAPES) project, Bangalore
Lele S, Srinivasan V (2013) Dissaggregated economic impact analysis incorporating ecological and social trade-offs and techno-institutional context: a case from the western ghats of India. Ecol Econ 91:98–112
Martinez-Alier J (2002) The environmentalism of the poor: a study of ecological conflicts and valuation. Elgar, Cheltanham
Murty MN, Panda M (2012) Generating National Income accounts for Measuring Green GDP in India A report submitted to the Central Statistical Office, Government of India, February
Norgaard RB (2008) Finding hope in the millennium assessment. Conserv Biol 22(4):862–69
Rockstorm J, Steffen W et al. (2009) A safe operating space for humanity. Nature 461:472–475
Simon H (1972) Theories of bounded rationality. In: Mcguire CB, Rodner R (ed) Decision and organization (chap 8). North Holland Publishing Company, Amsterdam
Stern N (2006) The economics of climate change: the stern review. Cambridge University Press, Cambridge
UNEP (2011) Towards a green economy: pathways to sustainable development and poverty eradication. UNEP, Nairobi
Vatn A (2010) An institutional analysis of payments for environmental services, ecological economics. Ecol Econ 69(6):1245–1252
Walker B, Barrett S et al. (11 September 2009) Looming global scale failures and missing institutions. Science 325(5946):1345–1346
World Bank (2010 and 2011) World development reports. Washington DC, World Bank

Chapter 3
Social Metabolism and Environmental Conflicts in India

Joan Martinez-Alier, Leah Temper and Federico Demaria

3.1 The Standard of Living

The industrial economy works in practice by shifting costs to poor people, to future generations, and to other species. Could an industrial economy work otherwise? K. W. Kapp wrote in1950 that capitalism is an economy of unpaid costs, but the socio-environmental impacts are not due to capitalism as such, they would not be different in another system of industrial economy if there was one. The impacts occur at various temporal and geographical scales. They arise because of the increased social metabolism, and this article shows which are its main trends in India.

Sometimes, environmental liabilities appear in the public scene when there are sudden accidents as in Bhopal in 1984 or the Gulf of Mexico (with British Petroleum) in 2010 or Fukushima in 2011 and so many other cases. But here we shall look more at the smooth trends than at the (not very surprising) surprises.

Shrivastava's and Kothari's brilliant book of 2012, *Churning the Earth: the making of global India* is written in the spirit of Karl Polanyi's *Great Transformation* drawing also on the critique of uniform development brought forward since the

This chapter has been in the making for some time, with parts of it being presented at two INSEE conferences and also at the Nehru Memorial Museum and Library in New Delhi in December 2012. It contributes to the EJOLT project.

F. Demaria (✉)
Institute of Environmental Sciences and Technology (ICTA), Autonomous University of Barcelona (UAB), Edifici Z, ICTA-ICP, Carrer de les Columnes, Campus de la UAB, 08193 Bellaterra (Cerdanyola del Vallès), Barcelona, Spain
e-mail: federicodemaria@gmail.com

Centre for Studies in Science Policy (CSSP), Jawaharlal Nehru University (JNU), New Delhi, India

J. Martinez-Alier · L. Temper
Institut de Ciència i Tecnologia Ambientals ICTA, Universitat Autònoma de Barcelona, Barcelona, Spain
e-mail: joanmartinezalier@gmail.com

N. Ghosh et al. (eds.), *Nature, Economy and Society*,
DOI 10.1007/978-81-322-2404-4_3

1980s by Ashish Nandy, Shiv Visvanathan, Arturo Escobar, Gustavo Esteva, Wolfgang Sachs, etc. (Sachs 1991) and also by Norgaard's "Development betrayed" (Norgaard 1994).

Shrivastava and Kothari also drew attention on ecological economists such as K. W. Kapp and Herman Daly, and (in a footnote, Chap. 7, fn. 4) they criticize Amartya Sen's notion of "development as freedom." Sen insisted that development is not only growth of income per capita and the movement of low-productivity farmers into higher productivity occupations, together with industrialization and urbanization. Sen's canvass is broader, summarized in the idea of "capabilities." Development should mean to acquire the material circumstances and the mental and social abilities to choose as much as possible your own path in life. One can agree to all this, but there is still in Amartya Sen a positive view of economic development in contrast to the authors quoted above. From the current Indian experience, Shrisvastava and Kothari assert that "development as freedom" falls short of accounting for disappearing natural environments and human cultures, and they ask why "even as sophisticated a writer as Amartya Sen. . .omits any discussion of the loss of land and livelihood, human community and culture that is invariably involved in the displacement induced by development." In Sen's writings, the natural environment has been seen, if at all, as amenities to be enjoyed once you are well-off, although in fact using Sen's own conceptual framework, it could be claimed that what development achieves is the loss of traditional "entitlements" to products and services formerly available outside the market. This is taken up again below in the section on "The Gross Domestic Product of the Poor."

There is still a persistent trend among economists to see the environment as a luxury good and to consider that the poor are "too poor to be green" (Alier 1995; Temper and Alier 2007). This view is vigorously opposed by Shrivastava and Khotari whose book is "dedicated to the many movements for ecological and social justice taking place today, in India and elsewhere."

In British historiography there was a debate on whether the "standard of living" for the common people increased between 1760 and 1850, and on what the "standard of living" meant. In India, a country that at present is compressing into a short time socioeconomic historical periods that could be represented in the West by Charles Dickens, Henry Ford, and Bill Gates all at once, there could be a similar debate. Moreover, sensitivity to environmental values and the diversity of languages and cultures has increased even further since E.P. Thompson and Eric Hobsbawm argued on the "pessimistic" side in Britain. In a country like India as in the world in general development means in practice environmental and cultural losses on a much larger scale than British industrialization meant internally for Britain.

In the British "standard of living" debate, the issues were not only the enclosures and the application of the "Poor Laws" but also the rate of increase in real wages of rural laborers and industrial workers. The point was that the quality of life had perhaps deteriorated (in terms incommensurate with money wages) because of overcrowded urban housing conditions, pollution, loss of access to land, and loss of status of independent skilled workers.

In due course, the condition of the English working class improved. This was due to the application of increasing amounts of more efficient power of coal to move the machines of the "thermo-industrial" revolution and to the creation of markets for the textile industry all over the world including India. It was also due to the slowly increasing union power of the new working class and the political power of Britain to exploit other territories, not only the colonies but also the Southern USA that until the 1860s exported cheap slave labor-time and cheap soil services transformed into cotton. The "ghost acres" and slave labor-time in sugar production in the Caribbean also helped in this (Hornborg 1998; 2007).

Although they tried, the British were far from being able to literally strip the world bare like locusts because there were too few of them (about 27 million by 1870) and many were poor and also because coal was extracted from the island itself. In 1870 (when Charles Dickens died) coal extraction in Britain was 3–4 t per person, and it still went up per capita until 1914.[1]

Poverty itself had been created by enclosures and dispossession. It decreased in Britain after the first decades of the thermo-industrial revolution, but there were many losses unaccounted for. Then, something unexpected happened internationally that should have dampened the positive views on the thermo-industrial revolution. In 1896 Svante Arrhenius published the first article showing that temperatures would increase because of increased carbon dioxide concentration in the atmosphere by burning coal. Nevertheless, climate change was not a spanner thrown into doctrines of economic growth until 1985, 90 years later, with the creation of the Intergovernmental Panel on Climate Change (IPCC; chaired since 2002 by an Indian scientist, Dr R. K. Pachauri).

There was early awareness that the economy was increasingly relying on nonrenewable sources of energy. From Jevons in 1865 to Patrick Geddes from the 1880s to the 1920s, and Frederick Soddy in the 1910s and 1920s, it was repeatedly pointed out that economic growth was based on fossil energy stocks which were being burnt and irreversibly dissipated. Later, in the late 1940s, calculations of "peak oil" began to appear, and in the 1970s estimates of a decreasing energy return on energy input (EROI) in agriculture and in the commercial energy sector were made (Pimentel et al. 1973; Hall et al. 1986; Martinez-Alier 1987, 2011). Energy cannot be recycled; therefore, even an economy that would not grow but would use large amounts of fossil fuels would need "fresh" supplies coming from the commodity frontiers. The same applies to materials, which in practice are recycled only to some extent (copper, aluminium, steel, or paper). Water is recycled in nature by sun energy, but we use groundwater and sometimes also surface water quicker than it is replenished. When the economy grows, the search for water and other materials and energy sources is of course even greater.

[1] In India coal extraction is now only about 0.5 t/c/year, mainly for electricity production (a process somewhat more efficient than 1870 steam engines).

3.2 The Gross Domestic Product of the Poor

When economic historians reconstruct gross domestic product (GDP) series, they would have to balance gains (in monetary and nonmonetary terms) with losses. A notion developed in India that points to such unaccounted losses is the "GDP of the poor." It was popularized by The Economics of Ecosystems and Biodiversity (TEEB) reports (TEEB 2008; TEEB Foundations 2010). Authors Haripriya Gundimeda, Pushpam Kumar, and Pavan Sukhdev in the first TEEB report "found that the most significant beneficiaries of forest biodiversity and ecosystem services are the poor, and the predominant economic impact of a loss or denial of these inputs is to the income security and well-being of the poor." Here "economic" and "income" are not or should not be meant in a chrematistic sense.

Assume a woman making a living by collecting shells in a mangrove forest with a husband who sustainably produces charcoal for the family and for the local market. Assume that a local shrimp farming corporation or an urban developer encloses the mangrove forest and destroys it (legally or illegally). The loss in the standard of living because of displacement, the increased fear because of threats of security guards, lack of access to food and domestic energy, are not well measured in money terms. That family is likely to lack money to compensate for the losses through buying alternative accommodation, food, fuels, and other sources of livelihood. The notion of equivalent compensation itself is in question. Moreover, the surrounding populations are now in danger because they lack protection against storms or tsunamis, and there has also been a loss of biodiversity in itself, beyond products or services for humans.

Assume (as it happens in India, driven by the increased extraction of materials) that another poor rural woman working at home and also outside the home for subsistence or for wages notices that the water in the river or the well is polluted because of mining nearby. The water was free before if social institutions of caste allowed access to it. Now it is polluted; if the woman gets a National Rural Employment Guarantee Act 2005 (NREGA) wage, she is unlikely to be able to incur the extra cost of buying water in plastic bottles. If she buys water, she cannot buy food or wood or clothes.

When people see their access to nature's products or services destroyed by deforestation, mining, tree plantations, dams or transport infrastructures, they often complain. They might ask for economic compensation (to "internalize the externalities") or, very likely, they will eschew chrematistic accounting and appeal instead to a language of rights.

3.3 Ecological Distribution Conflicts and the Defence of the Commons

The use of fossil fuels and minerals, the human appropriation of the available biomass, the diversion of water to industrial use, cause conflicts on the access to environmental products and services and on the distribution of the burdens of pollution. Poor people are not always on the side of conservation. However, in many conflicts of resource extraction, transport, or pollution, the local poor people (indigenous or not) are often on the side of conservation not because they are card-carrying environmentalists but because of their livelihood needs and (often) their cultural values (Guha and Martinez-Alier 1997; Martinez-Alier 2005).

Therefore, there are movements of environmental justice or an environmentalism of the poor and the indigenous often appealing to nonmonetary values such as livelihood, territorial rights, or sacredness of the land or the rivers. These movements combine livelihood, social, economic, and environmental issues. They set their "moral economy" in opposition to the logic of extraction of oil, minerals, wood, or agrofuels at the commodity frontiers, defending biodiversity and their own livelihood. In many instances they draw on a sense of "place" and local identity, but they could also connect easily with the politics of the Left. However, the traditional left in southern countries (as in West Bengal in 2007) still tends to see environmentalism as a "luxury of the rich." The same applies to the nationalist-popular movements in Latin America.

In India there are many conflicts on mining of coal and lignite, bauxite, iron ore (usually by domestic corporations, public or private). There are not so many foreign corporations as in Latin America, Indonesia, and Africa (from Canada, China, the USA, Australia, Europe) engaging in opencast mining or fossil fuel extraction. There are no foreign land grabs in rural areas of India, on the contrary there are some Indian land grabs outside India. There are many conflicts on illegal sand and gravel extraction, which do not happen so often in other countries. There are also biomass conflicts (deforestation, tree plantations), on water use (dams, excessive consumption of underground water, pollution by mining and by industry) in common to other countries. There are renewed conflicts on the uncertain risks from nuclear electricity, as in Maharashtra (Ranatgiri) against the French Jaitapur grandiose nuclear projects, and in Tamil Nadu with the world famous Russian-built Kudankulam power plant of Russian design where in December 2012 "fishermen from the coastal villages set out into the sea in more than 100 boats towards the Kudankulam nuclear power plant while women and children stayed back at the protest site in front of St Lourdes Church in Idinthakarai, the epicenter of the 400-day-long protest."[2] At the other extreme of the commodity chain, there have been persistent conflicts on uranium mining in Jharkhand.

Until 2011, it was foreseen that in Haripur in West Bengal six Russian-built nuclear power plants would come up.[3] This would be a second nuclear park, after Kudankulam. Haripur was to come up after all the six reactors at Kudankulam became operational. There was opposition. After peasant opposition had prevented, in 2007, the setting up of Special Economic Zones (SEZ) at Nandigran and Singhur (for the chemical industry and for a Tata car factory), the Communist Party government of West Bengal, which also favored Haripur, was defeated in elections.

---

[2] Times of India, 11 December 2012.

[3] http://www.thehindu.com/news/national/haripur-site-cancellation-is-one-more-wrinkle-in-india-russia-ties/article4026380.ece.

Its general secretary, Prakash Karat, complained in 2007 against "the modern-day Narodniks who claim to champion the cause of the peasantry" while neglecting the historic task of industrialization. He mentioned "the likes of Medha Patkar" among the Narodniks.[4] It would be funny if it were not so sad.

Anywhere one goes in India, one can find small civil society organizations (Gandhian and, in an Indian sense, Narodnik) documenting and involved in conflicts on land grabbing, water grabbing, and expoliation of other natural resources. They all follow roughly a similar line, although there is no politbureau lying it down. Thus, one such group in Allahabad claims that defense of the people's "communitarian ownership of natural resources is the underlying idea behind the mass upsurge, assertion and activism" in states like Jharkhand and Chhatisgargh against coal mining, in Uttarakhand and other Himalayan states against dams, in Tamil Nadu, Maharashtra, Haryana against nuclear energy threats. Thus, in Jharkhand, in the Karapura valley of Hazaribagh, "indigenous and agrarian populations of 205 villages have not allowed the entry of 35 corporations to start mining of coal and build large thermal power stations," while in Haryana "rural people in and around the village of Gorakhpur of Fatehabad district are opposing the proposed nuclear plant there. They have been sitting in a continuous dharna since July 20, 2010. Three persons have died while sitting dharna."[5]

The primary causes of such ubiquitous movements of resistance are the increase in the social metabolism and the defence of livelihoods against resource extraction. Meanwhile, there are also new urban waste disposal conflicts. Which of these types of conflicts existed already in 1947, which ones are new? Which are the trends? Are there historical studies of political ecology following the pioneering work of the 1980s and 1990s by Ramachandra Guha, Rohan d'Souza, and other authors on forests and water management conflicts since colonial times?[6]

3.4 Methods for the Study of Social Metabolism

The economy may be described in terms of economic indicators such as growth of GDP, savings ratio, budget deficit as percentage of GDP, current account balance in the external sector, etc. Social factors may be taken into account, as in

---

[4] http://pd.cpim.org/2007/0128/01282007_prakash.htm.

[5] Nai Azadi, monthly journal of Azadi Bachao Andolan, October–December 2012, special issue in memory of Banwari Lal Sharma, issued by the Swaraj Vidyapith in Allahabad.

[6] There is an overview of *India's Environmental History: Colonialism, Modernity and the Nation* edited by Mahesh Rangajaran and K. Sivaramakrishnan (Permanent Back, Ranikhet, 2012, 614 p.). This formidable collection does not trace changes in the growing human use of energy, materials and water, and the resulting ecological distribution conflicts in rural and urban areas. There is more research on social metabolic trends (particularly energy, including biomass) and on environmental conflicts in the EPW essays edited by Rohan d'Souza, *Environment, Technology and Development. Critical and Subversive Essays*, Orient Black Swan, New Delhi, 2012.

the Human Development Index which nevertheless correlates closely with GDP per capita leaving aside (as Shrivastava and Kothari emphasize) environmental and cultural loses.

The economy may also be described in terms of physical indicators. Economic, social, and physical indicators are nonequivalent descriptions. An economy may provide 260 GJ (gigajoules) of energy per person/year, its human appropriation of net primary production (HANPP) is 35 %, material flow amounts to 16 t per person/year of which fossil fuels account for 5 t. Of the material flows, 5 t are imported, 1 t is exported. Income per capita is US$34,000. It ranks 10th in the Human Development Index (HDI).

Of another economy, we say that it provides only 35 GJ person/year, its materials flow amounts to only 5 t person/year, its HANPP is 65 % (a heavily populated country, relying on biomass, with little external trade). Foreign trade is less than 0.3 t per capita/year of exports or imports. Income per capita is US$3000 (at PPP). It ranks 127th in the HDI. Different regions and different classes of people in such countries could be classified by their metabolic profiles.

Materials and energy flows accounting (MEFA) is a set of methods for describing and analysing socioeconomic metabolism. It examines economies as systems that reproduce themselves not only socially and culturally but also physically through a continuous exchange of energy and matter with their natural environments and with other socioeconomic systems. Material flow accounts are published at national level by Eurostat and now also by the United Nations Environment Programme (UNEP), drawing on methodologies established by the research group led by Marina Fischer-Kowalski in Vienna and other groups over the past 20 years. We need material flow accounting also at regional (state) level. This should be done in India because all-India accounts show levels of domestic extraction per capita and trends very different from those that would be shown by individual states, some of which have very large Physical Trade Deficits as they perform the role of suppliers of cheap materials.

In the Material Flows we calculate first the domestic extraction (in tons per year) divided into biomass, minerals for building materials, mineral ores for metals, and fossil fuels. They show different levels and trends in different countries. The domestic extraction is denoted as DE. The domestic material consumption (DMC) is equal to domestic extraction plus imports minus exports.[7] Physical imports and physical exports measure all imported or exported commodities in tonnes. Physical trade balance (PTB) equals physical imports minus physical exports. So countries like Brazil or Russia (among the BRICs) have large Physical Trade Deficits, but not India as a whole.

Such accounts (including carbon or energy "rucksacks," "virtual" water, and "embodied HANPP") are relevant for historical and current debates on ecologically unequal exchange and the ecological debt.

---

[7] The Material Flows of India have been calculated in Singh et al. 2012.

Energy flow accounting (EFA) is an integral part of the analysis of social metabolism. Primary and final energy delivered are usually classified in the statistics according to the source. Such energy flows (including hydroelectricity) are also unequally distributed, and in India they are creating not only coal mining conflicts and the new nuclear conflicts but also many conflicts in the Himalaya and in the North East on hydroelectricity. Notice that energy accounts are separate from the Material Flows. The idea of linking economic history to the use of energy goes back to Wilhelm Ostwald and later to Leslie White and other authors, but it was only in the 1980s when several histories of the use of energy in the economy were published. The most interesting EFA indicator is that of EROI. EROI is a useful coefficient for assessing the increasing costs of obtaining energy in developing tar sands or heavy oil in Alberta in Canada or the Orinoco Delta, Venezuela, or when taking oil from the bottom of the sea (and in the Brazil's *pre*-sal) or in the new gas fracking, or for agrofuels such as those derived from sugarcane or *Jatropha curcas* (with low EROI).

An economic-ecological history would establish the changes in the EROIs in a country such as India over the years noticing an improvement as biomass energy (fuelwood, charcoal) is substituted or supplemented by fossil fuels and, later, indicating perhaps a decline because getting energy while going down the Hubbert curve (after peak oil) will require (it seems) increasing amounts of energy.

3.5 The HANPP

The HANPPof biomass is calculated in three steps. First, the potential net primary production (NPP, in the natural ecosystems of a given region or country) is calculated. Then, the actual NPP (normally, less than potential NPP, because of agricultural simplification and soil sealing) is calculated. The part of actual NPP used by humans and associate beings (cattle, etc.) relative to potential NPP is the HANPP, meant to be an index of pressure on the biodiversity (because the higher the HANPP, less biomass is available for "wild" species). So, an increasing HANPP is an indicator of increasing pressure on biodiversity. This should be relevant to do a history of India's conservation areas and threats to its wildlife.

In India, due to high-population density and land conversion, and due to also a relatively high use of biomass per capita (which would still be larger if the Indian population ate more meat), the HANPP is very high (as it is also in Bangladesh). Simron Singh et al. (2012b) put it at 73 % compared to about 40 % in the European Union (EU, with comparable population densities), and only 24 % in Japan (that imports much biomass). The higher the HANPP, the stronger the pressure on biodiversity. We can also ask the question of who gets the HANPP among groups of humans, as when a commercial tree plantation is planted in a former forest used sparsely by adivasi groups.

Table 3.1 A classification of socio-environmental conflicts

Geographical scope stage	Local	National and regional	Global
Extraction	Resource conflicts in tribal areas, such as bauxite mining in Odisha, coal and uranium mining in Jharkhand, oil extraction in the Amazon of Ecuador or Peru	Mangrove uprooting. Tree plantations for wood or paper pulp. Collapses of fisheries	Worldwide search for minerals and fossil fuels at the "commodity" frontiers. Biopiracy. Attempts at regulation of "corporate accountability"
Transport and trade	Complaints against urban motorways or heavy traffic in rural mining areas because of noise, pollution, landscape loss	Interbasin water transport oil/gas pipelines (e.g., Burma to Thailand) no TAV movement (Italy)	Oil spills at sea "Ecologically unequal exchange" because of large South to North material flows
Waste disposal and pollution, post-consumption	Conflicts on incinerators (dioxins) or VOCs, NOx, ozone, particulate matter	Acid rain from sulfur dioxide Nuclear waste disposal (Yucca Mountain, Nevada, the USA). Ship dismantling (Alang-Sosiya, Chittagong)	CO2, CFC: causes of climate change/ozone layer destruction/ocean acidification. POPs even in remote pristine areas. Claims for a "carbon debt," climate justice

3.6 From Social Metabolism to Valuation Languages

Table 3.1 classifies environmental conflicts depending on the stage of the commodity chains where they occur, and depending on their geographical reach. One could fill in Table 3.1 mostly with Indian examples (in India's territory or where Indian companies are active overseas—for example, the controversial Aratirí iron ore mining project in Uruguay belonging to Zamin Ferrous Metals). Sometimes local conflicts become "glocal", that is, well known also outside the territory in question, as with the Dongria Kondh vs. Vedanta in Odisha or in Latin America, the Chevron-Texaco case in Ecuador.

In such conflicts, a variety of Valuation Languages are deployed. Some of them were perhaps more powerful in the past (livelihood, sacredness) than in this era of the generalized market system where even "the fetishism of fictitious commodities" is in the ascendant in schemes for payment for environmental services. One wonders for instance how effective it is still in India to oppose the sacredness of a "sacred grove" against a mining project or a dam (Kososy and Corbera 2010).

Notice in India the attempt at compensation of damages in resource extraction conflicts by ascertaining the so-called net present value (in money terms) of the foregone products and services from forests. This is a peculiarity of Indian state administration (Temper and Martinez-Alier 2013). Other valuation languages apart from money compensation (human rights, indigenous territorial rights, environmental justice against "environmental racism," and even the Rights of Nature as in art 71 of the 2008 Constitution of Ecuador, are perhaps gaining in strength across the world.

There is much scope for historical and comparative work on the valuation languages deployed in environmental conflicts. In the Environmental Justice Organizations, Liabilities and Trade (EJOLT) project (www.ejolt.org), we collect socio-environmental conflicts. We hope to collect and map about 2000 by the end of 2014. We aim at having the atlas of country and thematic maps. We classify the conflicts according to the commodity in question, as in the following list:

- Nuclear: uranium extraction, nuclear power plants, nuclear waste storage
- Ore and building materials: mineral extraction; mineral processing; tailings
- Waste Management: e-waste and other waste import zones, ship-breaking, waste privatization, waste-pickers, incinerators, landfills, uncontrolled dump sites, industrial, municipal waste
- Biomass: land grabbing, tree plantations, logging, non-timber products, deforestation, agro-toxics, GMOs, agrofuels, mangroves versus shrimps, biopiracy and bioprospection, intensive food production (monoculture and livestock), fisheries
- Fossil fuels and climate justice/energy: oil and gas extraction, oil spill, gas flaring, coal extraction, climate-change-related conflicts (glaciers and small islands), Reducing Emissions from Deforestation and Forest Degradation (REDD)/Clean Development Mechanism (CDM), windmills, gas fracking
- Transportation and infrastructure: megaprojects, high-speed trains
- Water management and Hydric justice: dams, water transfers, aquifers, hydroways, desalination
- Biodiversity: invasive species, damage to nature, conservation conflicts
- Industrial and utilities conflicts: factory emissions, industrial pollution

It would be easy to trace in India hundreds of examples of each of these types of conflicts, listing the specific commodities in question, the social actors involved, tracing their geographical distribution patterns over time, accounting for the reasons for the occasional successful outcomes (stopping projects).

3.7 India's Social Metabolism

Following the example of Japan for 20 years at least until 2013, economic growth has stopped in many rich countries since 2008 (less by design than by the economic crisis), while in the BRICs but also Peru, Indonesia, Colombia, Turkey, and many other countries there has been sustained growth even after 2009. Poverty in terms of income per capita is declining in all such countries, including India.

This growth is achieved at great environmental and social costs. Peasants are squeezed out of the land, tribals in India and elsewhere are being displaced because they happen to live at the commodity frontiers. Biodiversity is being rapidly lost, and the concentration of carbon dioxide in the atmosphere is still increasing at 2 ppm per year. It was 300 ppm when Arrhenius first wrote on the enhanced greenhouse effect, and it is now almost 400 ppm. There is no international agreement on reduction of emissions. The United Nations (UN) gave up after the Copenhagen meeting of 2009.

This is the background to the study of India's Material Flows from 1961 to 2008 carried out by Singh et al. and published in *Ecological Economics* in 2012. India per capita consumes less fossil fuels, less building materials, and less mineral ores than most other countries. In the 1960s, about three quarters of the total material consumption consisted of biomass while construction materials were second in importance. Fossil fuels and industrial minerals and ores were insignificant in relation to the total flows. In the course of the almost 50 years under study, this changed in quantity and composition. The use of biomass only doubled. Fossil fuel consumption multiplied by a factor of 12.2, industrial minerals and ores by a factor of 8.6, and construction materials by a factor of 9.1.

Until the 1980s the population grew at a slightly faster pace than material throughput. Throughout the 1960s and 1970s, material use remained at a low and slowly declining level of less than 3 t/cap/year. Only since the early 1980s (10 years before Dr. Manmohan Singh became finance minister in 1991) a sustained growth in per capita material consumption set in, growing to 4.3 t/cap/year, accelerating in the period since 2004. Taking into account further growth from 2008 to 2012, India is probably at a level of 5 t/cap/year. In comparison, per capita material consumption in EU countries is about 15 t per person/year. (Notice moreover that in the EU imports are very significant, and they exceed exports in tons by a factor of 4).

The trends in total and per capita material domestic consumption in India (extraction plus imports minus exports), on material intensity, and on physical trade are summarized in Fig. 3.1.

In regard to the material intensity (or its reverse, resource productivity), India's GDP (in constant US$2000) increased by a factor of 12.4 between 1961 and 2008. The monetary economy grew faster than the physical economy. The material intensity of the Indian economy, measured as the ratio of DMC per GDP, declined by 69 %, from almost 20 kg of DMC per US dollar GDP to only 6 kg per US dollar. This decline can be attributed to the slow growth of biomass consumption that only doubled. In contrast, the use of minerals and fossil fuels grew at about the same pace as GDP.

In Singh et al. (2012) we wrote that, in general, construction minerals are abundant and scarcity and extraction conflicts are usually only regional phenomena. But as we shall see, almost all regions of India suffer from the phenomenon of "sand mafias."

As regards trade, India has no Physical Trade Deficit. In the case of India, unequal trade has to be analysed at state level. For instance, as we shall see, Odisha is a large net exporter. Its exports cause locally large environmental and social damages.

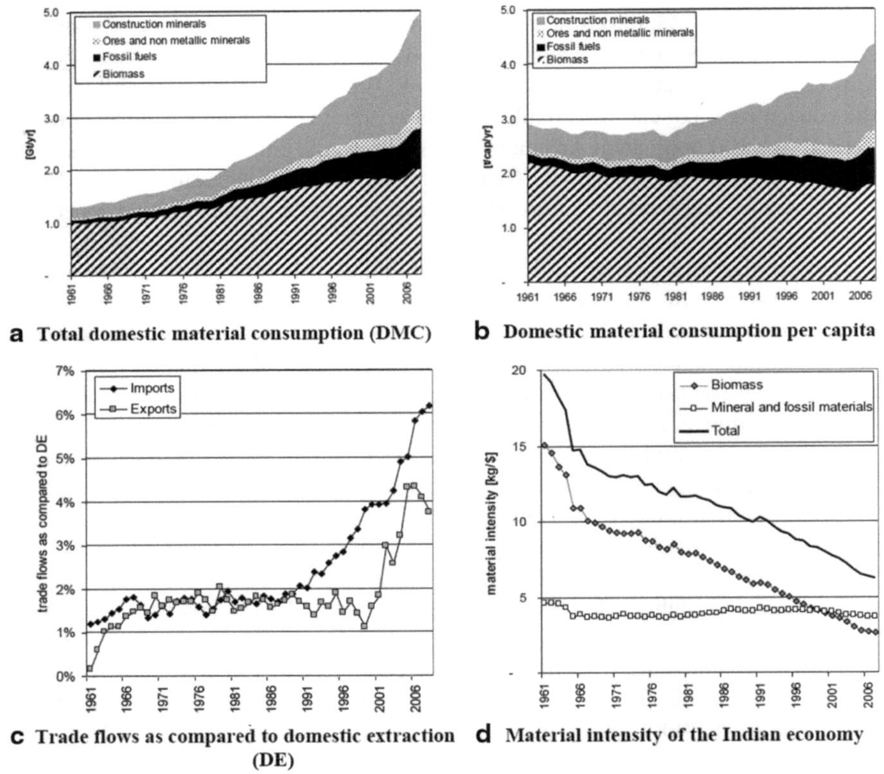

Fig. 3.1 Material flows accounts of India 1961–2008. (Source: Singh et al. 2012, op. cit)

Could Odisha tax exports substantially for its own benefit? The same applies to other states or regions of India (including Goa).

Table 3.2 shows again the domestic material consumption (domestic extraction + imports − exports) per capita from 1961 to 2008, excluding the biomass, and comparing to GDP and population growth.

3.8 Regional Variations in Social Metabolism and Resources Extraction Conflicts: The Case of Odisha

In this section (inspired by Felix Padel and Samarendra Das over the years), we account for a historical novelty of the past 30 years in the very long recorded history of Odisha: the clashes because of plans for mineral extraction (as in Maikanch) or because of industrial land grabbing (as in Kalinganar; Padel and Das 2010).

The mining and quarrying sector has been the fastest growing sector in Odisha at above 10 % per annum growth since 1980–19 81 to 20.09.2008 and beyond. Yet

Table 3.2 DMC (t/cap/year) of India for the three main groups of mineral and fossil materials and their average annual growth rates (%) in comparison to population and GDP. (Source: Singh et al. 2012)

	1961	1980	2008	1961–1980 (%)	1980–2008 (%)
GDP (billion US dollar at constant 2000)	66	156	812	3.5	6.0
Population (million)	444	687	1140	2.3	1.8
Fossil energy carriers (DMC t/cap/year)	0.1	0.2	0.6	4.7	6.0
Ores and industrial minerals (DMC t/cap/year)	0.1	0.1	0.3	3.4	5.6
Construction minerals (DMC t/cap/year)	0.4	0.5	1.6	2.0	6.2

DMC domestic material consumption, *GDP* gross domestic product

at the same time, virtually the whole of Odisha, including Kashipur in Rayagada, Lanjigarh in Kalahandi, Lower Suktel area in Balangir, Kotagarh in Phulbani, the mining-industrial belt in Jharsuguda, Kalinganagar and Rourkela, has turned into a battleground on the issue of development and displacement. The issue at hand in these conflicts is sometimes demanding better compensation packages, but often communities are raising serious objections to the notion of development itself that is being promoted, and championing a different system of values and an alternative vision of development to that advocated by the state. The Niyamgiri-Lanjigarh case as well as the struggle over the investments of the Korean steel company POSCO have become emblematic of such clashes, also found in neighbouring states.

While other growing economies such as Brazil, Argentina, Russia, and South Africa have large physical trade deficits in relation to the size of their economies, India's exports and imports (in tonnes) are small relative to the physical size of her economy (Fig. 3.1). However, within the subcontinent, there are resource-rich regions with very large net material exports to foreign countries and to the rest of India. As such, looking at material flows only at the national level obscures large variations between per capita (and per hectare) material extraction between states. Particularly in the eastern part of the country in states such as Jharkhand, Chhattisgarh, and Odisha, the ecological distribution conflicts associated with the growing social metabolism can most clearly be evidenced, but if we look at the energy flows, we would also understand the resistance to hydro-electric developments in other areas. It is always the same story: from increased social metabolism in terms of materials and energy, to socio-environmental conflicts expressed in different valuation languages.

Odisha is the 9th largest state by area in India and the 11th largest by population. It holds many minerals. Despite (or because of) increasing exports of this wealth to other regions, it remains one of the poorest states in India with 40 % inhabitants below poverty line (Economic Survey, Government of Orissa, 2003–2004 and

Table 3.3 Extraction of important minerals in Odisha, 1960–2000. (Source: Statistical Abstracts of Orissa 1961, 1969, 1979 and Economic Survey 2003 and 2006 and Directorate of Mines, Government of Orissa: http://www.orissaminerals.gov.in/Mines/MineralProduction.aspx?GL=ming&PL=3)

Year	Iron ore	Coal	Bauxite	All minerals	Tonnes per capita
	(million t/year)	(million t/year)	(million t/year)	(million t/year)	
1960	3.5	0.9	0	8.9	0.5
1970	nd	nd	0	12.3	0.6
1980	6.6	3	0	14.6	0.6
1988–1989	7.3	nd	nd	31	1.0
1995–1996	9.3	32.6	2.4	51.3	1.5
2000	14.3	47.8	2.9	87.3	2.4
2005–2006	48.0	70.5	4.9	138.7	3.6
2008–2009	77.1	97.7	4.7	189.0	4.7

2008–2009). Recent years have seen unprecedented levels of investment from both domestic and international companies hoping to profit from this mineral bounty. In 2009, Odisha was the Indian state with the 2nd most Foreign Direct Investment (FDI), after industrial Gujarat.

Thus, in states like Odisha, the incidence of mining is much larger than in the country as a whole. Iron ores, coal, bauxite together with other minerals extracted reach in Odisha 4.7 t per capita per year (Table 3.3), with rapid increase since 1980. The incidence of conflicts depends on the growth in extraction, but there is no rule that the amount of extraction in tons be proportional to the number and intensity of conflicts. The type of mineral and the rate of growth are relevant together with other variables such as population densities, water scarcities, indigenous presence, local political activism, etc.

Odisha holds one third of the country's iron ore reserves, a quarter of its coal, half its bauxite and more than 90 % of its nickel and chromite. The state, under the firm government of Naveen Patnaik who has been relected several times, has attracted large investment proposals from Tata, Jindal, Posco, ArcelorMittal, and Vedanta. Some large investment projects include building new harbours. There are problems over land acquisition, accusations of corruption, and violations of environmental regulations. There is increasing Naxalite presence in the state. But surmounting all obstacles, the increase in mineral extraction is undeniable, to be slowed down in 2013 because Odisha (as Karnataka and Goa) is undergoing a review of illegal iron ore leases and because of the Vedanta (temporary?) fiasco in Lanjigarh.[8]

Compare with Table 3.4, for the whole country, showing rapid growth and then stagnation (for metallic minerals) due sometimes (as acknowledged by the Ministry

[8] Games Vadanta Plays, EPW, 22 December 2012, http://www.epw.in/system/files/pdf/2012_47/51/Games_Vedanta_Plays.pdf.

Table 3.4 Extraction of some key minerals, years 1997–1998, 2008–2009, 2010–2011 (000 t per year). (Source: Shrivastava and Kothari, Churning the Earth, 2012, p. 125 from Ministry and Coal and Mines, Annual Report, 2001–2002, Ministry of Mines, Annual Report, 2008–2009; last column has been added from Ministry of Mines, Annual Report, 2011–2012)

Bauxite	6108	15,250	13,172
Coal	297,000	493,000	533,000
Iron ore	75,723	225,544	191,522
Chromite	1515	3976	3900

of Mines' Annual Report of 2011–2012) to "temporary discontinuance of mining for want of environmental clearance."[9]

Shrivastava and Kothari do not provide a systematic Political Ecology of resource extraction, infrastructures, and waste disposal conflicts in India linking social metabolic flows to such conflicts, but they (2012, p. 125) state that the socio-environmental impacts of mining are horrifying and they are far from being randomly distributed. "The blasted limestone and marble hills of the Aravallis and Shivaliks; the cratered iron ore or bauxite plateau of Goa, Madhya Pradesh, and Odisha; the charred coal landscapes of eastern India; and the radioactive uranium belt in Jharkhand are all witness to the worst that economic "development" can do." The worst affected, they conclude, are possibly the adivasis of central and eastern India.

While Odisha had only two iron and steel plants in 1995, today there are 14 steel plants and four pig-iron plants in the state. Yet this was only the tip of the expansion, with over 43 Memoranda of Understanding (MoUs) to set up steel plants already signed by 2005, including a US$12 billion plant that South Korean company POSCO aims to establish near Paradip Port, plans from Arcelor-Mittal to invest in a mega-steel plant worth US$10 billion, and Russian Magnitogorsk Iron and Steel Company (MMK) plans to set up a 10 MT steel plant. The state is also attracting record investment in aluminum, coal-based power plants, petro-chemicals. Proposed investments would see an annual production of 76 million t of steel, 5 million t of cement, 4 million t of aluminum and 25,000 MW of electricity to fuel this production in the coming years (Mishra 2010).

Examining the relevant material flow data from 1960 to 2006 (Table 3.3), Odisha's role in providing itself and the rest of the country with minerals can be seen. In 2008–2009, over 4.7 t of minerals (including coal) were mined per capita in Odisha—equivalent to the per capita consumption of all materials including biomass for an average Indian. The growth of the mining sector was vertiginous from 1995 onwards, when the deregulation of the mining sector allowed increased foreign investment and more state-level control over mining concessions (Asher 2009).

[9] http://mines.nic.in/writereaddata%5CContentlinks%5 C1ed4a15b370646d7be2c6defb2ecf6c9.pdf).

The need for land acquisition in a wide wave of enclosures to accommodate Special Economic Zones and mines, often forestland, has induced local level conflicts between the state and those slated to be displaced. The population density in Odisha is close to the average for India (approximately 300/km^2) leading to land scarcity in a state where 85 % of the population is rural and almost entirely dependent on agricultural land and forest resources for their livelihoods. From 1950 to 1995 over 250,000 people were displaced in Odisha, half of these were adivasi. Of these, only 25 % were ever resettled (Fernandes and Ashif 1997).

We have here a clear link from increasing social metabolism to resource extraction conflicts. In the state, hydroelectric dams (at the service of the mining industry), eucalyptus plantations and shrimp farms have proved no less controversial than open cast mines. Conflicts include the killing of three young men in Maikanch protesting the UTKAL Aluminum plant in Kashipur on the 16 December 2000; in Kalinganagar, 12 people were shot dead by police in 2006 resisting displacement by industrial conglomerate TATA. The South Korean multinational POSCO's attempts to acquire land for a steel plant that would displace many people growing crops have led to violent clashes, including a protest in 2006 where 11 persons were injured (Padel and Das 2010). As it is well known in India, this is a type of violence different from that of Naxalite (or counter-Naxalite) origin. There is a geographical overlap in some areas between Naxalite violence and resource extraction conflicts though not yet in Odisha at the time that all such incidents took place.

3.9 Illegal Sand Mining in India

From a factsheet for EJOLT prepared by the Centre for Studies in Science Policy at Jawaharlal Nehru University (JNU) and from other sources, we take evidence and interpretations regarding this practice (Singh 2012a) which is not unique to India. In the Observatorio de Conflictos Mineros de América Latina (OCMAL) archive of mining conflicts in Latin America, we find similar cases of illegal or contested extraction of "aggregates" (sand and gravel) in several countries, including the very visible case of the Tunjuelo river in the outskirts of Bogota where Holcim, Cemex, and the Catholic Church are involved.

Sand mining refers in India to the extraction of sand and gravel from riverbeds and seashores for construction activities and for minerals such as gold, silver, silicates. Mining sand for silica or for metals in coastal areas (such as ilmenite for titanium) is different from "mining" sand and gravel from river beds or beaches as building materials. What in India are called the "sand mafias" are small contractors serving the building industry. There are cases of violence due to sand mining that all too often make the national news.

Sand mining is legal when authorized or permitted by relevant governments, and when existing regulations are followed. Illegal sand mining on the other hand includes all other unauthorized extraction activities. In India, the Ministry of Forests and the Environment (MoEF) has formulated guidelines on sand mining for both riverbeds and coastal areas (Coastal Regulation Zone 2010; MoEF Notification 2010). The Ministry of Mines is also supposed to govern sand mining.

This is a type of conflict where the actors can be, on the one side, poor people worried above all by livelihood issues, honest administrators and policemen, holy men defending sacred rivers, and also conservationists. On the other side are the building industry and unscrupulous providers of raw materials. In a case in Kerala[10] in which turtles were threatened by sand mining, an alliance between conservationists and local fishermen (who saw their beach destroyed) was able to stop sand mining.

The increasing demand for materials for booming real estate and infrastructure projects together with weak governance and rampant corruption, lead to illegal mining of sand and gravel in the rivers of India. The illegality arises because the practice is forbidden as damaging to the environment. It is reported in Madhya Pradesh, Bihar, Haryana, Karnataka, Goa, Andhra Pradesh, Rajasthan, Chattisgarh, Odisha, and West Bengal (12 March 2012, Express News Service). It is also practiced in Kerala, Tamil Nadu, Maharastra, Gujarat, Uttar Pradesh, Uttarkhand, covering the whole country. The Centre for Science and Environment has published several reports on sand mining.

The use of excavators to remove sand causes riverbeds to erode, banks to collapse, damages infrastructure like bridges and transmission lines, causes problems in drinking water systems. Uncontrolled, illegal sand mining has caused depletion of groundwater tables and degradation of groundwater quality.

Along the Konkan coast of Maharashtra, illegal sand mining is taking a toll, causing landslides, destroying large tracts of mangroves and the natural habitats of turtles and crocodiles. As various reports indicate, illegal sand mining generates huge revenues that are shared among different stakeholders. The actors involved in conflicts related to both legal and illegal sand mining in India are usually the local people, nongovernmental organizations (NGOs), contractors, private companies, bureaucrats, and politicians; but others are also involved. The term "sand mafia," as used by the Indian media, refers to groups composed of local politicians, contractors, and bureaucrats, who pursue illegal extraction and resort to physical violence to do so.

On the same coast, in tiny Goa, south of Maharasthra, it was reported in 2011 that sand extraction meant an ecological disaster for villages along the eroded river banks.[11] The boom in real estate fuelled the activity. Activist Rajendra Kerkar said the Mandovi river was already facing the brunt of incessant iron ore transportation. Now, sand extraction at the mouth of a river could let saline water from the ocean rush into its basin while banks of the river would collapse as extraction went on.

Colvale, a village 40 kms away from Panaji, was known for extraction. State Tourism Minister Nilkant Halarnkar, who formerly owned a licence, said that there

---

[10] Kartik Shanker in Mahesh Rangajaran and Ghazala Shahabuddin, eds. Making Conservation Work, Securing Biodiversity in this new Century, Permanent Black, Delhi 2007.

[11] http://zeenews.india.com/news/eco-news/sand-mining-threatening-three-goan-rivers_686300.html.

is hardly any sand left for mining in Colvale. Meanwhile, it was reported that 500 trucks from Goa and 1000 from Karnataka were left without work due to the current ban on sand mining. Cranes are also lying idle, said South Goa Sand Transportation Association treasurer Moti Desai. Before the ban Goa had issued 59 licences for sand extraction but officials conceded their lack of manpower to check flouting of extraction norms.

There is still illegal sand mining in Goa, but also pressure is mounting to import sand from Kawar in Karnataka. In late December 2012, advocate B S Pai alleged that "if sand mining resumes to Goa, sand miners will excavate huge quantities of sand and this will damage fishing activity in the Kali River, threatening the livelihood of thousands of families"[12].

Three other conflicts are briefly presented below showing how different people complain against sand mining, including an environmentalist, a religious priest, and a high-ranking police officer:

1. Awaaz Foundation vs. the Sand Mafia Awaaz Foundation is an environmental NGO based in Mumbai working extensively on raising awareness about the vulnerability of the environment through educational projects in different states of India. Ms. Sumaira Abdulali, Founder of Awaaz Foundation, was physically assaulted on 17 March 2010 by the son and employees of a local politician, who are part of an extensive politically-controlled sand mafia in Maharashtra (17 March 2010, *Times of India*). Awaaz Foundation filed a case at the Bombay High Court through Public Interest Litigation, demanding a ban on sand mining activities along the Konkan coast of Maharashtra. The Bombay High Court banned mining in the Coastal Regulation Zone (CRZ). Moreover, the court ordered the state government to implement the alternative measures mentioned in the report prepared by the prestigious IIT Mumbai (Indian Institute of Technology), which includes reusing sand from building debris and using of environmentally sound techniques for sand extraction.
2. Swami vs. the Uttarakhand State In June 2011, Hindu priest Swami Nigamananda Saraswati died after a 4-month fast in protest of reckless state-sponsored sand mining and stone crushing on the banks of the Ganga, near Haridwar in Uttarakhand. Millions of pilgrims visit this holy place to dip in the river during Kumbh Mela to wash away their sins. A few days before Swami Nigamanand died, the Uttarakhand government ordered a ban on mining activities in the region considered sacred. The ban also followed a directive of the Uttarakhand High Court on 26 May 2011 that expressed concerns over the degradation of the river's ecology and in general the area used for Kumbh celebrations (15 July 2011 Down to Earth).
3. Narendra Kumar (IPS Officer) vs. the Mining Mafia In Madhya Pradesh, Narendra Kumar (Indian Police Service, a high-ranking officer) was brutally crushed

[12] The Herald, 27 December 2012.

to death by a tractor loaded with illegally-mined stones, allegedly by the "mining mafia" in Morena on 8 March 2012. The Chhatarpur district administration ordered the cancellation of all sand mining contracts in the district after the media outrage over the IPS officer's killing, and a second attack on a sub-divisional magistrate and police officials in Panna. Later, the Madhya Pradesh chief minister announced they were handing the murder case of the IPS officer over to the Central Bureau of Investigation (13 March 2012, *Times of India*).

Finally, we consider a case of the people against the sand Mafia in Tamil Nadu already over 10 years ago. The Cauvery River has been seriously impacted by indiscriminate sand mining. The groundwater table has been depleted, rendering the water scarce. Decrease in soil fertility has led to a sharp decline in agricultural productivity, forcing farmers to sell off their lands and allowing miners to dredge the precious sand lying beneath their fields. People who realized their very livelihood were at stake due to mining, took to the streets at the call of the Association for Rural Education and Development (AREDS) on several occasions. Since 1991, AREDS, together with the local people, women's organizations and activists, organized several road blockades. AREDS also filed a case through Public Interest Litigation (5762/90) to the Madras High Court in 1990. As a result, mining was banned in the Cauvery River by the High Court on 25 January 1999.

In conclusion, in comparison with older times, we realize that rapid economic growth coupled with the drive to industrialize, has significantly increased the demand for materials, including sand. Legal sand mining, in line with existing regulations, was not enough to meet the demand generated by booming real estate and infrastructure projects. Riverbed and seashore ecosystems are severely impacted due to sand mining. In our small sample of cases, different stakeholders have drawn attention to the issue, for very different reasons. In Goa and in Maharashtra, environmentalists defended the ecosystem. In Uttarakhand, a religious priest attempted to protect a holy place. In Madhya Pradesh, a high-ranking police officer tried to enforce the law; and in Tamil Nadu local people mobilized to defend their livelihoods. The growing number of conflicts, the court cases against sand mining, and even the number of deaths testify to the dimensions of the problem.[13]

3.9.1 The Ban on Iron Mining in Goa

On 5 October 2012, iron mining was banned by the Supreme Court in the state of Goa. The former Portuguese colony is divided into a prosperous coastal strip with many tourist visitors, and a mountainous area, belonging to the biologically rich Western Ghats from where the water courses come down. Iron mining is done in

[13] There is a recent Master thesis by Kiran Pereira on sand mining in India, at the Department of Geography, King's College, London (2012).

the area between the coast and the Goan portion of the Western Ghats, protected by a series of small contiguous national parks. There were and still are some tribal populations in that area.

The iron ban in Goa is remarkable because open cast production was above 50 million t of low-grade iron per year in 2010 and 2011 (worth above US$4000 million), transported to jetties by a fleet of perhaps 20,000 trucks (owned by small contractors, often employing drivers from outside the state). The iron was loaded on barges at jetties and taken to the ships outside. Most of the iron went to China. The huge dumps of mining overburden invade cultivated fields and forests, and the tailings are deposited in pits with water. Mining produces large amounts of wastewater for washing the mineral that then runs into rivers or into excavated pits. Such scars are very visible when one travels up to the Western Ghats through the mining belt, although most tourists in Goa's beaches (or importers of iron) have never seen them. Although total rainfall may reach 4000 mm per year, there is a long rainless period with high temperatures before the monsoon, with scarcity of water for the coastal area and for the farmers. Therefore, the overuse and pollution of water by the mines are important issues.

The mining companies are of different sizes. At the end of 2012, we visited the main SESA Goa mine in Codli that produced 7 million t of iron per year. Its 800 employees are still paid their salaries, but the large mining area with its tailings pits and overburden dumps, and factory where "fines" and "lumps" were produced, are totally silent. The trucks are parked in the villages. SESA Goa had belonged initially to Mitsui, from Japan, and it was bought by Vedanta from London a few years ago. This is a famous company because of the clash with the Dongria Kondh over bauxite mining at the other extreme of India, in the Niyamgiri Hill in Odisha. Almost all other mines in Goa are Indian-owned and private.

Goa's population is about 1.5 million living in 3000 km^2. The birth rate is low, but there is much immigration into this state. There is a vociferous mining lobby (*The Herald,* 22 December 2012) that sees in iron exports the economic mainstay. There are many open questions: how much of the revenue from mining remains in Goa, what the local fiscal dues should be, who pays for the socio-environmental liabilities from mining (groundwater overuse and surface water pollution, encroachment on farming or protected areas), who would pay for rehabilitation?

The Congress party, the truck and barge owners, the antienvironmental Left, the Unions of bargemen, truckers and miners, all support mining. The companies themselves are somewhat divided. Some hope that they will be declared "legal" or at least less illegal than others, and will be able to resume mining. The present chief minister and the local government belong to the BJP, and some mining companies such as Vedanta itself in Bicholim are building new Hindu temples for the people. Companies claim with that they have invested a lot in other forms of social capital, like schools and hospitals.

The chief minister of the state of Goa, Manohar Parrikar, accepted without complaint the ban imposed by the Supreme Court after the reports of the Justice Shah Commission came out, and he sits on the fence. Manohar Parrikar is in his third term in office, interrupted by spells in opposition. While in opposition he chaired the Public Accounts Committee, and gave to the press its provisional conclusions claiming that Goa government ministers were involved in illegal mining (*Frontline*, 28 (23), November 2011). Afterwards he came back to power on a platform of cleaning up the environment, attacking the corruption nexus, and supporting the tourist sector. Meanwhile activists who coined the slogan *Pani khay Khani* (water or mines) are enjoying their triumph.

The mining ban will have measurable positive effects on traffic accidents and on other aspects of health. In the period leading to the mining ban, there were complaints from tribals and poor farmers. In Cavorem in 2011, they mobilized against atrocities by mine owners and police and against the destruction of fields by mining. But this is not only an "environmentalism of the poor and indigenous." Indeed, the tourist and building interests in the coastal area and an influential part of the media strongly support the ban on iron mining.

In the politics of Goa and in India as a whole, mining and environmental issues have had no priority in elections, despite the many conflicts. While electoral politics is rather silent on environmental issues, there is, however, a lot of "judicial activism" by concerned citizens and environmental groups all over India. In the case of Goa, given the public scandals, the Government of India had set up the Justice M. B. Shah Commission of Inquiry for Illegal Mining of Iron Ore and Manganese in Goa in November 2010. Its task was to determine the fiscal losses from illegal mining, also its negative effects on forest wealth, the damage to environment including water pollution, and the prejudice to livelihood and other rights of tribal peoples and other persons in the mining areas.

Then on 5 October 2012 the Supreme Court of India (its Forest Bench), relying on the findings of the Shah Commission, stopped the mining operations and transport in all iron ore mine in Goa after a petition had been submitted by the Goa Foundation (an environmental action group led by Claude Alvares).[14] The Goa Foundation says that *all* 90 operating mines were to some extent "illegal," if nothing else at least because of lack of compliance with rules regarding forest and wildlife conservation.

The ban on iron mining in Goa followed that in Bellary, in the neighboring state of Karnataka. The Supreme Court had a decisive role, together with environmental activists. The question is now whether the ban will last forever or only for a few years. Perhaps iron mining will start again (although world market prices came down in 2012). Goa's low-grade ores have demand in China but not in other countries or in India itself. Should the ban be permanent, or should there be a "cap" at 10

[14] http://www.thehindu.com/news/national/supreme-court-halts-mining-in-goa/article3968492.ece.

million t per year or perhaps at 20 million t together with an Environmental Rehabilitation Plan? Notice that 10 million t still implies an extraordinary rate of Physical Exports of 7 t per person per year.[15]

3.9.2 *Delhi Waste Wars: From Cradle to Grave*

The industrial capitalist system prospers not only by appropriating as cheaply as possible energy and material resources (in a process of accumulation of profits and capital by dispossession, as David Harvey calls it) but it also needs to dispose of waste as cheaply as possible. The largest waste disposal conflict in terms of volume is that arising from the climate change. Hence the movements for Climate Justice. There are also injustices in the appropriation of water and the environmental services done by water, and therefore there are movements which have been identified internationally as movements for Hydric Justice.

There are problems with urban solid waste disposal. The connection to climate change comes through the methane escaping from non-recycled organic waste going to dumps. There are schemes to collect and burn the methane (which is a powerful greenhouse gas), gaining "carbon equivalent credits."

In common with so many other cities in the world, Delhi and other cities of India are producing more and more solid waste and have policy debates and social disputes on how to manage this waste. Research by Federico Demaria and colleagues looks at the customary "property rights" that recyclers had on the waste. Now being threatened the recyclers form unions. There are unequally distributed advantages and disadvantages in changing from a system of informal recycling by poor people to a formal system of collecting waste and then burning most of this waste in incinerators.

Authorities in Delhi proclaim that waste management is in a state of crisis. They say that waste is commonly dumped in the open illegally and the existing landfills are overcapacity. This narrative portrays the crisis as a failure of management. What should be done, and who gets the costs and the benefits of this change made necessary by the increase in the social metabolism in such a large city?

A modern waste management system, based on subsidizing and supporting the recyclers' unions (as has been done in Porto Alegre in Brazil and other cities) could consist in separating organic waste from the other waste, and then composting it for fertilizer, while the rest of the waste would be separated (even more than at present by "traditional" recycling) into glass, paper, plastics, to be used as raw materials again. This should be done in hygienic conditions. Such alternatives are being tried in other places. In Pune, for example, waste workers are organized in a union, the Kagad Kach Patra Kashtakari Panchayat (KKPKP) (6000 members),

---

[15] Our gratitude to UAB students Aida Vila, Clara Solé, Eloi Puigdollers, Mireia Planell, Miriam Pablos and Xavier Llavina for information on Goa in December 2012.

that has promoted a waste management cooperative, the Solid Waste Collection and Handling (SWaCHCoop) authorized by the Pune Municipal Corporation to provide door-to-door collection.

Instead, what is proposed in Delhi is something unprecedented in the long history of the city, namely, to dismiss the recyclers "expropriating" or "dispossessing" them from their customary rights to collect and make some money from the waste, bringing most of it to incinerators that would produce some electricity. The critics point out, in India as in Europe, that large-scale incineration (even if disguised as "energy recovery") is an incentive to produce more waste instead of moving towards a "zero waste" objective. Out of sight and into the fire also means out of mind, at the risk, however, of dioxin production if the process of incineration is not done well. Also, after incineration, there remains still about one third of ashes as waste that needs to be disposed of. Besides, incineration is technically difficult (apart from unjust and wasteful) when the organic fraction is large (over 50 %), as it is often the case in poor cities. This is because incineration generally requires a moisture content of less than 30 % (food waste is generally 75 % water). The higher is the moisture content, the more smoke and pollution will be produced due to incomplete combustion.

The social metabolism starts with resources but ends with waste. The conflicts on waste management in Delhi are between, on the one side, the "traditional" recyclers and the neighbors who distrust the Ghazipur and Okhla incinerators, and on the other side, the city administration and the private sectors interested in making profits from the various stages of waste management. Federico Demaria, Shashi B. Pandit, and Seth Schindler argue that the informal sector should be incorporated into an efficient and equitable waste management system that is also environmentally sustainable. Their article is an example of action research, coauthored by the secretary of the "traditional" recyclers union, the All India Kabadi Mazdoor Mahasangh (AIKMM). The incineration alternative (called "waste to energy facility") poses a major threat to the livelihoods of waste workers because they must increasingly compete with private firms for ownership and control over recyclable waste at multiple stages. There are approximately 150,000-waste workers in Delhi, who belong to unprivileged communities and cannot easily find alternative livelihoods. These workers provide environmental services in the form of high level of recycling in working conditions that are extremely hazardous, that could be made more just and safer (Schindler et al. 2012).

3.9.3 Shipbreaking in Alang

Finally, another type of conflict that takes place in India (as also in Bangladesh) arises from one form of waste disposal that has directly to do not so much with the internal metabolism of the Indian economy (although it contributes to it) as with what we call "Lawrence Summers' Principle." The well-known economist, when he worked at the World Bank, wrote a memorandum that was leaked to the press. *The*

Economist (8 February 1992) titled the story as "Let them eat pollution." The memo recommended putting polluting industries in areas without people or where people were poor, because the costs of illness or mortality of poor people were lower than those of richer people. Any insurance company would agree. Lawrence Summers argued "from a strictly economic perspective."

The shipbreaking yards at Alang-Sosiya are practical applications of Lawrence Summers' principle. Shipbreaking is a successful case of cost shifting, or in other words, profit accumulation by contamination.

This business shows the ugly face of globalization, although in terms of economic value added and in terms of raw materials recovered, it is not so important. Over 500 big ships per year (the quantity depending on the world economic cycles) reach the beaches of Gujarat, are grounded at high tide, and they are dismantled by a legion of manual workers. In 2012 Alang was again in the news when the notorious Exxon Valdez after many changes of flag (its last name was "Oriental Nicety"), reached Alang for final demolition.

Federico Demaria's work[16] written with activist Gopal Krishna, explains that more than 80 % of international trade in goods by volume is carried by sea. The shipping industry constitutes a key element in the infrastructure of the world's social metabolism. Ocean-going ships are owned and used for their trade by developed countries but are often demolished, together with their toxic materials, in relatively poor countries. Ship breaking is the process of dismantling an obsolete vessel's structure for scrapping or disposal. Ship owners and ship breakers obtain large profits shifting the environmental costs to workers, local farmers and fishermen, and their families.

The international and national uneven distribution of power has led to an ecological distribution conflict. The valuation languages deployed can be analyzed. The Supreme Court of India has been called on more than one occasion to decide on the costs and benefits of ship dismantling in Alang-Sosiya because of appeals through public interest litigation. Are the benefits of ship dismantling (the jobs, the recycled steel) larger than the costs to the local environment and to human health, since the dismantled ship carry asbestos, heavy metals? Are such costs and benefits commensurable? For instance, there was a controversy at the Indian Supreme Court in 2006 over the dismantling of the ocean liner "Blue Lady," showing how the different languages of valuation expressed by different social groups clashed and how the language that expresses sustainability as monetary benefit at the national scale, dominated.

Here we see not only that "the poor sell cheap," and that capitalism is an economy of unpaid social and environmental costs. We also see that at a small scale how pertinent are the questions that Kothari and Shrivastava ask about the costs of economic growth in India as a whole. Another growing stream of waste from North to South is electronic waste. The environmental justice organization, Toxic Links, is trying to keep track of these flows in India.

---

[16] Shipbreaking at Alang-Sosiya: an ecological distribution conflict, Ecological Economics, 70(2) 2010). See also http://www.shipbreakingplatform.org/shipbrea_wp2011/wp-content/uploads/2012/05/120410_Ejolt-1_Low2.pdf.

After Alang-Sosiya, the second largest ship-breaking yards are in Chittagong in Bangladesh. Similar controversies have arisen. The Bangladesh Environmental Lawyers Association (BELA) convinced the Supreme Court in 2009 to ban all ship breaking not meeting certain environmental standards. The industry stopped in 2010, but then pressure from the government and from the Bangladesh Ship Breakers Association led to about 150 ships being dismantled there in 2011.[17]

3.10 Sustainability Indicators

In India, with the exception of the HANPP, which is very bad, many other indicators are still good, *per capita*. Based on Singh et al. 2012, we foresee that biomass is unlikely to increase very much. But it is also unlikely to decrease as other demand for biomass substitutes for decreased fuelwood and decrease fodder for cattle. Biomass might increase because of more wood, paper, and meat consumption, although at a slow rate. Agrofuels will remain marginal. Historically, it is interesting to compare the slowly moving trends in biomass from the rapid moving trends in other materials.

Material flows per capita, driven by building materials (hence so many conflict on sand mining), mineral ores for metals, and fossil fuels, will increase with income, tending towards 10 t per capita in 15 years at current rates of growth (still below EU average), with important internal and international impacts.

This chapter has explained the methods used to count the material flows in the economy, giving the main results for the economy of India between 1961 and 2008. Drawing on work done by colleagues of the EJOLT project, brief illustrations of the links between social metabolism and ecological distribution conflicts have been given, looking at clashes over illegal sand mining in India, responses in Odisha to bauxite mining, the iron mining ban in Goa, social disputes on waste incineration in Delhi, and valuation contests on ship dismantling in Alang, Gujarat.

Instead of anecdotal evidence as in the present chapter, what is needed is a historical reconstruction of hundreds or indeed thousands of environmental conflicts in India (classified as biomass conflicts, mining, and fossil fuels conflicts, waste disposal conflicts, etc.), showing whether they increase in number and intensity with the growth of the social metabolism, what the outcomes have been, how they have been solved by technological modernization or by repression, by displacement, by criminalization of activists, or by monetary compensation for the "externalities," etc. The historical significance of movements of environmental resistance has been analyzed by historians of India on many occasions already (conflicts on tree plantations or on deforestation, which are therefore conflicts on the HANPP; conflicts on fisheries; conflicts on water use; conflicts on urban space and amenities; conflicts on mining; conflicts on pesticides and on transgenic crops, etc.).

---

[17] "Ship breaking in Bangladesh. Hard to break up," The Economist, 27 October 2012.

In a pioneering article in 1988 linking ecological economics and political ecology, Jayanta Bandyopahdyay and Vandana Shiva, who had already published remarkable articles against eucalyptus plantations, provided a theory of what they called the Political Economy of Environmental Movements. Citing repeatedly Nicholas Georgescu-Roegen, they showed the incapacity of economic theory to deal with resource exhaustion and with pervasive externalities, they made fun of Solow's phrase of 1974 that the economy could get along without natural resources, they pointed out that environmental movements made the externalities visible (as Enrique Leff had written in 1986 in *Ecología y Capital*), they asserted that care of the environment was not a "luxury of the rich" and explained (following N.S. Jodha) that human survival in India was directly dependant for many people on the direct utilization of natural resources held in common. They listed a number of historical and current movements such as the Chipko and Appiko movements against deforestation and tree plantations, the movements against limestone quarries in the Doon Valley and in Almora and Pithoragath in Uttarakhand, the early successful movements against bauxite mining in the Gandhamarardan hills in Odisha against the state company BALCO (Bharat Aluminium Company), the conflicts on coal mining in Singrauli ("the energy capital of the country," a terrible place), and also the movements attempting to stop dams whether in the Silent Valley in Kerala (whose motivation was biodiversity conservation rather than the survival of the people), the Tehri Dam in the Himalaya, and other movements fighting submersion in Bedthi, Inchampalli, Bhopalpatnam, Narmada, Koel-Karo, Bodhghat. Among the authors quoted, there was already Medha Patkar (Bandyopadhyay and Shiva 1988).

Current conflicting coal mining cases are Jharia in Jharkhand where another state company, Bharat Coking Coal (BCCL) operates the collieries in a landscape of underground fires and land subsidence, or the Mahanadi Coal Fields in Odisha. In Jharkhand, Uranium Corporation of India (UCIL) mines uranium. An outstanding documentary, "Buddha weeps at Jadugoda," showed birth malformations negated by the company and corroborated by Xavier Dias. Iron mining in Bailadila in Chhatisgarh is another socio-environmental disaster, narrated as the preceding cases in one of the chapters of *The Caterpillar and the Mahua Flower* edited in 2007 by Rakesh Kanshian.[18] Note the following title: the mahua tree of the central and eastern zone of India provides a nice drink for tribal peoples while the caterpillar is here not a life form but the fossil-fuel driven machine.

There is much accumulated knowledge on socio-environmental conflicts coming from "activist knowledge" since the 1980s in the Centre for Science and Environment's citizens' reports of 1982 and 1985. There must be also an enormous variety of regional sources. What would be now required are studies at state or regional levels of the trends in social metabolism (at least, the domestic material extraction and the physical trade balances), including also historical statistics on the increase in the HANPP, on water use, on energy flows. And then see the connections to changing

[18] Available at http://www.panossouthasia.org/pdf/Caterpillar%20and%20the%20Mahua%20Flower.pdf.

patterns of environmental conflicts. For instance, we are aware of the conflict around Bangalore between sustainable farming and the brick-making industry (Shrivastava and Kothari 2012, p. 143). Farmers are enticed or forced to sell off the top soil for brick-making, destroying the productive capacity of their farms—"over 4000 truck loads from farms and irrigation tank-beds are removed every day for the building industry." To what extent does this Bangalore specialty go on in other places?

3.10.1 Climate Change

Per capita carbon emissions and (therefore) the ecological footprint, amount in India to only one third of the global mean, and about an eighth of industrialized economies. The feeling of injustice expressed by Anil Agarwal and Sunita Narain in their influential booklet of 1991, *Global warming in an unequal world: a case of environmental colonialism;* the subsequent movements claiming the "ecological debt" (and Climate Justice) had strong resonance in other countries of the South, although they have not succeeded in shaming the rich countries in paying back such liabilities, not even in stopping their growth.

Internationally, India's government and citizens claim with reason that the country has, by its low per capita consumption, made a contribution to world sustainability. With a population that is almost one fifth of the global total, India currently uses only 10 % of the global supply of material resources (in terms of tonnage) and 6 % of global primary energy supply. Even though India's per capita level of resource use and emissions is strikingly modest, India's requirements are not negligible already at present. Internationally, an India whose carbon dioxide emissions per capita would increase from 2 t before 1990 to a European average of 10 t would have (ceteris paribus) a most significant impact on world climate.

Internally, there is need to discuss what is happening to the "standard of living" of the citizens of India in the midst of their "thermo-industrial revolution." The GDP per capita is growing fast and therefore the HDI must be improving rapidly. How do they compare to other indicators—rapid loss of biodiversity, increasing ecological footprint? A task of historians with training in ecological economics (and also a task for the Planning Comission) would be to do Multi-Criteria Assessments over long periods of the economy of India.

Given the fact that India might hold nearly one fifth of humanity when world population (optimistically) peaks by 2045 or 2050 at 8.5 billion people, one can easily foresee the non-negligible impact that the social metabolism of the growing Indian economy (largely fuelled by coal) will have on world environmental pressures. To give room for India, China, and the rest of the world, the rich countries should decrease their social metabolism, this could be hopefully achieved by moving towards a steady-state economy preceded by a period of moderate degrowth in material and energy use. There is a possible alliance between the small *décroissance, post-Wachstum,* "prosperity without growth" movements in some Northern countries (they ask themselves, how much should a person consume?), and the large and growing world movements for environmental justice (Martinez-Alier 2012).

3.11 Conclusion

The planet is being plundered through economic growth, the search for profits, and high levels of consumption by some of the world increasing population. Meanwhile, since the planet is limited in size; the frontiers of resource extraction and waste disposal are reaching the farthest corners. The movement to impose market values and increase profits by expanding the frontiers of capitalism is resisted by counter-movements (as Karl Polanyi already explained in *The Great Transformation* in 1944) aiming to protect nature and humans. One can see India today according to Polanyi's view. One could also appeal to James O'Connor theory of a "second contradiction" of capitalism put forward in 1988. Or to K. W. Kapp's view in 1950 of externalities not so much as market failures as cost-shifting "successes" that nevertheless sometimes cause social complaints. Or to David Harvey's 2003 notion of "accumulation by dispossession" to which one could add "accumulation by contamination" thinking of Alang-Sosiya and of e-waste imports into India. In Marx's theory, primitive or original accumulation of profits and capitals by colonial robbery and enclosures would be substituted in due course by exploitation of labor as the capital/labor ratio in the economy increased, and new technologies enhanced labor productivity together with new methods of disciplining wage labor. Harvey's footnote to Marx emphasizes that primitive or original accumulation has never been so important as today.

Movements of resistance are, on their side, rarely successful. Democracy in India is systematically undermined by the very processes of internal colonialism in an unavoidable type of economic growth which is intensive in energy and materials use. There is corruption and violence in large-scale land grabbing. Fortunes are made by changing land zoning in urban areas and also in auctions of coal leases (and also in other less-materially intensive cases such as corrupt auctions of communications spectrum).

In the countryside, in areas with strong Naxalite presence (a rebellion of Maoist origin that predates the economic boom), like Chhatisgarh, there are allegations (Shrivastava and Kothari 2012, p. 224) that extractive companies Tata and Essar have helped to pay for brutal counter-insurgency forces (the Salwa Judum) recruited from local populations.

Ramachandra Guha is on record strongly criticizing the Salwa Judum. He would agree also to the fact that in regard to mineral and fossil fuels there is no "dematerialization" in the economy of India (even relative to GDP) while biomass extraction decreases relative to GDP but still increases in absolute terms. We know well and share Ramachandra Guha's enthusiastic defence of the largely successful efforts made since 1948 in the building of the world's largest democracy with regular elections, no officially imposed religion, little military political power, respect for minority linguistic rights, a free press, etc. Would he agree that the population grew too much over that period, this being a failure not only of public policies but also of collective resolve? Women's freedom has been less that desired in a democracy. When Gandhi asked in 1927 what would happen to the world if India, with a population of 300 million, would follow industrialization on Britain's steps, he did not foresee an India (plus Bangladesh and Pakistan) that would exceed 1500 million in 100 years time. Did Gandhi make a mistake by not joining E.K. Ramaswamy's and Margaret Sanger's neo-malthusianisms of the 1920s?

India should take pride in the fact that its material throughput per person is still at 5 t per year while the EU is at 15 t. But such numbers are politically meaningless and such pride is largely absent. On the contrary there is emphasis in many circles in India on the glories of economic growth, leaving aside the metabolic implications. However, the links from social metabolism to resource extraction and waste disposal conflicts at different scales are a reality that we have only superficially explored in this article. The growing social metabolism causes internally great environmental and human livelihood losses, and also, increasingly, negative external effects at world level.

One should look at the historically changing social actors and types of resistance in such conflicts, as has been done in historical work on ecological distribution conflicts (as when Ramachandra Guha compared Garhwal and Kumaun in *The Unquiet Woods*), looking also at the changing valuation languages deployed. For instance, when and why did the Kanchan Chopra committee at the prompting of the Supreme Court establish methodologies to count the Net Present Value (NPV) of destroyed forests, and what have been its effects? Indeed, has the Supreme Court the power to impose the single language of valuation of the NPV, and at which discount rate? On the contrary, which are the regional and historical patterns of development projects stopped in India by the use of "sacredness"? Or, for instance, has the use of adivasi rights been increasingly effective to stop development projects, as the use of Convention 169 of International Labour Organization (ILO) has been in Latin America? Such studies would be important contributions to the history of the global environmental justice movements.

Since the 1970s and 1980s authors and groups in India have taken part and have sometimes led movements for environmental justice at local, national, and international levels. There is now a global Climate Justice movement, there is a growing Hydric Justice (Water Justice) movement, there are networks with strong Indian presence sharing strategies in the combat against biopiracy or against particularly aggressive international firms (Coca Cola in water extraction). Such environmental justice movements contribute to democracy and they constitute the strongest forces for environmental sustainability.

Instead, there is too much emphasis among policy makers on hypothetical economic valuations of environmental damages and on economic instruments, and too little on this great tide of environmental justice. When the stakeholders in such conflicts do not insist so much on economic compensation for externalities as on different local alternatives (as explained in the second part of Shrivastava's and Kothari's book 2012) they join those in Latin America searching for a *buen vivir* or *Sumak Kawsay,* perhaps translatable as *aparigraha,* a voluntary simplicity rooted in local social values, and they also join and support those few in rich countries who preach a moderate *décroissance* (french for degrowth) leading to a "steady state economy" or a "prosperity without growth."

References

Asher M (2009) Striking while the iron is hot: a case study of the Pohang Steel Company (POSCO) proposed project in Orissa. National Centre for Advocacy Studies, Pune, India. https://stoposco.files.wordpress.com/2009/09/striking-while-the-iron-is-hot_posco-case-study.pdf

Bandyopadhyay J, Shiva V (1988) Political economy of ecology movements. EPW 23(24): 1223–1232

Fernandes W, Ashif JV (1997) Development induced displacement and rehabilitation in Orissa, 1951 to 1995: a database on its extent and nature. Indian Social Institute, New Delhi, pp 72–80

Guha R, Martinez-Alier J (1997) Varieties of environmentalism. Essays north and south. Oxford, UP

Hall CAS, Cleveland CJ, Kaufmann RK (1986) Energy and resource quality. The ecology of the economic process. New York, Wiley Interscience

Hornborg A (1998) Towards an ecological theory of unequal exchange: articulating world system theory and ecological economics. Ecol Econ 25:127–136.

Hornborg A (2007) Footprints in the cotton fields: the industrial revolution as time-space appropriation and environmental load displacement. In: Hornborg A, McNeill J, Martinez-Alier J (eds) Rethinking environmental history. World-System history and global environmental change. Altamira Press, Lanham, MD

Kosoy N, Corbera E (2010) Payments for ecosystem services as commodity fetishism. Ecol Econ 69:1228–1236

Martinez-Alier J (1987) Ecological economics: energy, environment and society. Blackwell, Oxford

Martinez-Alier J (1995) The environment as a luxury good or "too poor to be green"? Ecol Econ 13(1):1–10

Martinez-Alier J (2005) The environmentalism of the poor. A study of ecological conflicts and valuation. Oxford, Delhi

Martinez-Alier J (2011) The EROI of agriculture and its use by via campesina. J Peasant Stud 38(1):145–160

Martinez-Alier J (2012) Environmental justice and economic degrowth: an alliance between two movements. Capital Nat Soc 23:51–73

Mishra B 2010 Agriculture, Industry and mining in Orissa in the post-liberalization era: an inter-district and inter-panel analysis. EPW XLV(20) May 15. http://www.epw.in/special-articles/agriculture-industry-and-mining-orissa-post-liberalisation-era-inter-district-and-i

Norgaard RB (1994) Development betrayed: the end of progress and a coevolutionary revisioning of the future. Routledge, London

Padel F, Das S (2010) Out of this earth: East India adivasis and the aluminium cartel (foreword by Arundhati Roy). Orient Black Swan, Hyderabad

Pimentel D et al (1973) Food production and the energy crisis. Science 182(4111):443–449

Sachs W (ed) (1991) The development dictionary: a guide to knowledge as power. Zed Books, London

Schindler, Seth, Demaria, Federico, Pandit, Shashi B (2012) Delhi's waste wars. EPW XLVII(42): 18–2. http://www.ejolt.org/2012/02/watch-a-gaia-ejolt-video-on-waste-wars-in-delhi/. Accessed Feb 2013

Singh AA, Mahongnao M, Demaria F, Krishna VV (2012) Illegal sand mining conflicts in India, EJOLT factsheet by CSPP, JNU. http://www.ejolt.org/wordpress/wp-content/uploads/2012/11/Mining-Report-low.pdf. Accessed Feb 2013

Singh SJ, Krausmann F., Gingrich S, Haberl H, Erb K-H, Lanz P, Martinez-Alier J, Temper L (2012) India's biophysical economy, 1961–2008: sustainability in a national and global context. Ecol Econ 72:60–69

Temper L, Martinez-Alier J (2007) Is India too poor to be green? EPW XLII(17). http://www.epw.in/commentary/india-too-poor-be-green.html

Temper L, Martinez-Alier J (2013) The god of the mountain and Godavaram: net present vale, indigenous territorial rights and sacredness in one bauxite mining conflict in Odisha, submitted to Ecological Economics. Ecological Economics. 96, December 2013 (Pages 79–87). http://www.sciencedirect.com/science/article/pii/S0921800913002978

TEEB (2008) The economics of ecosystems and biodiversity: an interim report. European Commission, Brussels

TEEB Foundations (2010) The economics of ecosystems and biodiversity: ecological and economic foundations. Kumar P (ed) Earthscan, London

Chapter 4
Current Status of Environmental and Economic Accounting: Review of Some Countries Experiences and Way Forward for India

M. N. Murty and Manoj Panda

4.1 Introduction

Linking natural resources and economy, and efforts to find out contribution of natural resources to economic development can be found in the economic policies pursued by some countries even from the start of the second half of the last century. It all started with an attempt to correct shortcomings of the United Nations (UN) framework System of National Accounts (SNA) released in 1968 in the context of treatment of income from unsustainable use of natural resources (Peskin 1989, Peskin and Lutz 1990; Repetto 1989; Dasgupta 1990; Bartelmus et al. 1991; Dasgupta and Maler 1991). The developed countries taking a lead in natural resource accounting include Norway, Sweden, Netherlands, France, Canada, Germany and the USA. Norway, Netherlands, France and Canada were early users of natural resource accounting, developing country-specific methods during the 1970s and 1980s to meet their own economic and environmental policy requirements. Germany has taken a lead on material flow accounts (MFA) and land accounting. Attempts of natural resource accounting can also be found earlier in some developing countries like the Philippines, Indonesia and Namibia.

The methodology of integrated environmental and economic accounting first appeared as *Handbook of National Accounting: Integrated Environmental and Economic Accounting* published by the UN in 1993. The most recent version of the System of Environmental–Economic Accounting (SEEA) appeared in 2003 as a joint publication of the UN, the International Monetary Fund (IMF), the Organization of Economic Cooperation and Development (OECD) and the Statistical Office of the European Communities. The next version of the SEEA may appear soon with further revisions being attempted by the experts from different countries. This methodology has been developed out of decades of rich experience of several countries and international organizations in natural resource accounting. Even after almost two decades of its existence, it is now found that no country in the world has

M. N. Murty (✉) · M. Panda
Institute of Economic Growth, New Delhi 110007, India
e-mail: mn.murty71@gmail.com

N. Ghosh et al. (eds.), *Nature, Economy and Society*,
DOI 10.1007/978-81-322-2404-4_4

made full use of this methodology for estimating its green gross domestic product (GDP). Given that the full implementation of this methodology in national income accounting could provide an estimate of the comprehensive measure of a country's green GDP, it is therefore hard to find as of today reliable green GDP estimates even for a few countries in the world.

The remaining chapter is planned as follows. Section 4.2 reviews past practices and current initiatives in natural resource accounting by some developed and developing countries. Section 4.3 discusses attempts of some developed countries to integrate their current practices of natural resource accounting with the SEEA. Section 4.4 presents a brief account of some resource-specific studies attempted for implementing the SEEA methodology. Section 4.5 provides some suggestions for implementing the SEEA methodology in India. Finally Sect. 4.6 contains the conclusions.

4.2 Country Experiences of Resource Accounting: Lessons from Past Practices and Some Current Initiatives

4.2.1 Developed Countries

Norway is among the first countries to develop a system of natural resource accounting. Its work began in the 1970s in response to the Club of Rome's publication of *Limits to Growth* and a growing environmental movement. The Norwegian system of resource accounting (SRA) was developed with twin objectives: to develop regular reviews of the volume, quality and use of natural resources in the form of "resource accounts"; and to coordinate and present proposals regarding the future use of such resources in the form of "resource budgets". The resource-budgeting provision is peculiar to the Norwegian SRA, especially given that the accounts were developed entirely in physical units. The exclusion of monetary resource accounting grew from a Norwegian perception that monetary accounting may require the exclusion of vital resources from the framework and the use of unreliable valuation methodology.

In the early phase of natural resources and environmental accounting in Norway, accounts were prepared for a large number of natural resources. These included energy, minerals, sand and gravel, forests, fish, land use, fresh water, air pollution and waste. The accounts were published in annual reports from 1981 (Central Bureau of Statistics 1981). The material resources accounts were kept in physical units and consisted of three parts covering (1) reserves or capital accounts; (2) extraction, conversion and trade accounts; and (3) end use accounts of the resources. These accounts designed much earlier, provide a structure of information of natural resource accounting similar to the SEEA methodology which was developed much later. The salient features of these accounts are: (1) They provide information of stocks or reserves of the resources as well as their end uses; (2) they are provided in physical units with information of market prices, if available; (3) they are classified such that natural resource accounts and SNA accounts are interlinked; (4) they are prepared such that specific characteristics of resources in terms of reserves and end uses are taken into account[1] and (5) most of the natural resources and environmental accounts were constructed using data from secondary sources.

A close look at natural resource accounting practices in Norway during the last two decades of the twentieth century reveals that significant efforts were made to integrate SNA and natural resource accounts as we currently see in the SEEA for estimating the green GDP. Norway has been making some limited efforts to implement the SEEA, especially in the context of developing a national accounting matrix including environmental accounts (NAMEA) as in other European countries like the Netherlands.

Sweden has been another European Union (EU) country very rigorously pursuing environmental accounting starting in the 1990s. Energy and air pollution accounts and climate change models constitute the core of environmental accounting of Sweden. The general equilibrium model of Swedish economy has been extended to include an environmental module. This helps to link emissions to productive sectors of the economy and assess the economic effects of different environmental objectives including Kyoto Protocol targets for all economic activities.

Attempts have been also made to estimate the green GDP for Sweden. Some work has been done towards it, essentially following the SEEA approach. Monetary values for the depreciation of natural resources and the costs of preventing further pollution by Swedish households and industry have been estimated on a regular basis. These estimates have been presented as shares of the GDP. Environmental valuation studies have been done in Sweden for assessing the costs imposed by pollution. These studies provide estimates of costs due to forest loss, crop loss, health impacts and declines in real estate values.

The Netherlands is pioneer in considering the extension of national input–output tables (IOT) for linking economic and natural resource accounts. It has developed the NAMEA. The NAMEA builds on the IOT of the national income accounts by introducing additional columns containing physical data on air pollutant emissions by sector. The tables also include imports of pollution from outside the national boundaries and exports beyond national boundaries. Using the NAMEA, some environmental indicators are created pertaining to international effects (greenhouse effect and ozone layer depletion) and domestic effects (acid rain, eutrophication and waste) of pollution. The NAMEA accounting system has been adopted by the EU, and the European Commission has been providing financial support to all countries to develop their own NAMEA systems.

[1] A biotic resource like fish required a relatively detailed reserve or stock account with specification of age structure and localization of the different fish stocks. The end use part of the accounts is, however, quite simple since relatively few sectors use fish as an input factor in their production. For other resources, like energy, the situation is different since energy is an important input factor in almost all sectors of the economy requiring detailed end use accounts, while the reserve account could be kept relatively simple.

In the scheme of the NAMEA mainly two types of extensions could be distinguished (see European Commission 2006):

a. Natural resources covering mineral and energy resources, water and biological resources flow from the natural environment into the national economy.
b. Residuals consisting of emissions to air, water and soil are the incidental and undesired outputs from the economy discharged into the environment.

In the environmentally extended input–output table (EEIOT), environmental extensions are considered in the form of satellite accounts. The conventional monetary IOT remains as it is and non-monetary environmental extensions are attached in the form of separate accounts underneath the monetary accounts. The satellite accounts of environmental extensions consist of an input matrix of environmental extensions and an output matrix. The inputs are primary natural resources (gifts from nature) and the outputs are emissions. The EU countries could use EEIOT for environmental policy analysis in three ways:

1. Environmental problem analysis: Analysis of the nature and causes of environmental problems, as related to resource use and emissions relevant for policy. For example, it includes the analysis of life cycle environmental impacts of product groups (cars, meat, houses, etc.), consumer groups (urban vs rural dwellers) and impacts related to primary resources used (oil, iron ore, coal, wood, etc.).
2. Prospective effect analysis of policies: ex ante prediction of effects of policy measures. These include economy-wise environmental implications of changes in lifestyle and consumption expenditure patterns, technological changes of products and processes, emission reduction measures, price effects of environmental taxation and other ways of internalizing the environmental externalities.
3. Monitoring and ex post effect analysis of policies: ex post analysis of impacts and effectiveness of policy measures. It involves the analysis of the relationship between environmental impact (emissions, material requirement) and economic output.

France has also been another country adapting environmental accounting as early as the 1980s. It developed its own method of environmental accounting resulting in what are known as the patrimony accounts. Three distinct accounts known as resource, place and agent accounts are prepared which are linked for policy analysis. Resource accounts in the form of stock and flow accounts are measured in physical terms. Apart from natural resource accounts, patrimony accounts also include cultural resources and any other assets. Geographical accounts provide physical data about assets organized by location and by ecological and land characteristics. The agent accounts show identity and actions of agents/people and institutions using the resources.

Germany is one of the EU countries which started working on environmental accounting as early as the 1980s. The main focus is on developing physical accounts of materials and energy flows, flows of air and water pollution, solid waste and land use by industry. Germany is a leader in the development of MFA and provides physical data of flows of materials. It has implemented a large part of the SEEA placing a main emphasis on physical accounts. This has been done by extending physical input–output tables (PIOT) for integrating natural resource accounts with production accounts and by constructing the NAMEA. The PIOT describe the flows of materials and energy within the economic system, and between the economic system and the natural environment. These flows include changes in the natural environment caused by human activities like using natural assets as source of raw materials and as sink for residuals. It consists of input tables to show sector-wise use of different materials, output tables to show different sectors producing goods and materials as residuals and material integration tables showing the material flows between the sectors. All the flows are measured in physical units. NAMEA-type flow accounts have been prepared for different sectors like energy and emissions, materials, land and environmental protection expenditures. Germany publishes annually a *Report of the German Environmental–Economic Accounts* with analysis and detailed tables.

Canada has initiated in the late 1970s a framework of action in order to coordinate the collection, storage and manipulation of environmental statistics (Rapport and Friend 1979). The framework was called STRESS[2] with the ultimate goal to establish quantitative links between environmental systems and economic systems. Subsequently, the Environment and Natural Resources Division at Statistics Canada has started to implement a separate framework for natural resource accounting which is much more limited than the previous one with two main goals: to produce satellite accounts for natural resources and to value these natural assets for inclusion in the national balance sheets of the SNA. In this framework, only the economic reserves of commercial resources (resources which could be exploited at current market prices with available technology) form the core of the resource accounting system, at least in so far as the connection between the resource and economic accounts are concerned (Hamilton 1989). The natural resource accounting programme of Canada is named "Econnections: Linking the Environment and the Economy". This programme includes work on natural resource accounts, environmental protection expenditures, the environmental protection industry, MFA, and so on. Canada publishes all of its environmental accounts data, along with a comprehensive description of the methods used. Its publication entitled *Human Activity and Environment* provides both statistical data and syntheses of the links between economy and environment.

The *USA*, unlike the EU countries, has not made any efforts for formally constructing natural resource accounts. This has happened in spite of the existence of a rich data system in the country linking natural resources and various economic activities. The Environmental Protection Agency (EPA) provides databases on pollution control and emissions that could provide basic data for some parts of environmental accounting. Attempts have been made in the USA to integrate the EPA databases on emissions into a system analogous in some ways both to the NAMEA systems of many European countries and to the micro-level use of MFA. Attempts have also

[2] It is so named because the structure of the system is based upon the process of anthropogenic environmental stress and response and the implications of this process for human populations.

been made to build national MFA similar to the accounts in Germany described above. Even in the absence of an environmental accounting framework, considerable work has been done in the to assess the costs and macroeconomic impacts of greenhouse gases, estimate the physical stocks and flows of a wide variety of minerals and provide physical data of forests.

4.2.2 Developing Countries

Developing countries had not started working on developing databases linking natural resources with economic activities until the turn of the twenty-first century. However, one can find some case studies of selected sectors in some countries which were attempted independently of the government. A study by Repetto et al. (1989) for Indonesia was one of such attempts. This study looks into the problem of how some measure of natural resource depletion might influence estimates of the national income of Indonesia. It compiles accounts of stocks and flows of resources over time for timber, petroleum and soil resources. A measure of the stock at the beginning of the accounting period was listed together with the unit price and an estimate of the total value for each resource. Changes in the resource stocks are recorded together with changes in the unit cost over the accounting period. At the end of this period, final stock, closing unit price and final total value of the resource stock are recorded. This study assessed the effects of natural resource depletion on macroeconomic indicators of the Indonesian economy during 1971–1984. It is found that making corrections for resource depletion reduces the annual rate of growth of the GDP of Indonesia from 7 to 4 % during this period.

A study by Gilbert (1990) provides another case study of resource accounting in the context of developing countries. She suggested a framework consisting of three accounts: resource user accounts, ecological or stock accounts and socio-economic accounts and attempted a case study of natural resource accounting for Botswana. The system of resource user accounts is similar to an input–output approach describing the input of natural resources and environmental services to various economic sectors and the output of materials from these sectors to the natural environment. User accounts pertain to fisheries, livestock, crops, forestry, mining, conservation, recreation, water storage, urban transport and waste disposal. In these accounts, inputs are described as stocks of resources, effort, infrastructure, investment and government policy, while the outputs are yield, income, value added and environmental impact. The resource user accounts are prepared as both physical and monetary accounts. The ecological or stock accounts are classified as abiotic consisting of accounts of air, land, water and subsoil resources and biotic having accounts of ecosystems and ecosystem components. These accounts are only physical measures of resource stocks and changes. The socio-economic accounts have three subcomponents relating to economic, demographic and policy aspects of the economy. The economic accounts are derived from the SNA with reclassification of certain sectors. The demographic accounts describe aspects of societal interaction with the environment, as for example information on direct relationships between local populations and natural resources. The policy accounts provide a description of existing government environmental policy.

The methodology of resource accounting described in Gilbert's and other early studies (Repetto et al.1989; Peskin 1989; Theys 1989) on the subject have paved the way for gradual development of more comprehensive and practical approaches of natural resource accounting during the last decade of the twentieth century and the first decade of the twenty-first century. The country-specific experiences of resource accounting in developed countries briefly described above and independent studies attempted by researchers during the end of the past century have helped to understand more clearly the relationship between economy and natural resources. This has led to undertake a number of studies in different countries and international organizations (UN; EU) to attempt necessary changes and extensions in conventional methods of measuring the GDP of a country in order to properly account for the contribution of natural resources to the well-being of the people. The currently developing methodologies of extended input–output tables for accounting of environmental externalities of the European Commission and the SEEA are offshoots of some of the past experiences in this area of research.

The *Philippines* is one of the first among developing countries that tried to implement the SEEA in national income accounting. In the mid-1990s, it received financial and technical support from the UN for implementing the SEEA. New methodologies of "green accounting", proposed as a satellite of the conventional SNA, were implemented by the National Statistical Coordination Board (NSCB) in collaboration with other national agencies and with the technical support of the UN Statistics Division (UNSD). This project has produced asset accounts of natural resources and estimates of the GDP for the Philippines. Prior to this work, there were also attempts publishing overall macroeconomic accounts for the years 1988 and 1992 including environmentally adjusted GDP estimates prepared, of course, using the Peskin approach which differs from the SEEA.

Estimates of physical and monetary accounts of SEEA asset accounts are prepared for a comprehensive list of natural resources mostly for the period covering 1988–1999. The list of resources includes land, minerals, coal, water, forests and fisheries. Also estimates of environmental protection expenditures of private and government sectors are obtained for the period of 1994–1998. These accounts show only the commercial or economic value of natural resource stocks as per the SNA and do not consider non-marketable environmental values of use and non-use values and option value. In the case of accounting for pollution, the SEEA application in the Philippines uses the maintenance cost method of valuation. The SEEA defines maintenance costs as those that could have been avoided if appropriate technologies or protection measures would have been applied during the accounting period. As explained in this report, these are hypothetical costs.

In *China,* measurement of the green GDP has been a topic of interest to policy makers and researchers for as long as the past three decades. This is evident from the following initiatives taken up in China towards measuring the green GDP since the 1980s (Yu et al. 2006): (1) The National Environmental Pollution Cost and

Ecological Damage Valuation was conducted by the Chinese Research Academy of Environmental Sciences in the 1980s; (2) the project team for resources accounting and integrating it into the National Economic Accounting System was set up by the Development and Research Centre of the State Council in 1988; (3) calculation of genuine saving was jointly conducted by the State Environmental Protection Administration (SEPA) and the World Bank in 1998; (4) at the beginning of the year 2000, the SEPA and the World Bank began to conduct research on valuation approaches for Chinese environmental pollution cost; (5) during 1998–2001, an energy account covering 25 kinds of major energy sources was developed jointly by the National Bureau of Statistics of China (NBSC) and Statistics Norway; (6) in 2002, the Program of Environmental Physical Quantification Calculation in the Reform of the National Economic Accounting System, namely the Environmental Satellite Account Program, was carried out by the SEPA at the request of the NBSC; (7) in 2004, the research on environmental index systems for building a well-off society in an all-round way and for assessment of leaders' performance was conducted by the SEPA with the aid of the NBSC and other agencies in China; (8) in 2004, the Framework Research on the Green National Economic Accounting System, one of the key technology R & D programmes during the Tenth Five-Year Plan period, was conducted by Renmin University of China in cooperation with the Chinese Academy for Environmental Planning (CAEP).

In the light of various initiatives of natural resource accounting as mentioned above, some attempts have been made in 1994 to estimate the gross ecological cost of China, and it was found to be 2000 billion yuan (Liu and Guo 2008). There have been a number of pilot studies focusing on accounting of pollution cost without paying much attention on extending the SNA framework or implementing the SEEA in China. However, things have changed in China during the start of this century. In a major development for implementing the SEEA in China, the research project of Integrated Environmental and Economic Accounting (Green GDP Accounting) of China was initiated in March 2004 by the SEPA and the National Bureau of Statistics (NBS). The report of this project, which was completed in 2006, contains physical and monetary accounts of water and air pollution and solid waste for different industrial sectors and 31 provinces and municipalities in China and calculations of an environmentally adjusted GDP. This study subscribes to only partial implementation of the SEEA in China because it studies only accounting of pollution but not other natural resources' depletion. Physical accounts of pollution consist of pollution generated (influent), pollution abated and pollution discharged (effluent) for various pollutants. Both methods of environmental valuation, maintenance costs and damage costs, are considered for developing monetary accounts of pollution.

In the case of maintenance cost estimates for the year 2004, there is a large difference between imputed or hypothetical costs and actual abatement costs implying that the investments in pollution abatement are much lower than required in China. The GDP of China was 15,987.8 billion yuan in the year 2004. The total imputed abatement cost or hypothetical cost required to reduce the pollution to a safe level is 287.44 billion yuan, which constitutes 1.8 % of the GDP. Therefore, the pollution-adjusted GDP is 1.8 % less than the conventional GDP in China. The estimated damage costs of water pollution and air pollution in China were 286.28 and 219.8 billion yuan, respectively, in the year 2004. These costs together constituted 3.02 % of China's GDP in the year 2004.

There have been also attempts in China to go beyond air, water and solid waste pollution accounting described above in implementing the SEEA. There was research jointly conducted by the NBSC and Statistics Canada with emphasis on the accounting system for mineral resources. The Forest Resources Accounting Program was jointly conducted by the NBSC and the State Forestry Bureau. The Marine and Fishery Resources Accounting Program was launched by the State Oceanic and Fishery Administration. The detailed sector level studies of natural resource accounting initiated so far in China could ultimately provide a comprehensive database of linking economy and natural resources for estimating the green GDP using the SEEA methodology.

4.3 SEEA: The Current Status

Many countries as evident from the review of some country experience in the earlier section have started using some parts of the SEEA that could be easily integrated into natural resource accounting methods historically used by them. Also, countries may differ with respect to the importance of specific natural resources to their economies. For example, forestry or fisheries may not be a significant natural resource in some countries. Therefore, some countries have started implementing the SEEA for some selected sectors of the economy. Also, country experiences in using the SEEA differ with respect to developing both physical and monetary accounts of natural resources. Some countries like Germany have attempted to develop physical accounts (both asset and flow accounts) of natural resources recognizing the limitations on the monetary valuation of resources, especially non-marketable or non-commercial services offered by them. However, countries like Sweden, the Philippines and China have been trying to develop monetary accounts of specific environmental resources in an effort to estimate the green GDP. Some countries in Europe, for example the Netherlands, have tried to integrate the SEEA into the input–output matrix of the country that has been historically used for economic policy analysis. It could be useful here to attempt a brief discussion of recent experiences of implementing the SEEA in part or full. An attempt is made to describe the efforts made by some countries, especially EU countries, to integrate SEEA methodology with the methods of natural resource accounting they have been using historically.

Attempts have been made at *EU* level for developing a natural resource accounting system integrating some immediately implementable components of the SEEA with the methods of resource accounting currently used by many member countries. The commission of the EU suggested in 1994 a number of actions to be taken by member countries for developing environmental accounting in view of availability of the SEEA 1993 handbook. As a follow up, the European Strategy for Environmental Accounting (ESEA) is defined by the Statistical Programme Committee of Eurostat. The contents of the ESEA, though similar to the accounting framework of the SEEA, are more practical. The ESEA is developed taking into account (a) the consideration of the user needs and of the actual use of environmental accounts, (b) the need to harmonize environmental accounting among EU member countries and (c) the identification within the SEEA 2003 of the accounts deemed to be particularly relevant in the European context.

Eurostat has developed handbooks and a standard set of tables for collection of data for environmental accounting in Europe which are consistent with both the SEEA 2003 and the ESEA. The handbooks are available for the MFA, NAMEA, environmental protection expenditures accounts (EPEA), and forest, water and subsoil asset accounts. Table 4.1 presents the main features of the accounting system of the EU.

Most countries in Europe now produce on a regular basis NAMEA, MFA and EPEA accounts. Table 4.2 presents a prototype of a NAMEA prepared and presented by many EU countries. These accounts describe the physical flows of environmental resources to the economy and from the economy back to environment media as residuals. These flows form part of production accounts of various economic activities, agriculture, industry, and services and consumption accounts of households.

A discussion of natural resource accounts of the UK, which is one of the major EU countries, could be useful to know about integration of current accounting practices with the SEEA in some of the EU countries. Environmental accounts of the UK provide natural resource accounts of land cover, forestry, fishing, oil and gas extraction and reserves. These accounts provide information of physical flows of fossil fuel and energy consumption, atmospheric emissions with a breakdown of greenhouse gas and acid rain precursor emissions by industry, material flows, solid waste and water. Monetary accounts of environment-related information are provided for environmental protection expenditures and environmental taxes. These accounts are developed as satellite accounts to the main national accounts. They use similar concepts and classifications of industry to those employed in the national accounts, and they reflect frameworks recommended by the EU and the UN.

Most of the environment accounts of the UK are physical flow accounts except the accounts of environmental protection expenditures and environmental taxes. The physical accounts contain a large part of information required by physical accounts of natural resources prescribed by the SEEA, particularly for the relatively more implementable parts of it. But the SEEA requires a full set of corresponding monetary accounts which current UK and many other European country environmental accounts do not provide. The SEEA requires the monetary evaluation of all environmental and material resource flows. It prescribes the valuation methods of maintenance cost or non-market valuation methods of revealed and stated preferences. The monetary accounts of environment protection expenditures provide information about the actual cost incurred by different economic activities to

Table 4.1 Key features of the environmental accounting framework adopted in the environmental stress screening (ESS)

Main types of accounts	Main objectives
Economy-wide MFA	To construct an economy-wide balance sheet inclusive of all material flows between the economic system and the natural system (in both directions) in order to quantify the extent to which the economy makes use—according to its own trends—of natural resources and environmental media, including those located abroad
NAMEA-type flow accounts	To account for the physical flows taking place between the economic system and the natural system (atmospheric emissions, intake of natural resources, etc.) in correspondence with the specific economic activities that generate them, in particular—for a given activity—side by side to distinct economic flows such as production, value added, etc.
Economic accounts for the environment (EPEA)	To account for the economic transactions connected with the environment (environmental protection expenditures, resource use and management expenditures, environmental taxes, etc.); to describe the economic activities that produce goods and services for the environment ("eco-industries")
Asset accounts in physical terms	To construct an asset account in physical terms for a given natural resource (initial stock, increases and decreases during the accounting period—both natural and anthropogenic—closing stock); quality aspects are taken into account as appropriate by means of indicators or by breaking down the account by quality classes
Integrated environmental and economic accounting for natural resources (NRIEEA)	To construct—for a given natural resource of interest (e.g. forests, subsoil assets, etc.)—an accounting system made up of NAMEA-type flow accounts, economic accounts for the environment (EPEA) and asset accounts in physical terms

MFA material flow accounts, *NAMEA* national accounting matrix including environmental accounts, *EPEA* environmental protection expenditures accounts

achieve the objective of sustainable use of environmental resources. If these expenditures fall short of expenditures required for sustainable use of the resources, the SEEA maintenance cost method requires the estimation of hypothetical additional costs to be incurred for achieving the sustainable resource use objective.

Many EU countries, with the exception of Sweden, do not aim to estimate the green GDP using the environmental accounts they have been currently preparing. Estimating the green GDP requires the information of a full set of monetary accounts corresponding to the physical accounts they have been publishing.

Table 4.2 Structure of a national accounting matrix including environmental accounts (NAMEA)

Economic activities and household consumption	Economic aggregates				Environmental pressures: pollution		Environmental pressures: intake of natural resources		
	Production	Value added	Employment	Final consumption	Atmospheric emissions	Water emissions	Fossil fuels	Minerals	Biomass
Agriculture									
Industry									
Services									
Household consumption Transport Heating Others									

Table 4.3 Implementation of the SEEA. (Source: UN 2000)

Adaptation of the national accounts for environmental analysis
Step 1: Compilation of the supply and use accounts
Step 2: Identification and compilation of environmental protection expenditures
Step 3: Compilation of produced natural asset accounts
Natural resource accounting
Step 4: Compilation of physical natural resource accounts
Step 5: Valuation of natural resources—compiling the monetary accounts
Accounting for environmental assets
Step 6: Compilation of physical environmental asset accounts
Emission Accounts
Step 7: Compilation of emissions by economic sector
Step 8: Maintenance cost of emissions
Presentation and analysis
Step 9: Aggregation and tabulation
Step 10: Comparison of conventional and environmentally adjusted indicators

They perhaps believe that monetary valuation of non-market environmental services requires a lot of data and models with a host of assumptions. However, the SEEA-prescribed method of maintenance cost to account for degradation of environmental resources in measuring the green GDP could be a feasible method for developing monetary accountings of environmental resource flows.[3] In this context, the monetary accounts of environmental protection expenditures of industries and environmental taxes currently prepared by many EU countries already provide useful information for making use of the maintenance cost method of valuation.

4.4 SEEA Accounts of Different Sectors

4.4.1 Implementation of SEEA

The *Handbook of National Accounting: Integrated Environmental and Economic Accounting—An Operational Manual* (UN 2000) explains step by step the implementation of the SEEA and provides methodologies and case studies of preparing accounts for specific resources like forests, subsoil resources, renewable aquatic resources, soil degradation and air emissions. Table 4.3 explains these steps.[4]

[3] As explained in the earlier section, China has used the maintenance cost method to account for air and water pollution and solid waste in estimating the green GDP.

[4] Table 4.4 provides a list of worksheets for preparing SEEA accounts.

A number of case studies using EU framework for natural resource accounting and the SEEA are being attempted by some countries and institutions to prepare natural resource accounts for different resources/sectors. Physical and monetary accounts of forest resources have been prepared by some EU countries on a regular basis. In the case of land resources, physical accounts of land use changes are prepared by many of these countries. The Food and Agriculture Organization (FAO) has discussed SEEA methodology for preparing fisheries accounts and provided case studies of some countries. The UNSD has discussed the methodology of integrated environmental and economic accounting for water resources and attempted some country case studies.

4.4.2 Forestry Accounts

EU countries adopted some resolutions at a ministerial conference held in Helsinki 1993 on the protection of forests in Europe. These resolutions provide general guidelines for sustainable management of forests and the conservation of biodiversity in Europe. As a follow-up pan-European criteria and indicators were defined for gathering and assessing information to monitor the development of European forests. These are:

(i) Maintenance and appropriate enhancement of forest resources and their contribution to global carbon cycles
(ii) Maintenance of forest ecosystem health and vitality
(iii) Maintenance and encouragement of productive functions of forests
(iv) Maintenance, conservation and appropriate enhancement of biological diversity in forest ecosystems
(v) Maintenance and appropriate enhancement of protective functions in forest management (notably soil and water)
(vi) Maintenance of other socio-economic functions and conditions

Consequently, proposals for a "European Framework for Integrated Environmental and Economic Accounting for Forests" were developed and tested by the Eurostat Task Force on Forest Accounting. The objective of this framework is to consistently link forest balance sheets and flow accounts, forest-related economic activities and the supply and use of wood within the economy in physical and monetary terms. Moreover, the forest accounting framework contributes to the issues of classification and valuation of forest-related assets within the European System of Accounts (ESA), the SNA, the new Economic Accounts for Forestry and the SEEA. In order to implement the forest accounting framework, accounting tables were drafted covering physical stocks and flows of forests, monetary accounts of stocks and flows, balance sheets for land and standing timber, economic accounts of forestry and supply–use tables (see European Communities 1999). All these tables together provide a large part of forestry accounts required as per SEEA methodology and NAMEA accounts.

Physical stocks and flows accounts of forests provide information of forest cover and volume of standing timber. This account provides stocks and changes in stocks due to human activities, natural or accidental processes as well as changes in classification of land and standing timber. Monetary stock and flow accounts describe in monetary terms the stocks and changes in stocks of land and standing timber. Land and biological assets of forests, say for example commercial timber, differ from an economic point of view and therefore require separate valuations. These accounts present only economic or commercial value of forest stocks using standard valuation methods prescribed in the SNA and SEEA. However, they fall short of SEEA forest accounts for not presenting environmental values of forest resource stocks.

Economic accounts for forestry and logging link the value of forest assets with the economic benefits of different activities related to forests. These accounts are dedicated to the establishment of economic accounts for all industries which use "wooded land" as a basis for their activity. For example they present detailed production accounts of forestry and logging describing detailed output related to woodland, and primary and intermediate inputs used including land, inventories and value added.

Supply and use of wood accounts describe in physical and monetary units the transformation of wood from the stage "output of forestry" to the final products and establish a link between the supply and use of wood, the forest balances and the economic accounts. The use tables show the uses of wood products as intermediate input uses of industries and final uses (final consumption, changes in inventories and exports). The supply tables show the supply of wood products (output of industries and imports). A complementary use table serves to record residuals (wood waste, paper waste and other residuals containing wood, e.g. black liquors) which are not accounted for as intermediate consumption of industries and waste treated by external recycling activities. A complementary supply table records production of waste not counted as output of industries. The supply and use tables are drawn up both in monetary and physical terms, leading to physical material balances.

Mass balance accounts integrate wood and wood products material balances into the forest accounting framework. Mass balances are disaggregated into two tables: use tables and make tables. The use table shows the wood content of the intermediate input use of industries in selected wood products. The make table shows the wood content of the output in selected industries. A complementary table shows the wood content of waste and residuals not accounted for in the output of industries. Therefore these accounts are useful to know the total storage of carbon dioxide (CO2) in wood products and standing timber. They facilitate linking of mass balances to balances of CO2 and activities in the NAMEA-type of accounting. These balances could be further allocated to global environmental problems like greenhouse effect.

The ESA accounts of forest resources described above consider only economic assets as in the SNA. The SNA defines economic assets as assets over which "ownership rights are enforced and from which economic benefits are derived by their owner(s) by holding them or using them over a period of time". However the SEEA extends the asset boundary of forest resources by including also non-economic assets. It accounts for forest land and related ecosystems, biological assets (plants, animals, etc.) in the forest and other assets related to forests. The *Handbook of National Accounting* of the UN (2000) describes the methods of preparation of a country's forestry accounts as per the SEEA. It defines non-economic environmental forest land covering land of both protected and non-exploitable forests as corresponding to forests that are not exploitable for economic reasons including virgin forests, and to forests where the exploitation of biological resources is severely restricted by virtue of the protection status. The UN Handbook describes the compilation of forestry accounts in the following eight steps:

Step 1: Compilation of the Supply and Use Accounts
These accounts identify and separate, within the general supply–use tables, the transactions that are relevant for the description of forest assets and forest-related activities. The main activities are forestry and logging, gathering of non-wood forest products, hunting, etc. and also activities that result in deforestation (e.g. agriculture, construction, etc.).

Step 2: Identification and Compilation of Environmental Protection Expenditures Related to Forests
Environmental protection expenditures of forests consist of costs of fire protection, afforestation, the improvement of forest soils and protection against game, insect attacks, etc. These also include costs of sustainable use of forests such as forest protection costs.

Step 3: Compilation of Produced Forest Asset Accounts
Produced forest assets accounts consist of the value of standing timber located on land cultivated for wood production.

Step 4: Compilation of Physical Forest Accounts
Two main categories of non-produced economic assets are detailed here: (a) economically used land (specifying forest land) and (b) standing timber in economic non-cultivated (native) forest.

Step 5: Valuation of Forests: Compiling the Monetary Accounts
A monetary value is given to stocks and flows related to non-produced economic assets.

Step 6: Compilation of Physical Environmental Forest Accounts
This step is necessary to complete the description of all stocks of forest land and standing timber and all changes that affect these stocks. It describes accounts of (a) all non-economic land, in other words, all the land that is not described in step 4, except forest land and associated ecosystems; (b) all non-economic forests (areas of forest land and volume of standing timber) and associated ecosystems; (c) other environmental, in other words non-economic, assets—fauna, flora, water and air. The comprehensive description of non-economic forests necessitates a classification of ecosystems (type of forest, etc.) and state of forests from an environmental and ecological point of view.

Table 4.4 List of worksheets of SEEA implementation

WS 2, 2A Environmental protection expenditures
WS 3, 3A Monetary asset accounts: produced assets, including natural assets
WS 4, 4A Physical asset accounts: non-produced economic assets
WS 5 Monetary asset accounts: non-produced economic assets
WS 5A Market valuation of non-produced economic assets
WS 5B Monetary asset accounts: non-produced economic assets
WS 5C Allocation of depletion costs to economic activities
WS 6 Physical asset accounts: non-produced

Step 7: Compilation of Emissions by Economic Sector
Accounts are prepared for three main categories of emissions: (a) emissions by forestry and related industries, (b) emissions that affect forests, and (c) absorption of CO2.

Step 8: Maintenance Cost of Environmental Degradation
Maintenance cost accounting requires the assessment of the extra cost for the maintenance of the state of forests from a quantitative and qualitative point of view. Maintaining the state of forests requires restricting forest-related activities like logging to sustainable levels and reducing the impacts of non-forest-related activities like agriculture and construction to the sustainable levels.

The natural resource accounts of EU countries as described above (natural resource accounts of the UK and forestry accounts of the European framework) have come close to the implementation of the SEEA in these countries. However, these accounts and the SEEA differ mainly with respect to the definition of the asset boundary of natural resource stocks and monetary valuation of especially stocks and flows of environmental resources. The EU countries' accounts consider only economic assets as defined in the SNA, while the SEEA considers both economic and non-economic assets of natural resources by extending the asset boundary. EU countries mainly focus on physical accounts, especially of environmental resources. They do not aim at measuring the green GDP given the limitations on measuring non-marketable environmental services. However, the SEEA provides for developing monetary accounts of environmental resource stocks and flows, especially in the context of sustainable use of these resource stocks (Table 4.4).

4.4.3 Some Other Sector-Specific Studies

4.4.3.1 Fisheries

The UN Handbook 2000 and the FAO *Handbook of National Accounting: Integrated Environmental and Economic Accounting for Fisheries* (FAO 2004) discuss the methodology of preparing natural resource accounting for fisheries resources. The case of preparing natural resource accounting for fisheries using the procedure prescribed in the SEEA given in Tables 4.3 and 4.4 requires to note specific characteristics of this resource. The approaches to be adopted for defining production and asset boundaries, and developing accounts of physical asset and flow accounts will be different from those for other resources.

In defining the production boundary of resources, natural growth of fish stocks in open seas is not counted as production, as the process is not fully managed as required by the SNA. On the other hand, the growth of fish in fish farms and fish harvested in open seas from commercial or recreational fishing are treated as processes of production. The SEEA for fisheries accounting considers all fish stocks within the exclusive economic zone (EEZ) of a country as an economic asset and therefore included in environmental assets, for example, marine and freshwater ecosystems. Physical accounts of fisheries have to be developed using the concept of sustainable use or harvesting and using different approaches for cultivated and wild fish stocks. The production and use tables showing the relationship between fisheries and fishery-dependent activities and ecological resources and accounts of environmental protection expenditures require specific approaches to develop them.

4.4.3.2 Water Resources

Integrated Environmental and Economic Accounting for Water Resources (United Nations Statistics Division 2005) and the UN Handbook 2000 provide detailed methodology and description of data requirements for preparing natural resource accounts for water resources. This report first discusses how to develop physical supply and use accounts of water resources and provides some case studies at a country as well as river basin level. It also describes monetary accounts of supply and use of water, water protection and other water-related activities. Asset accounts of water resources are developed by extending the asset boundary of the SNA to account for aquatic ecological resources. This also provides a description of emission accounts for a river basin with a case study.

4.5 A Way Forward for Measuring Green GDP of India

4.5.1 Some Studies Funded by CSO in India

An attempt to develop a comprehensive database for measuring the green GDP of India requires a review of currently available data sources and methods giving the information about the natural resource stocks and their links to various economic activities in India. In this context, the Central Statistical Organization (CSO) has initiated some studies for developing databases and methodologies for specific sectors in India. Table 4.5 provides a brief description of CSO-funded projects.

Table 4.5 Summary features of CSO-funded studies

Authors	Sectors covered	Valuation methods	Data	Observations
Murty and Gulati (2006)	Air and water pollution	Maintenance cost valuation methods using production function and cost function methodologies Non-market valuation method of hedonic prices	Both primary and secondary sources	Methods of developing physical and monetary accounts of water and air pollution are described with the help of case studies from Andhra Pradesh and Himachal Pradesh Methodologies for estimating maintenance costs of air and water pollution are described, and estimates are provided for thermal power generation, industry and transport sectors Estimates of benefits from urban air quality improvements are provided A method for the generalization of industry and farm production accounts in India for estimating environmentally corrected net value added is described with the help of case studies
Roy et al. (2008)	Air and water pollution	Maintenance cost valuation method using cost functions Non-market valuation method of household health production function	Both primary and secondary sources	Methods of developing physical accounts of water and air pollution are described for West Bengal Maintenance cost estimates are made for water pollution of paper and pulp industry Estimates of benefits from reduced arsenic pollution of groundwater and air pollution are obtained
Parikh et al. (1992, 2008)	Solid waste, forests and land quality	Review of valuation methods Surveys used to collect data Maintenance cost method used	Both primary and secondary	Physical and monetary accounts of forestry and land sectors are described for Goa Information about the status water quality in Goa is provided Industrial, vehicular and indoor air pollution valued using abatement cost method Physical and monetary accounts of municipal solid waste are developed using surveys and abatement cost method Environmentally adjusted SDP for Goa presented

Table 4.5 (continued)

Authors	Sectors covered	Valuation methods	Data	Observations
Singh et al. (2008)	Forests and land quality	Benefit transfer methods	Secondary	Compiled data on natural resources, particularly land and forests of Meghalaya, which could be used for developing physical and monetary accounts Information about expenditures made for soil and forest conservation in Meghalaya is provided
Haripriya et al. (2007)	Forests and land quality	Maintenance cost and non-market valuation methods of travel cost and benefit transfer methods	Secondary and primary	Physical and monetary accounts of forests and land quality for Tamil Nadu are developed Methods for estimating maintenance costs are discussed with case studies of land quality in Tamil Nadu
Verma and VijayKumar (2006)	Forests and land quality	Maintenance cost and non-market valuation methods of travel cost and contingent valuation	Secondary and primary	Physical and monetary accounts of forests and land quality for Madhya Pradesh and Himachal Pradesh are developed Methods for estimating maintenance costs are discussed with case studies of land quality in these states
Panchamukhi et al. (2008)	Forest and land quality	Maintenance cost and non-market valuation methods of travel cost and contingent valuation	Secondary and primary	Information for developing physical and monetary accounts of forests and land in Karnataka is provided A case study of valuation of sacred groves is attempted
Datt et al.	Exhaustible resource coal	User cost, net price and net present value methods	Secondary and primary	Physical and monetary accounts of coal resources for Madhya Pradesh and West Bengal Environmental costs of mining coal are also estimated

SDP state domestic product

4.5.2 *Possible Methods for Measuring Green GDP in India*

The recent literature on the measurement of sustainable income has developed in two important ways for accounting of contribution of natural resource stocks. One set of studies directly addresses the problem of measuring genuine savings or extended wealth formation including changes in human resource capital and natural capital. The second set of studies use the extended conventional national income accounting methods for accounting of changes in natural resource stocks (SEEA and extended IOT). The international experience in integrating natural resource accounts with the SNA has been progress in three directions: (a) attempts to estimate genuine savings for different countries, (b) implementation of practical aspects of the SEEA and their integration into the national accounts and (c) using SEEA methodology for preparing natural resource accounts of specific resources and sectors. In India, an attempt could be now made for integrating practical aspects of the SEEA with current national income accounting practices.

4.5.3 *Developing MFA and NAMEA for India*

Develop a strategy plan as the European Strategy for Environmental Accounting (ESEA) for integrating currently used national income accounting methods with the SEEA. Unlike some EU countries, India had not made any advancement in natural resource accounting prior to the publication of the SEEA methodology. European countries currently prepare and publish regularly material flow- and NAMEA-type of accounts. These accounts are still short of information for estimating the green GDP since they avoid putting values on environmental changes reported in these accounts.

It is surprising that a major developing country like India has not yet attempted to develop databases for preparing these accounts. The NAMEA-type of accounts given in Table 4.2 could be prepared using information from extended databases of Annual Survey of Industries (ASI) accounts, input–output accounts, cost of cultivation accounts of agriculture and National Service Scheme (NSS) consumer expenditure accounts in India. In addition, preparation of emission and MFA is needed to have environment-related information of various economic activities. India has been preparing a very detailed input–output matrix for the Indian economy on a regular basis. Environmental extensions of this matrix could be attempted to produce NAMEA-type of accounts.

In extending IOT, mainly two types of extensions could be distinguished:

a. Natural resources covering mineral and energy resources, and water and biological resources flow from the natural environment into the national economy.
b. Residuals consisting of emissions of air, water and soil are the incidental and undesired outputs from the production activities discharged into the environment.

The conventional monetary IOT remains as it is, and non-monetary environmental extensions are attached in the form of separate accounts underneath the monetary accounts. The satellite accounts of environmental extensions consist of an input matrix of environmental extensions and an output matrix. The inputs are primary natural resources (gifts from nature) and outputs are emissions. The ongoing project of the European Commission (EXIOPOL Project) in which many institutions from Europe are collaborating is going to come out soon with an environmentally extended input–output matrix for Europe and some EU countries.

4.5.4 Preparing Monetary Accounts of Environmental Changes

To start with, India should aim at developing an NAMEA reporting physical flows of environmental resources and emissions along with flows of goods and services in the economy. The SEEA mentions maintenance cost and non-market methods of valuation for estimating monetary accounts of environmental changes. The CSO-funded projects mentioned above and some recent studies in South Asia (see Haque et al. 2011) provide good case studies using both maintenance cost and non-market valuation methods. The SEEA recommends the maintenance cost method of valuation for developing monetary accounts recognizing the limitations on the non-market valuation methods.

In India, and for that matter in any country thinking of using SEEA methodology, it is useful to start with the maintenance cost method of valuation for developing monetary accounts. This method of valuation is based on the concept of strong sustainability. Maintenance cost is a more objective and quantifiable method of valuation, while the methods of non-market valuation are more subjective and normative. Maintenance cost depends upon the available technologies of pollution abatement, afforestation and methods of resource conservation. Maintenance cost studies have to be done for different sectors: industry, agriculture, forestry and households. Maintenance cost could vary across sectors depending upon sector-specific environmental and resource depletion problems. Using the maintenance cost method, monetary accounts of changes in environmental resource stocks could be unambiguously constructed for different sectors at national level.

Monetary valuation of environmental changes could be region specific as could be seen in some case studies of non-market valuation done in India and South Asia (Murty 2010; Haque et al. 2011). Learning from a number of non-market valuation studies already done in India and the South Asian region, it is also important to have more similar studies sponsored by the CSO and other relevant government and non-government agencies in future as a long-run strategy for having more accurate estimates of the green GDP of India. These studies could be specific to households and regions and could be used to develop region-specific monetary accounts of environmental changes.

4.5.5 Using SEEA in India

Implementation of SEEA methodology for estimating the green GDP for India requires development of physical and monetary accounts of natural resource stocks at national level. A top-down approach of developing these accounts may not be feasible in big countries like India, the USA and China. Physical changes in environmental resource stocks could be valued differently in different regions, especially in the context of non-market valuation. A bottom-up approach in developing these accounts could ensure to take into account region-specific values of environmental changes.

The SEEA could be applied at subnational or regional level in big countries for developing region-specific physical and monetary accounts of environmental changes. National accounts could be developed by aggregating over the regional accounts. In the Indian context, attempting to apply the SEEA at state level could be a right approach. In this context, the CSO-funded projects addressing the problem of developing natural resource accounts at state level are a good beginning in using the SEEA for measuring the environmentally corrected net national product in India.

4.5.6 Resource-Specific Studies

4.5.6.1 Forestry

Forestry accounts similar to the accounts of EU countries described above have to be developed in India in the near future. Attempts already made by some CSO-funded projects and other studies using SEEA methodology for developing forestry accounts could provide some leads for developing forestry accounts at subnational or state level in India. Given the difficulties involved in monetary valuation of non-marketable services of forest resources and the absence of a comprehensive set of valuation studies for forests even while using maintenance cost methods in India, it is feasible to develop detailed physical accounts of forests in the near short run. Data from secondary sources could be used to develop physical accounts of changes in forest resource stocks at some level of species-wise disaggregation. CSO-funded studies of Verma and Kumar, Haripriya and Panchamukhi et al. have shown how this data could be used for developing physical accounts of forest resource stocks. These studies also show the risk of having unreliable estimates in preparing monetary accounts in the absence of information from more needed valuation studies.

Preparation of monetary accounts of forests has to be treated as long-run objective in India waiting for more comprehensive valuation studies of forest services. The CSO-funded studies, while providing some insights for using both maintenance cost and non-market valuation methods in preparing forestry accounts, underscore the need for more valuation studies in India. In the medium run, priority should be given in India to having more valuation studies of using maintenance cost methods.

Monetary valuation of changes of forest stocks requires disaggregated physical accounts by types of forests, which could differ with respect to the ecological services they offer (biodiversity, carbon sequestration, recreation, etc.). This is especially so in the context of non-market valuation of forest services.

4.5.6.2 Soil Quality

The UN Handbook 2004 provides a detailed discussion of preparing accounts of soil degradation. Soil quality is an environmental good affected by anthropogenic activities and natural causes. Measurement of physical changes in land quality requires information from land quality modelling finding the relationship between soil quality and changes in land use patterns. The SEEA classifies land area as (a) soil; (b) area of land under economic uses such as land underlying buildings, land under cultivation, recreational land and forest land; and (c) non-economic land areas with connected ecosystems. In an attempt to prepare physical accounts of land quality changes using the SEEA in India, it is important to have studies of land quality modelling for different geographical regions. The discussion of land use changes in agriculture and attempting to relate these to soil quality changes in some CSO-funded studies underscore the need for having these studies. Studies of land quality modelling and preparation of physical accounts of land quality changes for different agro-climatic regions should be one of the short-run objectives of natural resource accounting in India.

An attempt to develop monetary accounts of soil quality changes using the SEEA in India should be treated as a medium-term objective. The UN Handbook considers expenditures on improvement in land quality as gross fixed capital formation. This cost covers land reclamation from the sea and forest clearance, and expenditures on soil conservation measures. The maintenance cost method is recommended for compiling comprehensive estimates of cost of soil degradation for a given area. This cost could be estimated as cost of replacement and restoration. Some of the CSO studies discuss and provide some information about replenishment and restoration costs.

Alternatively, monetary accounts of land degradation could be developed using the productivity change method. This is the value farmers place on land quality which has to be measured by estimating production functions considering land quality as one of the inputs in agriculture. There are some available studies in India estimating loss of land productivity due to increased land salinity, soil erosion and nutrient loss. A number of similar studies have to be done on a regular basis in India for estimating hypothetical agricultural production losses from land degradation.

4.5.6.3 Water Resources

Integrated Environmental and Economic Accounting for Water Resources (United Nations Statistics Division 2005) and the UN Handbook 2004 provide detailed methodology and description of data requirements for preparing natural resource accounts for water resources. It provides case studies of preparing accounts at country as well as river basin level. It is therefore useful in India to make immediate attempts to prepare water resource accounts for all its major river basins in a step towards implementing the SEEA.

Preparing physical accounts of water quality requires studies of water quality modelling for each river basin relating water pollution from sources with ambient quality of surface and groundwater. The information from these studies is important for developing physical accounts of ambient water quality and water quality changes attributed to specific sources, and also for designing an environmental policy for sustainable water resource development. Preparing physical accounts of water resources could be again one of the short-run objectives of natural resource accounting in India.

Some of the CSO-funded studies provided some insights in preparing monetary accounts of water quality using both valuation methods of maintenance cost and non-market valuation. However, preparation of monetary accounts could be a long-run objective for India waiting for information from more valuation studies of water quality and quantity in India.

4.5.6.4 Air

The UN Handbook 2004 explains how physical and monetary accounts of emissions have to be prepared. For preparing monetary accounts of air quality, information from studies of air quality modelling at regional or subregional level is required. Industry-level physical accounts of air quality could be prepared, and corresponding monetary accounts could be obtained using the valuation method of maintenance cost. Again, preparation of physical accounts of emissions could be one of the short-run objectives in India waiting for more valuation studies to prepare monetary accounts in the long run.

Some of the CSO-funded studies have done case studies of air pollution at sources as well as ambient pollution for preparing physical and monetary accounts at sector and regional levels. Data from both primary and secondary sources are used for this purpose. These studies have prepared monetary accounts using both maintenance cost and non-market valuation methods. Estimates of maintenance cost are obtained for the thermal-power-generating sector using the methodology of production function and for industrial water pollution using the methodology of cost function. A case study of estimating maintenance cost for the road transport sector is also provided by one of these studies. Estimates of benefits of air and water quality improvements are obtained using revealed preference methods of valuation. The hedonic property price method is used by one study for estimating monetary benefits of urban air quality improvements. Another study has used the household health production function method for estimating benefits of reducing arsenic pollution of groundwater.

4.5.6.5 Subsoil Resources

Physical accounts of exhaustible resources account for depletion of the resources during an accounting period. The depletion has to be measured with respect to proven and utilizable resource stocks. Valuation methods of net price, net present value and user cost are discussed in the literature for developing monetary accounts of exhaustible resources. The net price or resource rent is not an appropriate method of accounting for depletion for it cannot give credit to a country rich in exhaustible resources in relation to a country poor in these resources in measuring the GDP. Similarly, net present value is a measure of stock of a resource which cannot be used to measure the effect of resource depletion on the GDP, a flow.

Any method of valuation of resource depletion has to take into account the property rights of both present and future generations to an exhaustible resource stock and the problem of inter-temporal equity in resource use. The user cost could be an appropriate method for valuing the depletion of the resource because it is based on the concept of weak sustainability (manmade capital could be a substitute for natural capital) ensuring the same level of real income to the present as well as future generations from resource extraction.

In the context of preparing satellite accounts of exhaustible resources as per the SEEA in India, studies have to be done to estimate user cost of resource stocks of minerals, metals and fossil fuels. User cost depends on the life of proven reserves and rate of discount used to address the problem of inter-temporal equity. It decreases with the rate of discount and life of the resource stock making it resource specific. The rate of discount depends on the value judgements of the government on the property rights of present versus future generations to the resource stock. Studies have to be done on a regular basis for estimating user costs for all minable resource costs in India to implement SEEA methodology.

One of the CSO-funded projects has shown that the physical accounts of resource depletion in the case of coal reserves in India could be developed using data from secondary sources. The same could be true for other exhaustible resource stocks in India. Using data for a sample of mines, this study provides estimates of net price, net present value and user cost per tonne of coal extracted in India. It found that the user cost component for coal as percentage of mining sector state domestic product (SDP) in Madhya Pradesh is 2.5 % at a 6 % rate of discount during the years 2001–2002 at 1993–1994 prices.

A number of studies similar to this study have to be done in India for accounting the effects of mining of exhaustible resources in the measurement of the green GDP using SEEA methodology. These studies have to be done for all minable resource stocks in the country with a frequency of at least once in five years as proven reserves increase with new discoveries.

4.5.7 *Attempting to Make Estimates of Genuine Savings for India: An Approach to Measure Green GDP*

Methodology of genuine savings and estimates of genuine savings prepared currently for different countries by the World Bank provide a way forward to develop a database for preparing natural resource accounts of India. Data from secondary sources could be reorganized to develop this database. The World Bank group of economists has already developed some database for making genuine saving estimates for different countries including India on a regular basis. It could be useful if the Indian government made its own estimates of genuine savings with a more comprehensive database for India.

Estimates of adjusted net GDP or GDP adjusted for depletion and degradation of natural resources could be obtained given an estimate of the rate of genuine savings. If the difference between conventional net rate of savings and the rate of genuine savings is positive, the adjusted net GDP is lower than conventional net GDP for a country. As per the estimates of genuine savings in India made by the World Bank, the costs of rates of depletion of energy and mineral resources and degradation of forests and CO_2 emissions will be added up to 8.22 % of the GDP. Therefore, according to these estimates, the green GDP in India is 8.22 % lower than the conventional GDP. It is also important to know that the actual rate of adjustment for this purpose could be much higher than 8.22 % because the World Bank estimates do not account for the full cost of forest degradation, and the cost of soil erosion and domestic water and air pollution.

4.5.8 *Main Recommendations*

1. A feasible work plan for preparing natural resource accounts in India has to be implemented in the following three stages:

4.5.8.1 Stage I

Attempt to prepare physical asset and flow accounts for all the natural resources using secondary data sources in the near future. These data could be organized to generate physical accounts by either SEEA methodology or extended national IOT or both. Given that data from secondary sources are incomplete as of today to prepare these accounts for India, the accounts prepared could only provide rough estimates. However, in the case of preparing accounts of exhaustible resources, the secondary data sources could provide full information, and therefore more reliable estimates of these accounts could be obtained for India.

To have more accurate estimates of physical accounts of water, air, forests and land resources, new information has to be generated by having studies of air quality modelling of urban air sheds, water quality modelling of river basins and land quality modelling of agro-climatic regions. These have to be undertaken on priority basis.

4.5.8.2 Stage II

Attempt preparation of monetary accounts of natural resource stocks using the maintenance cost method of evaluation in the medium term. This requires the commissioning of a new set of sector- or industry-specific studies for having the estimated maintenance cost described in this report. Some available studies in India including some from CSO-funded studies have attempted to estimate the maintenance cost for some specific sectors in India such as road transport, coal-fired thermal power generation in the case of air pollution, some manufacturing industries in the case of water pollution and agricultural practices in the case of soil degradation.

4.5.8.3 Stage III

Attempt to use non-market valuation methods for preparing monetary accounts of natural resource stocks in the long run. This requires the monetary valuation of all ecological services provided by natural resources in India. A number of region-specific valuation studies for air and water and forest resources have to done in India for this purpose. There are already some environmental valuation studies using non-market valuation methods in India and the South Asian region. These could provide helpful insights for undertaking many future studies to have more reliable estimates of values of ecological services in India.

The tasks set for three stages of research work described above have to be concurrently undertaken with the priority to complete them in different stages. This is required especially because the empirical studies for obtaining reliable estimates of environmental values are more time-consuming and demanding of budgetary resources.

1. Prepare NAMEA-type of accounts using information from the extended database of ASI accounts, input–output accounts, cost of cultivation accounts of agriculture, and NSS consumer expenditure accounts in India. In addition, prepare emission and material flow accounts to have environment-related information of various economic activities. Attempts could be made to make environmental extensions of the input–output matrix for the Indian economy on a regular basis.
2. Attempt using SEEA methods and environmental extensions of IOT at subnational or regional level (state level) in India. Given that data of IOT are not available at the state level in India, environmental extension of these tables at the state level could be made with the assumption that the production technology is the same at national and regional level.
3. Use the maintenance cost method of valuation for developing monetary accounts to start with. Maintenance cost studies have to be done for different sectors: industry, agriculture, forestry and households. Maintenance cost could vary across sectors depending upon sector-specific environmental and resource

depletion problems. Using the maintenance cost method, monetary accounts of changes in environmental resource stocks could be unambiguously constructed for different sectors at national level.

4. Attempt to prepare forestry accounts similar to accounts of EU countries. Given the difficulties involved in monetary valuation of non-marketable services of forest resources and the absence of a comprehensive set of valuation studies for forests even while using maintenance cost methods in India, it is feasible to develop detailed physical accounts of forests in the near short run. Data from secondary sources could be used to develop physical accounts of changes in forest resource stocks at some level of species-wise disaggregation.
5. Attempt and commission studies of land quality modelling and prepare physical accounts of land quality changes for different agro-climatic regions.
6. Commission studies of water quality modelling for each river basin. The information from these studies is important for developing physical accounts of ambient water quality and water quality changes specific to sources and also for designing an environmental policy for sustainable water resource development. On priority basis, it is good to start with preparing physical and monetary accounts of the Ganges basin taking advantage of data generated by the national project Ganga Action Plan.
7. Attempt studies of air quality modelling at regional or subregional level. Many urban airsheds in India have been experiencing very serious air pollution problems imposing enormous health and other costs to urban households. The problem of preparing physical and monetary accounts of air pollution in major urban areas should be taken up on priority basis. This helps not only in measuring effects of urban air pollution on the green GDP of India but also to design an environmental policy for reducing urban air pollution to safe levels in India.
8. Commission studies to estimate user cost for subsoil resources on a regular basis probably with an interval of every five years. It depends on the life of proven reserves and the rate of discount used to address the problem of intertemporal equity. It decreases with the rate of discount and life of the resource stock making it resource specific. The rate of discount depends on the value judgements of the government on the property rights of present versus future generations to the resource stock.
9. The CSO and other concerned organizations could take advantage of information provided by genuine savings estimates by the World Bank and other agencies in preparing natural resource accounts for India, especially the accounts of exhaustible resources.
10. Given the difficulties in preparing natural resource accounts for India in the near future with the very weak database we are currently having, it is important to have a standing committee of a small group of dedicated experts to advice and monitor projects and programmes for generating data from primary sources and reorganizing data from the existing secondary sources. When data themselves are incomplete or absent, it is difficult for an expert group to make inroads into the complex problem of measuring the green GDP for India in a short time span. The literature provides well-developed and usable methodologies for measuring

the green GDP, but what we require in India is the development of a comprehensive database of natural resources and their links to the economic activities. Therefore, all efforts of concerned governmental agencies including the CSO have to be urgently diverted towards preparing a comprehensive database of natural resources.

4.6 Conclusion

Recognition of contribution of natural resources to economic development could be found in the economic policies pursued by some countries even from the start of second half of last century. It was found that the UN framework System of National Accounts (SNA) released in 1968 is inadequate for accounting of contribution of natural resources to the well-being of people in a country. Many developed countries and even some developing countries have started in the direction of preparing some form of natural resource accounts in addition to standard national income accounts of the SNA. These are mainly in the form of physical supply and use accounts linking natural resources with the natural resource-related economic activities and MFA and monetary accounts of environmental protection expenditures and taxes.

The methodology of the SEEA which first appeared as *Handbook of National Accounting: Integrated Environmental and Economic Accounting'* published by the UN in 1993 and its next version from 2003 is now available for the countries for integrating the conventional national income accounts with natural resource accounts. However, even after almost two decades of its existence, it is now found that no country in the world has made full use of this methodology for estimating its green GDP.

Many countries have started using some parts of the SEEA that could be easily integrated into natural resource accounting methods historically used by them. Also, countries differ with respect to the importance of specific natural resources to their economies. For example, forestry or fisheries may not be a significant natural resource in some countries. Therefore, some countries have started implementing the SEEA for some selected sectors of the economy. Also country experiences in using the SEEA differ with respect of developing both physical and monetary accounts of natural resources.

In the context of developed countries, attempts have been made, for example at EU level, for developing a natural resource accounting system integrating some immediately implementable components of the SEEA with the methods of resource accounting currently used by many member countries. Many of these countries, especially those from Europe, regularly prepare natural resource accounts consisting of MFA, NAMEA, environmental protection expenditures accounts (EPEA), and forest, water and subsoil asset accounts. These are mostly physical accounts with the exception of the EPEA accounts. Sweden has gone a step further in preparing monetary accounts with an objective of measuring the green GDP.

Most of the developing countries had not started working on developing databases linking natural resources with economic activities until the turn of this century. However, one could find some case studies of selected sectors in some countries which are attempted independently of the government. The Philippines is first among these countries trying to implement the SEEA in national income accounting even during the mid-nineties of the past century. Estimates of physical and monetary accounts of SEEA asset accounts are prepared for a comprehensive list of natural resources including land, minerals, coal, water, forests and fisheries. However, these accounts show only commercial or economic value of natural resource stocks as per the SNA and do not consider non-marketable environmental values of use and non-use values and option value. China and India have taken initiatives mostly during the past ten years for developing natural resource accounts. China now develops on a regular basis physical and monetary accounts of air pollution, solid waste and natural resource accounts of fisheries and forests.

The *Handbook of National Accounting: Integrated Environmental and Economic Accounting—An Operational Manual* (UN 2000) explains step by step the implementation of the SEEA and provides methodologies and case studies of preparing accounts for specific resources like forests, subsoil resources, renewable aquatic resources, soil degradation and air emissions. A number of case studies using the EU framework for natural resource accounting and the SEEA are now being attempted by some countries and institutions to prepare natural resource accounts for different resources and sectors.

In a feasible work plan for preparing natural resource accounts in near-future India, first an attempt has to be made to prepare physical asset and flow accounts for all the natural resources using secondary data sources. These data could be organized to generate physical accounts as per either SEEA methodology or extended national IOT or both. Second, prepare monetary accounts of natural resource stocks using the maintenance cost method of evaluation. Third, prepare NAMEA accounts using information from the extended database of ASI accounts, input–output accounts, cost of cultivation accounts of agriculture, and NSS consumer expenditure accounts in India. Fourth, attempt to use non-market valuation methods for preparing monetary accounts of natural resource stocks in the long run. This requires the monetary valuation of all ecological services provided by natural resources in India.

References

Bartelmus P (1989) Environmental accounting and the system of national accounts. In: Ahmad Y, El Serafy S, Lutz E (eds) Environmental accounting for sustainable development. World Bank, Washington, DC

Bartelmus P (1999) Green accounting for a sustainable economy: policy use and analysis of environmental accounts in the Philippines. Ecolog Econ 29(1999):155–170

Bartelmus P, Stahmer C, Van Tongeren J (1991) Integrated environmental and economic accounting: framework for a SNA satellite system. Rev Income Wealth 37:111–148

Bolt K, Matete M, Clemens M (2002) Manual for calculating adjusted net savings. Environment Department, World Bank

Costantino C An overview of Istat's environmental accounting in the light of EU methodological achievements. Istat, Direzione Centrale Contabilità Nazionale, Via Ravà 150, Roma, cecostan@istat.it

Chinese Academy for Environmental Planning et al (2004) A study report on china environmental and economic accounting in 2004

Chopra K (1993) The value of non-timber forest products: an estimation for tropical deciduous forests in India. Econ Bot 47(3):251–257

Chopra K, Kadekodi G (1997) Natural resource accounting in the Yamuna Basin accounting for forest resources. IEG Monograph, New Delhi

Chopra K, Bhattacharya BB, Kumar P (2002) Contribution of forestry sector to gross domestic product (GDP) in India. Institute for Economic Growth, Delhi

CSO (Central Statistical Organisation) (1989) National accounts statistics: sources and methods. Ministry of Planning and implementation, CSO, New Delhi

Dasgupta P (1990) The environment as a commodity. Oxford Rev Econ Pol 6:51–67

Dasgupta P (2000) Economic issues related to water contamination and waste disposal: a case analysis of urban development in Delhi. Ph.D Thesis, Jawaharlal Nehru University, new Delhi-110067, India

Dasgupta P (2006) Valuing health damages from water pollution in urban Delhi, India: a health production function approach. Paper forthcoming in environment and development economics. Forthcoming. [No volume or issue number?] 9:83–106

Dasgupta P, Mahler KG (1991) The environment and emerging development issues. In: Proceedings of the World Bank annual conference on development economies. World Bank, Washington, DC

Dasgupta P, Maler KG (1998) Decentralisation schemes, cost-benefit analysis, and net national product as measure of social well-being. The development economics discussion papers series no.12, LSE, STICERD

European Commission (2006) A new environmental accounting framework using externality data and input output tools for policy analysis (EXIOPOL). Ongoing Project

European Union (2000) Integrated environmental and economic accounts for forests: results and pilot applications

FAO (2004) Integrated environmental and economic accounting for fisheries. Handbook of national accounting

Forestry Statistics, India (2000) Indian council of forestry research and education. Forestry Statistics, India, Dehradun

FSI (Forest Survey of India). Annual reports: state of the forest report. FSI, Dehradun

Gilbert A (1990) Natural resource accounting: a case study of Botswana. In: Dixon JA, James DE, Sherman P (eds) Dryland management: economic case studies. Earthscan, London

Hamilton K (1994) Green adjustments to GDP. Resour Pol 20(3):1S5–68

Hamilton K (1996). Pollution and pollution abatement in the national accounts. Rev Income Wealth 42(1):13–33

Hamilton K (1997) Defining income and assessing sustainability. World Bank, Environment Department, Washington, DC (Processed)

Hamilton K, Clemens M (1999) Genuine savings rates in developing countries. World Bank Econ Rev 13(2):333–356

Hamilton K, Atkinson G, Pearce DW (1998) Savings rules and sustainability: selected extensions. Paper presented to the World Congress on environment and resource economics, Venice, June 25–27. Processed.

Haque AKE, Murty MN, Shyamsundar P (eds) (2010) Environmental valuation in South Asia, forthcoming. Cambridge University Press, Cambridge

Haque AKE, Murty MN, Shyamsundar P (eds) (2011) Environmental valuation in South Asia. Cambridge University Press, Cambridge

Haripriya GS (2000) Integrating forest resources into the system of national accounts in Maharashtra. Environ Devel Econ 5:143–156
Haripriya GS (2001) Accounting for the forest resources in the national accounts in India. Environ Resour Econ 19(1):73–95
Haripriya GS (2008) Environmental accounting of land and water resources in Tamilnadu. Final report submitted to the Central Statistical Organization, Government of India, Madras School of Economics, Chennai-600025, India
Haripriya G et al (2007) Estimating the value of educational capital formation in India. Monograph 5, GAISP (Green Accounting for Indian States Project)
Hartwick JM (1977) Intergenerational equity and the investing of rents from exhaustible resources. Am Econ Rev 66:972–974
Hartwick JM (1990) Natural resources, national accounting and economic depreciation. J Public Econ 43:291–304
Hartwick JM (1992) Deforestation and national accounting. Environ Resour Econ 2(5):513–521
Hecht JE (2000) Environmental accounting: findings from nine case studies, Green Accounting Initiative. The World Conservation Union, Washington DC, USA
ICFRE (Indian Council of Forestry Research and Education) (1995) Forestry statistics India. ICFRAE, Dehradun
ICFRE (Indian Council of Forestry Research and Education) (2000) Forestry statistics India. ICFRE, Dehradun
Lawn P (2007) A stock-take of green national accounting initiatives. Soc Indic Res 80:427–460
Lvovsky K et al (2002) India: household energy, indoor pollution, and health. ESMP, The Netherlands (BNPP Publications, World Bank)
Mäler KG (1991) National accounts and environmental resources. Environ Resour Econ 1:1–15
Markandya A, Murty MN (2000) Cleaning up Ganges: the cost benefit analysis, forthcoming. Oxford University Press, New Delhi
Markandya A, Murty MN (2004) The cost benefit analysis of clearing ganges: some emerging environmental and development issues. Environ Dev Econ
Murty MN (2010) Environment, sustainable development and well-being: valuation, taxes and incentives. Oxford University Press, Delhi
Murty MN, Gulati SC (2006) Natural resource accounts of air and water pollution: case studies of Andhra Pradesh and Himachal Pradesh states of India. Final report submitted to the Central Statistical Organization, Government of India, Institute of Economic Growth, University of Delhi Enclave, North Campus, University of Delhi, Delhi-110007
Murty MN, Kumar S (2004) Environmental and economic accounting for industry. Oxford University Press, New Delhi
Murty MN, James AJ, Mishra S (1999) Economics of industrial pollution: Indian experience. Oxford University Pess, New Delhi
Murty MN, Gulati SC, Chettri P (2003a) Valuation and accounting of urban air pollution in the Indian subcontinent. Monograph, South Asian Network of Economic Institutions (SANEI)
Murty MN, Gulati SC, Banerjee A (2003b) Health benefits from urban air pollution abatement in the Indian subcontinent. Working Paper No. E/236/2003, Institute of Economic Growth, Delhi-110007, India
Murty MN, Gulati SC, Banerjee A (2003c) Hedonic property prices and valuation of benefits from reducing urban air pollution in India. Working Paper No. E 237/2003, Delhi Institute of Economic Growth
Panchamukhi PR, Trivedi P et al (2008) Natural resource accounting in Karnataka: a study of land and forestry sectors. Final report submitted to the Central Statistical Organization, Government of India, Centre for Multi-displinary Development Research, Dharwad-580004, Karnataka, India
Parikh KS, Parikh JK, Sharma VK, Painuly JP (1992) Natural resource accounting: a framework for India. Indira Gandhi Institute of Development Research, Bombay
Parikh J et al (2008) Natural resource accounting in Goa Phase II. Final report submitted to the Central Statistical Organization, Government of India, Integrated Research and Action for Development (IRADe), New Delhi, India

Peskin HM (1989) Environmental and non-market accounting in developing countries. In: Ahmad Y, El Serafy S, Lutz E (oos) Environmental accounting for sustainable development. World Bank, Washington, DC
Peskin HM, Lutz E (1990) A review of resource and environmental accounting in industrialized countries. World Bank Environment Department Working Paper 37. World Bank, Washington, DC
Pezzey J (1997) Sustainability constraints versus "optimality" versus intertemporal concern and axioms versus data. Land Econ 73:448–466
Rapport D, Friend AM (1979) Towards a comprehensive framework for environmental statistics: a stress-response approach. Statistics Canada, Ottawa
Repetto R (1989) Wasting assets: natural resources in the national accounts. World Resources Institute, Washington
Roy J et al (2008) Natural resource accounting for West Bengal: sectors air and water. Final report submitted to the Central Statistical Organization, Government of India, Global Change Programme, Jodavpur University, Kolkata, India
Singh G (1997) Environmental impact assessment of mining projects. The Indian Mining & Engineering Journal, Special issue on Environment, pp 36–43
Singh OP et al (2008) Environmental accounting of natural resources of Meghalaya: land and forest resources. Final report submitted to the Central Statistical Organization, Government of India, Centre for Environmental studies, North-eastern Hill University, Shillang-793022, India
System of National Accounts (SNA) (1993) Commission of the European Communities. International Monetary Fund, Organisation for Economic Co-operation and Development, United Nations and World Bank, United Nations. http://unstats.un.org/unsd/sna1993/introduction.asp
Theys J (1989) Environmental accounting in development policy: the French experience. In: Ahmad YS, El Serafy S, Lutz E (eds) Environmental accounting for sustainable development. World Bank, Washington, DC
TERI (1994) Accounting for coal depletion in the national income accounts of India; natural resource accounting—a means to measure sustainable development in India. Energy and Resources Institute, New Delhi
UK Environmental Accounts Autumn (2005) A national statistics publication. www.nationalstatistics.gov.uk
United Nations (1997) Accounting and valuation of environment. A primer for developing countries, vol. 1 and 2, New York, ST/ESCAP/1810
UN (2000) Handbook of national accounting: integrated environmental and economic accounting—an operational manual. United Nations, New York
United Nations, European Commission, International Monetary Fund, Organisation for Economic Cooperation and Development, World Bank (2003) Handbook of national accounting, integrated environmental and economic accounting. Studies in Methods, Series F, No. 61, Rev.1 (ST/ESA/STAT/SER.F/61/Rev.1)
United Nations, SEEA (2003) Handbook of national accounting 'integrated environmental and economic accounting 2003'. European Commission/International Monetary Fund/Organisation for Economic Co-operation and Development/World Bank. http://unstats.un.org/unsd/envAccounting/seea2003.pdf
UN Statistics Division (2005) Integrated Environmental and Economic Accounting for Water Resources. United Nations Statistics Division in cooperation with the Subgroup of the London Group on Water Accounting
Verma M, Vijaykumar CVRS (2006) Natural resource accounting of land and forest sector for the states of Madhya Pradesh and Himachal Pradesh. Indian Institute of Forest Management, Bhopal
Wang J, et al (2005) A draft report on pollution-adjusted GDP accounting in China in 2004. Chinese Academy for Environmental Planning, Beijing, p 12
World Health Organization (2007) Indoor air pollution: national burden of disease estimates. WHO Press, World Health Organization

Yu F, Hongqiang J, Dong C, Jinnan W, Caizhon G (1989) A proposed environmental accounts framework. In: Ahmad Y, El Serafy S, Lutz E (eds) Environmental accounting for sustainable development. World Bank, Washington, DC

Yu F, Hongqiang J, Dong C, Jinnan W, Caizhon G (1990) Environmentally sound and sustainable development: a conceptual framework. Paper presented to the Canadian Environmental Advisory Council Workshop, Indicators for ecologically sustainable development economies, July

Yu F, Hongqiang J, Dong C, Jinnan W, Caizhon G (2006) Integrated environmental and economic accounting in China: proposed framework and preliminary findings. Department of Economic and Social Affairs Statistics Division United Nations

Chapter 5
Strengthening Forest Resource Valuation and Accounting System: A Case of Forest Resources of Kerala, India

Kiran P. Mali, Katar Singh, P. C. Kotwal and M. D. Omprakash

5.1 Introduction

Forests play an important role in India's economic development in terms of their contribution to gross domestic products (GDPs), employment and livelihoods of the poor people, who are mainly dependent on forests. Besides, they are also the main source of meeting food, fuel, fodder and timber requirements of the forest dwellers. In 2008–2009, the forestry sector contributed Rs[1] 888,230 million to India's GDP at the current prices, which was 1.7 % of the GDP (CSO 2009) estimated at base year 2004–2005. The contribution of forests to India's GDP at current prices has declined from 2.1 to 1.7 % over the 5-year period from 2004–2005 to 2008–2009. Though the contribution to the GDP has been declining over time, the value of its output had increased from Rs 737,830 million in the year 2004–2005 to Rs 1,052,410 million in the year 2008–2009, registering growth of 43 % over 5 years. In Kerala also, the relative contribution of forests has declined. For example, their contribution to the state's GDP has declined from 1.8 to 0.9 % over the 7-year period from 1999–2000 to 2006–2007 at current prices (DES 2007).

There are several reasons for the declining contribution of the forestry sector in India's economy and the state's economy. Underestimation of the value of all direct

[1] Indian national rupees. One US$ equal to Rs 44.838 as on 09.05.2011.

K. P. Mali (✉) · M. D. Omprakash
Indian Institute of Forest Management, P.Box-357, Nehru Nagar, 462 003 Bhopal,
Madhya Pradesh, India
e-mail: kiranpmali@gmail.com

K. Singh
Ex Director, Institute of Rural Management Anand, Gujarat, India

P. C. Kotwal
Ex-Faculty, Indian Institute of Forest Management, Bhopal,
Madhya Pradesh, India

N. Ghosh et al. (eds.), *Nature, Economy and Society*,
DOI 10.1007/978-81-322-2404-4_5

and indirect benefits is one of the major reasons for the decline in the contribution of the forestry sector to the GDP and consequently low budgetary allocation for the sector. Benefits from forests are at present grossly underestimated with the result that the actual benefits are several times higher than those reported and incorporated in the national income accounts.

In India, forests meet nearly 40 % of the energy needs of the country of which more than 80 % is utilized in the rural areas only. About 30 % of fodder needs are also met through forests. Forest products, especially non-wood forest products (NWFPs), provide sustenance to the rural poor, particularly tribals. For landless families and marginal farmers, forest-related activities often represent the primary source of income and livelihood. In view of this, sustainability of forest products is essential for sustaining their livelihoods. But, over the past few decades, forests have been under great biotic pressures. They have been subject to over-exploitation, unsustainable rates of felling, illegal removal of timber, unregulated grazing, fire hazards, encroachment and diversion of forestland to other development projects. Consequently, forest area, productivity and production have declined, and their sustainability has been threatened. Needless to say, meeting the requirements of wood for housing, furniture and allied activities in rural and urban areas result in unsustainable extraction of forest produce. The problems of degradation of natural resources and deterioration of environment pose a serious threat to the survival and livelihoods of mankind, particularly the poor in developing countries, who mainly depend on natural resources for their livelihood. In order to address the concerns mentioned above, there is greater need now than before to estimate the extent of damages caused to the natural resources and environment by the fast growing economic developmental activities and accounting for them while estimating the GDP of a country. In response to this need, a growing number of research studies have been conducted to assess the adverse environmental impacts of economic developmental activities and to account for them in the process of preparing national accounts including green GDP and green net domestic product (NDP). Green national accounts (green GDP accounts for short) refers to an accounting system that takes into account costs of depletion of natural resources and environmental degradation while assessing the quality of economic development in the real sense.

There have been numerous attempts in many countries of the world for preparing green national income accounts by incorporating environmental costs and benefits into the system of national accounts (SNA). Computation of the forestry sector's contribution[2] to economy is more common in developed countries than in the developing countries (Lange 2004). Vincent (2003) pointed out that one might expect countries with abundant natural resources to have a natural advantage in economic development, but the historical records suggest otherwise. Development experts in the early 1960s expected the resource-rich Asian countries like Burma, Philippines

[2] The contribution of any activity to the economy (e.g., GDP) is measured as the value added or produced by the sector.

and Sri Lanka to flourish, but they were vastly outperformed by the resource-poor 'tigers' (Hong Kong, Singapore, South Korea, Taiwan) over the following 30 years. Similarly, sub-Saharan Africa is rich in minerals, timber and land, but it has suffered negative economic growth rates since the mid-1960s. A cross-country econometric study by Sachs and Warner (1995) rigorously confirmed that per capita GDP has grown less rapidly throughout the world in resource-rich countries than in resource-poor countries.

In developing countries, most studies have examined the value of the subsistence use of forest products, rather than its broader social and environmental benefits. Although interest in forest valuation and payment for environmental services remains high, many developing countries are unable to use these techniques owing to the high costs of data collection, analysis and the establishment of markets for such payments. In addition, many developing countries face difficulty in collecting forest charges and taxes from the producers of forest products. This indicates that addressing the latter problem should be given a higher priority for immediate action than developing more sophisticated mechanisms such as payments for environmental services (FAO 2007). Unfortunately in India too, necessary action for preparing green national account has not yet been initiated at the national level. However, a few state level studies aimed at computing sectoral accounts particularly limited to land, water and forest resources have been conducted in the recent past (Singh and Mali 2005).

The National Forest Commission (NFC; 2006) has reviewed and examined various aspects of Indian forests including forest resource accounting (FRA). The NFC recommended, inter alia, that to reflect the real contribution of the forestry sector to India's economy, an appropriate system of FRA is necessary. More specifically, there are five recommendations in the report pertaining to FRA, that is, constitution of a national co-ordination committee, strengthening of the central statistical organisation (CSO), dissemination of information, preparation of a manual for capacity building and introduction of a new system of FRA (http://envfor.nic.in). FRA is an essential instrument for achieving the objectives of sustainable forest management (SFM). FRA promotes sustainable forestry by broadening the perspective of managing forests with the application of an economic approach. It not only helps identify the different forests benefits provided to local, regional and national economy but also helps identify different costs (social, environmental and economic)/threats to forests due to unsustainable management. The consolidated set of forest assets/capital, flow and an expenditure account depicts the impacts of forestry activities on environmental consequences, economic dimension and society.

Singh et al. (2009) pointed out that in the current FRA system, the revenue generated from the sale of only some of the forest products, such as timber, fuel wood, bamboo and a few nationalized non-timber forest products (NTFPs), such as *Diospyrous melanoxylon* (tendu) leaves, gum, etc., have been included. In the national accounts, the economic activities covered under the forestry sector include (i) forestry (e.g., planting and conservation of forests, gathering of forest products, charcoal burning carried out in the forests), (ii) logging (e.g., felling and rough cutting of trees, hewing or rough shaping of poles, blocks, etc.) and transportation of forest products to the sale depots/assembly centres, and (iii) farmyard wood (industrial wood and fuel wood collected by the primary producers from trees outside regular forests). Production of field crops (Jhum cultivation, etc.) and extraction of minor and major minerals from forests are included in agriculture and mining sectors, respectively (CSO 2007).

5.1.1 Concepts of FRA

FRA may be defined as a process of identifying and measuring various benefits and costs of forests, putting value tags on them and recording them in appropriate sets of accounts/statements. It comprises both physical accounts and monetary accounts. It also refers to the preparation of assets and liability statements of forest resources for the particular period of time together with the changes in the opening and closing balance (Fig. 5.1).

This chapter estimates the quantity and monetary value of unrecorded forest benefits in Kerala state and identifies some of the weaknesses of the existing system of FRA. FRA would facilitate the estimation of real contribution of the forestry sector to the state's as well as country's economy, which would provide a strong rationale for higher budgetary allocation of funds for the sector at both the state and national levels.

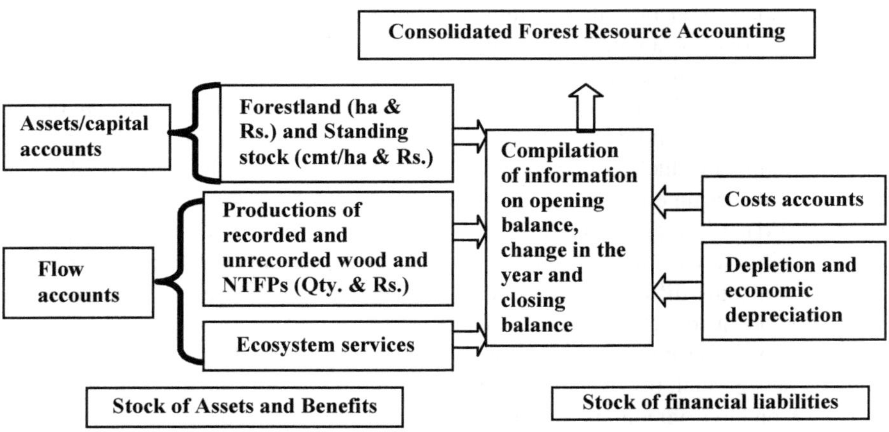

Fig. 5.1 Attribute of FRA. *NTFP* non-timber forest products, *cmt* cubic metre

5.1.2 Components of FRA

5.1.2.1 Forest Assets/Capital Account

The forestland and standing stock on it together form the forest assets. These two parts of forest assets are termed as forest estate. Forestland and standing stock (if not cultivated) are treated as non-produced assets, and if standing stock is cultivated, then it is treated as produced assets. The forestland may be categorized depending upon their legal status and land-use pattern, and the natural forests, plantations and privately managed forests constitute parts of standing stock accounts. There are two important changes related to forestland, viz., permanent loss in it and changes in it due to economic activity. The permanent loss in the forestland covers diversion of forestland to non-forest uses such as agriculture, hydropower generation and residential and industrial purposes. The changes due to economic activities, land encroachment, illegal logging, afforestation area, regeneration area, area subject to forest fires, area subject to floods, shifting cultivation areas and areas subject to grazing. The changes in the standing stock takes place due to activities such as logging/harvesting, illegal logging, damage due to logging, incidence of diseases and insect infestation, stand mortality, forest fires and natural calamities such as floods, thunderstorms, landslides, etc. that decrease (reduce) the stock of forests. The natural growth, regeneration and plantations increase (add) the stock of forests.

5.1.2.2 Flow Accounts

It comprises various kinds of wood and NTFPs provided by the forests. These forest products are in the form of timber, wood, fuel wood, bamboo, fodder, poles, NWFPs, Nistar supply and wastages in harvesting, storage and processing. The change in the flow of wood and NTFPs occurs due to forest departmental harvesting and the collection by the forest dwelling communities. Besides, forest also offers various ecosystem services like carbon sequestration, ecotourism, soil and water conservation, biodiversity, etc.

5.1.2.3 Costs

The cost accounts comprise plan and non-plan expenditure. Besides, it also covers environmental costs in term of damage caused to forest ecosystems due to fire, unsustainable harvesting and livestock grazing, impact of deforestations, diversion/encroachment of land for non-forestry purposes, shifting cultivation, etc. The estimation of environmental costs involves the assessment of loss in the ecological benefits and services of forests due to bad practices. It also covers the administrative and operational cost incurred to recoup/avoid damage for maintaining the forest ecosystem intact. Economic costs cover the costs of collection and marketing of forest products. The social costs cover damages which are caused by wild animals to the property of the people living in and around the forest areas.

5.1.2.4 Depletion and Economic Depreciation

Depreciation may be defined as the reduction in the quantity of assets due to economic uses, for example, timber harvesting. It should be noted here that even when the difference between the opening and closing stock in the physical account is positive or zero, depletion may still occur in monetary units as a result, for instance, of changes in the age structure of the standing timber. It is a thumb rule that for maintaining the sustainability, the quantity of timber harvested (cmt/ha) should not exceed the net growth rate/increment of timber. In economics and accounting, economic depreciation is the change in the market value of the capital over a given period. It is calculated as the market price of the capital at the beginning of the period minus its market price at the end of the period.

5.2 Materials and Methods

5.2.1 Selection of Study Area/Sites

The study was conducted in three forest divisions of Kerala, viz., Palakkad, Munnar and Thenmala. The selected forest divisions represent southern tropical wet evergreen forests, southern tropical semi evergreen forests, South India moist deciduous forests, southern subtropical broad leaved hill forests, southern montane wet temperate forests, west coast tropical evergreen forests, southern hilltop evergreen forests, west coast semi-evergreen forests, southern moist mixed deciduous forests (Fig. 5.2). Some of the salient features of the selected forest divisions are presented in Table 5.1.

5.2.2 Selection of Villages

For collection of primary data, five villages were purposively selected from each of the three selected forest divisions. Thus in all, a sample of 15 villages was selected. The criteria used for selection of the sample villages were: (a) location of the village within a radius of 5 km from the forest and (b) existence of a Van Samarakshana Samities (VSS) in the village. A list of selected villages in each of the three forest divisions is given in Table 5.2. The selected VSSs represent all major types of forests from tropical and temperate wet evergreen forests to tropical dry deciduous forests in the state.

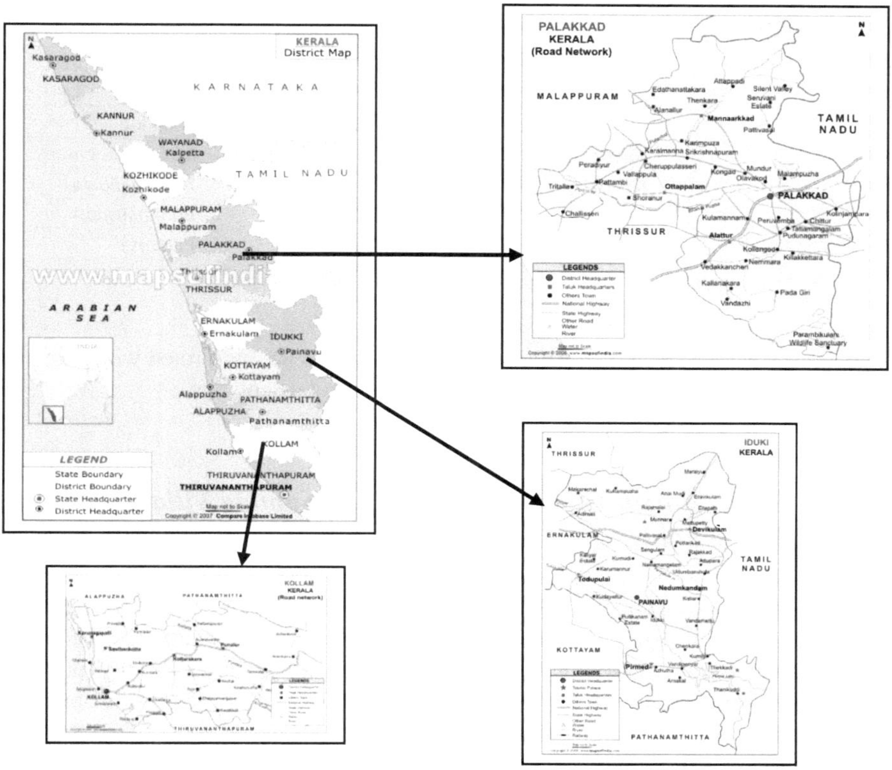

Fig. 5.2 Map of the study site. (Source: www.mapsofindia.com)

Table 5.1 Details of the selected forest divisions in Kerala. (Source: office records of the forest divisions concerned)

Name of forest division	Forest area (ha)	Particulars of villages within periphery of 5 km of forest division	
		Number of villages	Number of households
Palakkad	23,607.91	83	13,280
Munnar	38,666.20	68	11,016
Thenmala	14,075.00	12	1368

5.2.3 Selection of Households

From each of the five selected villages, a sample of five households was selected randomly. Thus, a sample of total 25 households was selected from each of the forest divisions and the whole sample comprised 75 households. Given the small size of the sample selected from each of the three forest divisions, we would like to

Table 5.2 List of villages selected from the sample forest divisions

Selected VSSs	Palakkad Forest Division	Munnar Forest Division	Thenmala Forest Division
	Vadyarchalla	Kurathilkudi	Ottakkal
	Nadupathy	Kanjiraveli	Malavedar colony
	Kottekkad	Elamplassery	Kadamanpara
	Thudikkode	5th Mile	Edappalayam
	Champan	Choorakettan	Palaruvi

VSS Van Samarakshana Samities

sound a caveat here that the results of this study should be interpreted with care and should be considered exploratory rather than being representative.

5.2.4 Some Salient Characteristics of the Sample Villages and Households

The level of literacy in the majority of the selected villages ranged from 30 to 100 %. Another important feature of the selected villages was the predominance of below poverty line (BPL) households, which accounted for, on average, 72 % of the total number of households. Besides, the size of the selected villages ranged from small to large in terms of human population, which varied from 198 to 1250 persons per village. Most of the households were partially dependent on the forest for their livelihood. The other major sources of income and employment were agriculture and wage labour. For a period of 5–6 months in a year, the villagers work as daily labourers in the adjoining tehsils, districts and neighbouring states, whereas some of the villagers earn their livelihood by working as black smiths, carpenters, truck drivers, tailors and masons. The labour work includes rubber tapping, reed cuttings and working in co-operative society and paper industry.

The members of VSSs were engaged by the forest department for carrying out plantations, assisted natural regeneration (ANR), rehabilitation of degraded forests (RDFs), tending operations, soil and water conservation work, construction of roads and community halls, raising plantation of cane, bamboo, *Emblica officinalis* (aonla), *Tecktona grandis* (teak), *Terminalia chebula, Pterocarpus morsupium* (vendek/white tree/karvena), *Aegle marmelos* (bael), *Santalum album* (sandalwood), *Caesalpinia sappan* (sappanwood/bukkum-wood), *Acacia catechu* (khair), *Albizia lebbeck* (siris), *Sorbus torminalis* (tenbau), *Swietenia mahagoni* (mahogany), *Mangifera indica* (mango tree), *Tamarindus indica* (tamarind), *Casia fistula* (kannikonna), *Eucalyptus spp., Artocarpus heterophyllus* (jackfruit), *Syzygium cumini, Terminalia panniculata, Dalbergia latifolia* (rosewood), *Terminalia tomentosa, Vitex altissima, Gmelina arborea,* (Akhil), Kambakam, Hersutica, Anjali, and other species. Some of the members of the VSSs were working in the ecotourism and rubber processing activities.

The local people were also actively involved in forest fire control, and for that purpose the forest department engaged some of them as fire-watchers during summer season. The members of the selected VSSs were also involved in the implementation of entry point activities (EPA) and forest protection. It was also observed that NTFPs were being collected from the forest areas of some of the VSSs by people living in the adjoining villages. Some of the VSSs were facing a problem created by elephants, which destroyed agricultural crops such as banana, tapioca, areca nut and coconut.

5.2.5 *Sources and Methods of Collection of Data*

5.2.5.1 Primary Data

Primary data were collected through personal interviews with the heads or representatives of selected households using a pre-structured questionnaire. The data collected included the quantity of forest products collected from the forests by the sample households, cost of collection of forest products and sources of income and employment. Besides the primary data collected from the sample households, some information about the number of villages within the radius of 5 km from the forest was also collected. This included the information about the number of households in each village, livestock population and requirement of wood and NTFPs.

5.2.5.2 Secondary Data

Secondary information was collected through interactive meetings and discussions held with the concerned officers of the Kerala Forest Department and selected forest divisions, the head and divisional offices of the Kerala State Federation of Scheduled Tribes & Scheduled Tribes Co-operatives Ltd, the Head Office of the Kerala Forest Development Corporation Ltd, Department of Economics and Statistics (DES) and selected VSSs of the sampled forest divisions.

Secondary data of the selected villages were collected from the working plans of the selected forest divisions. This included the information about the number of households, livestock population and wood and NTFPs consumption. Secondary data on revenue and costs of the sample forest divisions was compiled from the records available in the offices of the divisional forest officers (DFO) concerned and from the annual administrative reports, Nistar Patrikas (usufruct right records) and working plans of the forest divisions and the office of the Principal Chief Conservator of Forests (PCCF), Thiruvanathapuram. Some of the secondary data were also collected from the CSO and DES, Thiruvananthapuram. In case of *Nistar*[3] provided

[3] By Nistar, we mean a provision made under the Indian Forest Act, 1927, for providing some special privileges in the form of supply at concessional rate/free of charge of fuelwood, poles,

by the forest department to the communities residing within the radius of 5 km from the forest area, the concessional value of the goods provided was estimated by taking the difference between the revenue obtained at the Nistar price of forest products, viz., poles, bamboos and fuel wood stacks, and the corresponding market value of that particular forest product.

5.2.6 Methods of Pricing of Forest Produce

Prices of timber, poles, fuel-wood, grass and bamboos collected by persons are based on the market prices. The prices of forest products which are mainly NTFPs collected by households were ascertained on the basis of sale price of the products actually realized by the sample households. The actual prices of the products were further cross-checked or verified with the corresponding figures available with the forest department.

5.2.7 Valuation of Free Livestock Grazing

The valuation of free livestock grazing was done using the indirect substitute approach as depicted in Fig. 5.3.

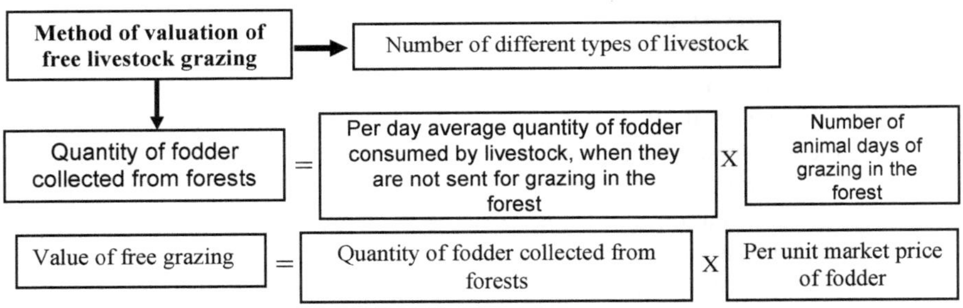

Fig. 5.3 Valuation of free livestock grazing

bamboo and other forest produce for their own use or consumption to the local people living within the 5-km radius of the forest. The Nistar benefits vary from year-to-year depending on the quantity of the forest produce available in the year which is distributed through village forest committees/forest protection committees/JFMCs/Gram Panchayats.

5.2.8 *Methods of Costing*

The cost of labour used in various forestry operations was based on the actual wages paid by the forest department to the labour. The cost of labour used by the villagers for collection of various forest products was based on the opportunity cost of labour. The annual administrative and operational costs incurred by the forest department were based on the actual expenditure incurred by the forest department.

The cost of collection of various forest products by local people was estimated using the opportunity cost method or shadow wage rate. This is based on the assumption that the person engaged in the collection has an opportunity to work elsewhere at the prevailing wage rate in the area for unskilled workers. This assumption is not totally correct, as there are not enough employment opportunities available in the area for all the persons engaged in collection of NTFPs. As a consequence, the use of current wage rate results in over-estimation of the cost of collection, which in turn led to a negative value added from the collection of some of the NTFPs such as gum.

5.2.9 *Method of Estimation of Unrecorded Value of Forest Benefits*

The following formula was used for estimating the value of unrecorded forest benefits while calculating the contribution of unrecorded forest benefits to the gross state domestic product (GSDP) (Fig. 5.4).

5.2.10 *Estimation of Standing Stock and Forestland*

For realistic assessment of the contribution of the forestry sector to the state's economy, it is necessary, among other things, to measure the growth of standing stock

$$\text{Percent contribution of value of unrecorded forest benefits to GSDP} = \frac{\text{Total value of unrecorded forest benefits* (Rs.)} \times 100}{\text{Average GSDP (Rs.)}}$$

$$\text{*Total value of unrecorded forest benefits (Rs.)} = \text{Value of unrecorded forest benefits (Rs./ha)} \times \text{Total forest area (ha)}$$

Fig. 5.4 Estimation of the value of unrecorded forest benefits. *GSDP* gross state domestic product

and the extent of forestland and put a value tag on them. There is some data available on the volume of standing stock and forest area with different types in the annual administrative reports of the Forest Department and Forest Survey of India (FSI) reports. Based on that information, we estimated the value of the standing stock using the stumpage value method. The value of forestland was estimated on the basis of the land prices fixed by the Supreme Court of India for pricing of forestland that is diverted for non-forestry uses.

5.3 Results and Discussion

The integrated System of Environmental-Economic Accounting (SEEA) framework (SEEA 2003) proposes a multi-step process for creating satellite environmental accounts for three types of natural assets, viz., produced assets, non-produced economic assets and non-produced environmental assets. Forests belong partly to produced economic assets (plantations) and partly to non-produced environmental assets (natural forests). In the first step, the SEEA involves the setting up of a physical account that contains information on opening stock, depletion, other volume changes and closing stock. The second step consists of valuation of opening stock by multiplying its items by the net price of the item at source at the beginning of the period. Similarly, it involves valuation of other volume changes by multiplying the estimates of volume by their average net price during the period and closing stock by net price at the end of the period. For known market environmental assets, the SEEA advocates the use of 'maintenance cost' method of computation. Finally, the SEEA proposes the computation of 'revaluation' (SEEA 2003). Revaluation is defined as the residual difference between the value of closing stock, value of depletion, other volume changes and opening stock (Bertelmus 1998). The major components of SEEA forest accounts are important in the Indian context also, the SEEA-FRA seems to be most appropriate as a basis for developing a new system of FRA for India. The proposed set of forest valuation and accounting could be maintained at the forest division level and state level with the help of existing staff and infrastructure with some training and additional funds.

5.3.1 Valuation of State's Forest Assets

The value of forestland and growing stock represents a type of stock benefit, which does not appear in the conventional system of FRA. While incorporating the value of forestland in the SNA, the annual incremental difference in the value of forestland should be considered. The estimates of flow of benefits from these two natural assets could be made by estimating the change during the year in the stocks in these two assets. If the changes are positive, their values are to be included in the forest

Table 5.3 The net present value of forestland in Kerala by class

Class of forest	Area[a] (ha)	Price (Rs lakh/ha)	Total value (Rs crore)
High conservation value (very dense forest)	144,300	9.18	132,467.40
Medium conservation value			
Moderately dense	941,000	8.28	779,336.20
Open forest	647,100	8.28	535,928.20
Low conservation value area (scrub and barren forest area)	5800	6.45	3738.70
Total	*1,738,200*	*8.05*[b]	*1,451,470.50*

[a]The forest cover includes plantations of coconut, areca nut and other miscellaneous species. It excludes tree cover (FSI 2009)
[b]Average price of forestland

resource accounts under the head 'Flow Benefits'. For example, in the case of forestland, if there is a diversion of forestland for non-forestry purposes or encroachments, the area diverted or encroached will represent the diminution in the stock of forestland, and its value would be considered as cost. In case of growing stock, the value of annual increment in the growing stock is to be considered as a benefit accruing in the particular year.

5.3.2 *Value of Forestland*

The net present value of forestland in the state was estimated by multiplying the area of forest under the categories of high conservation (evergreen and natural forest) value, medium conservation (plantations) value and low conservation (barren land) value by their respective prices, that is, Rs 9.18 lakh/ha, Rs 8.28 lakh/ha and Rs 6.45 lakh/ha as stipulated by the Supreme Court of India vide its order of 05.06.2008. The estimates of net present value thus computed are presented in Table 5.3. As shown in the table, the total net present value of forestland of all classes was Rs 1,451,470.50 crore.

The figure of Rs 1,451,470.50 crore estimated in Table 5.3 will form the opening stock of forestland. The increase in value (due to rise in prices of forestland)/decrease (due to diversion, encroachment and fall in prices, etc. of forestland) will constitute the 'change in the year'.

5.3.3 *Value of Growing Stock*

Like forestland, growing stock of forest trees is also a natural asset and therefore in any system of FRA; its value should also be estimated. There are many methods of

Table 5.4 Forest cover, growing stock, biomass and carbon store of Kerala forests. (Source: www.fsi.nic.in/fsi_projects)

Year	Forest cover (km^2)	Per ha growing stock (cmt/ha)	Growing stock (000, cmt)	Biomass per ha (Mt/ha)	Biomass (000, t)	Carbon per ha (Mt/ha)	Carbon (000, t)
1989	10,292	87	89,239.65	46	47,801.75	21	21,466.62
1994	10,334	87	89,532.17	46	47,959.28	21	21,537.31
2007	17,324[a]	87	150,718.80	46	79,690.40	21	36,380.40

[a]FSI (2009) reports forest cover with forest crown density more than 10 %. The forest cover includes plantations of coconut, areca nut and other miscellaneous species. It excludes tree cover

Table 5.5 Total growing stock, annual increment in it and its value in Kerala

Total forest cover (km^2)	Growing stock (000, cmt)	Asset value of growing stock at Rs 6000[a]/cmt (Rs)	Annual increment in growing stock (cmt/ha)	Total annual increment[b] (000, cmt)	Annual incremental value of growing stock at Rs 6000/cmt
17,324	150,718.80	904,312,800,000	1.14	1983	11,898,000,000

[a]Assuming one cmt of growing stock constitutes half cmt of biomass. The average price of industrial wood, plywood species, matchwood species, bobbin wood species, pencil wood species, packing case wood species, pulpwood species and miscellaneous species was Rs 10,839/cmt; therefore, the average stumpage price is taken as Rs 6000/cmt
[b]Source: Extent, Composition, Density, Growing Stock and Annual Increment of India's Forests. Forest Survey of India (1995), downloaded from http://www.envfor.nic.in/nfap/chap05.html

valuation of growing stock, but we have chosen the stumpage value method (value of standing timber) due to its practicability under the current situation with regard to availability of data. The findings of the study 'Assessment of Growing Stock, Biomass and Carbon in India's Forests' conducted by the Ministry of Environment and Forests, India, in 2001 have been used for estimating the growing stock of Kerala forests. The study estimated the growing stock of forest areas with a forest crown density of more than 10 %. The estimates of forest cover, growing stock, biomass and carbon for the year 1989, 1994 and 2005 are presented in Table 5.4. For the year 2005, the growing stock, biomass and carbon were estimated on the basis of findings of 2 years, that is, 1989 and 1994.

Using the above-mentioned method, we computed the total growing stock, annual increment in it and its value, as shown in Table 5.5.

5.3.4 *Carbon Sequestration*

One of the important benefits of forests is carbon sequestration, that is, their ability to store carbon. According to Manhas et al. (2006), the conifers of temperate

Table 5.6 Carbon sequestration by growing stock of Kerala's forest

Total forest cover (km^2)	Per ha carbon (t/ha)	Total carbon sequestration (000, t)	Asset value of carbon[a] at Rs 226/t (Rs)	Annual increment in carbon (t/ha)	Total annual increment in carbon (t)	Annual incremental value of carbon at Rs 226/t (Rs)
17,324	21	36,380.40	8,221,970,400	0.28	478,655[b]	108,176,030

[a]At US$5/t of carbon, 1 US$ = Rs 44.838 as on 09.05.2011

[b]Assessment based on the norm that 150718800 cmt of growing stock consist of 36380400 t of carbon (Table 5.4). Therefore, 1983000-cmt annual increments (Table 5.5) in growing stock sequester 478655 t of carbon

regions stored maximum carbon in their woods, 28.88–65. 21 t carbon/ha, followed by mangrove forests, 28.24 t carbon/ha, Dipterocarp forests, 28.00 t carbon/ha, and *Shorea robusta* forests, 24.07 t carbon/ha. *Boswellia serrata* with 0.22 million ha forest area stored only 3.91 t carbon/ha. Based on their estimates and using the data presented in Table 5.4, we computed the carbon sequestration of growing stock of Kerala's forest, as shown in Table 5.6.

5.3.5 *Flow Accounts*

5.3.5.1 Valuation and Accounting of Forest Goods and Services

This section presents a synthesis of the major findings of the study and extrapolates the results to the state level. The study revealed that the recorded net benefits from the forests were negative in the cases of Palakkad and Thenmala Forest Divisions. This could be justified because the main objective of the forest management now is to conserve and protect the forest resource. Furthermore, besides the recorded benefits, forests generate several benefits which are not valued and recorded. In fact, the value of unrecorded benefits including environmental goods, services and amenities is much more than the recorded benefits, and hence there is a need to take into consideration the values of unrecorded forest benefits to realistically estimate the value of all forest goods and services.

5.3.6 *Recorded Benefits and Costs*

Table 5.7 presents the aggregated estimates of recorded benefits and costs separately for each of the three selected forest divisions and for all three divisions as a whole and on per hectare basis. As one could see from the table, the net benefit was positive for the Munnar Forest Division. On the whole, the net benefit from the forest was

Table 5.7 Estimates of recorded benefits and costs at divisional level

S.N.	Name of forest division	Forest area (ha)	Recorded benefits[b]		Recorded costs[c]		Net benefits	
			Total (Rs lakh)	Per ha (Rs)	Total (Rs lakh)	Per ha (Rs)	Total (Rs lakh)	Per ha (Rs)
1	Palakkad[a]	23,607.91	56.89	241	321.75	1363	– 264.86	– 1122
2	Munnar	38,666.20	811.23	2098	380.97	985	430.26	1113
3	Thenmala	14,075.00	79.97	568	241.58	1716	– 161.61	– 1148
All		*76,349.11*	*948.09*	*1242*	*944.3*	*1237*	*3.79*	*5*

[a]In case of Palakkad Forest Division, average recorded benefits of last 5 years from 2002–2003 to 2004–2005 and from 2007–2008 to 2008–2009 and average recorded costs of last 2 years from 2007–2008 to 2008–2009, the value of benefits comprises the revenue obtained from the sale of timber, teak poles, billets, fuel wood, bamboo and reeds and NTFPs. The recorded cost was the average of 2 years from 2007–2008 to 2008–2009

[b]In case of Munnar and Thenmala Forest Divisions, the average of latest 5 years from 2003–2004 to 2007–2008, the value of benefits comprises the revenue obtained from the sale of timber, bamboo, reeds, ecotourism and NTFPs

[c]Average of latest 5 years from 2003–2004 to 2007–2008 in case of Munnar and Thenmala Forest Divisions, it comprises the plan and non-plan expenditure

positive and so also on per hectare basis. The average net benefit per hectare was almost negligible (Rs 5); it ranged from Rs (−)1148 to Rs 1113. However, if we factor in the unrecorded benefits in the form of timber, fuel wood, poles, bamboo, reed fodder and NTFPs that accrue to local people, the loss would not only be wiped off, but there would be a surplus of net revenue from the forests. The per hectare net benefit would go up further if we include the value of intangible benefits from the forests to the local people and society at large. This is done in the following section.

5.3.7 *Unrecorded Benefits and Costs*

In case of Palakkad Forest Division, the net benefit from unrecorded forest products accruing to the people living in and around the forest areas was Rs 282.35 lakh. It is important to note that almost all the unrecorded benefits are realized by the people. However, the forest division incurs the average cost of Rs 321.75 lakh annually in managing and protecting the forests. On an average, the value of net benefits from forests accruing to the local people was Rs 5315/household/year. Further, on an average, each of the households spent 30 person days in collection of forest products. At the current wage rate of Rs 120 per person per day, the average benefit in terms of employment generated from forests was Rs 1,55,37,600 or Rs 2925/household/year. Thus, the total benefit realized by local people was Rs 8240/household/year. Moreover, the forest department also engages on an average 25 persons/day for 50–60 days per VSSs per year for undertaking forestry improvement and EPA. This is an additional benefit accruing to the VSS from direct employment provided by the forest department.

In Munnar Forest Division, the net benefit from unrecorded forest products accruing to the people living in and around the forest within the periphery of 5 km was Rs 283.21 lakh. As we stated earlier, almost all the unrecorded benefits are realized by the people, while the forest department incurs the average cost of Rs 380.97 lakh annually on managing and protecting the forests. On average, the value of net benefits from forests in the Munnar Forest Division accruing to the local people was Rs 7791/household/year. Further, on average each household spent 70 person days in collection of forest products. At the current wage rate of Rs 125/person/day, the benefit in terms of employment generated from forests was Rs 442.00 lakh or Rs 4180/household/year. Thus, the total benefit realized by local people was Rs 11,971/household/year. Moreover, the forest department engages on an average 58 persons/day for 64 days per VSS per year for undertaking forestry improvement and EPA. This is an additional benefit from direct employment provided by the forest department.

In Thenmala Forest Division, the net benefit from unrecorded forest products accruing to the people living in and around the forest within the periphery of 5 km was Rs 17.47 lakh. As we stated earlier, almost all the unrecorded benefits are realized by the people while the forest department incurs the average cost of Rs 241.58 lakh annually on managing and protecting the forests. On average, the value of net benefits from forests in the Thenmala Forest Division accruing to the local people was Rs 5266/household/year. Further, on average each household spent 25 person days in collection of forest products. At the current wage rate of Rs 175/person/day, the benefit in terms of employment generated from forests was Rs 18.55 lakh or Rs 2712/household/year. Thus, the total benefit realized by local people was Rs 7978/household/year. Moreover, the forest department engages on an average 31 persons/day for 85 days per VSS per year for undertaking forestry improvement and EPA. This is an additional benefit from direct employment provided by the forest department.

Table 5.8 presents the aggregated estimates of unrecorded benefits and costs for each of the three selected Forest Management Unit (Forest Division in Indian context) (FMUs) and for all three FMUs as a whole and on a per hectare basis. As shown in the table, the average unrecorded net benefit was Rs 764/ha. The unrecorded net benefit per hectare was ranged from Rs 124 and Rs 1196. The average unrecorded net benefit per hectare of the order of Rs 764 seems to be quite good, as all these benefits go to the poor local people who mainly depend on forests for their livelihood. Further, the average net benefit per household was Rs 9396 per annum, which was a big sum by local standards.

5.3.8 *Projection of Benefits at the State Level*

Using per hectare averages, the recorded and unrecorded benefits were computed at the state level through extrapolation on the basis of four case studies. They are

Table 5.8 Estimates of unrecorded benefits and costs at FMU level

S. N.	Name of forest division	Forest area (ha)	Unrecorded benefits		Unrecorded costs		Net benefits	
			Total (Rs lakh)[a]	Per ha (Rs)	Total (Rs lakh)[b]	Per ha (Rs)	Total (Rs lakh)	Per ha (Rs)
1	Palakkad	23,607.91	437.73	1854	155.38	658	282.35	1196
2	Munnar	38,666.20	725.21	1876	442.00	1143	283.21	733
3	Thenmala	14,075.00	036.02	256	018.55	132	017.47	124
All		*76,349.11*	*1198.96*	*1570*	*615.93*	*807*	*583.03*	*764*

[a]Comprises wood for construction of new houses, repairing of houses, agricultural purpose, fuel wood, bamboo, reeds, grazing of livestock inside forests and NTFPs, viz., honey, honey wax, *Pseudarthria viscida* (moovila), *Desmodium gangeticum* (orila), *Acacia concioni* (cheenikka), *Vateria indica* (funthirikka), Honey Mizhaku, *Sida codifolia* (kurumthotty), *Mangifera indica* (mango), Padakizhangu, *Rauvolfia serpentina* (amalpori), Uruvanchi, *Emblica officinalis* (nellica), *Hydnocarpus laurifolia* (morottiki), tipali, karunji, parapula, arrowroot (kuwa) wild, camphor (telli), *Asparagus racemosus* (shatavari), *Stereospermum colais* (patri puwa/nut meg wild), *Garcinia malbica* (Puwa patrica), *Sapindus trifolius* (pulanchi/chinika/soap nut), edena (flower), *Garcinia gummigutta* (kudampuli), etc.
[b]Comprises the cost of collection of wood and NTFPs

Table 5.9 Projections of estimated recorded and unrecorded net benefit to the state level

Forest area[a] (ha)	Average recorded net benefit (Rs/ha)	Total recorded net benefit (Rs crore)	Average unrecorded net benefit (Rs/ha)	Total unrecorded net benefit (Rs crore)	Recorded and unrecorded benefit	
					Total (Rs crore)	Per ha (Rs)
1,126,500	5	0.56	764	86.06	86.62	769

[a]FSI 2009

presented in Table 5.9. It is amply clear from the table that the total net benefit from forests was Rs 86.62 crore. The corresponding per hectare estimates of total net benefits was Rs 769.

We would like to sound two caveats here. First, the unrecorded benefits of the three forest divisions need to be weighted by the density of the forests stocks and flows as these are not uniform across the sample and the state. An average of these three forest divisions as computed by us is not representative of the flow of forest products elsewhere. But due to non-availability of data about the density of forests stocks and flows for the state forests as a whole, we could not use weighted average net benefit to project net revenue at the state level. Second, forests provide a large variety of intangible goods, services and amenities such as biodiversity conservation, ecosystem resilience, soil and water conservation, recharge of groundwater and moderation of climatic aberrations which are prerequisites for sustainable development (Karl Goran Maler 2008). We have not been able to estimate and value all those intangible benefits of forests due mainly to the lack of availability of data and time.

Table 5.10 Extent of distortions in terms of revenue from the forestry sector at the state level

Official estimates of net revenue (Rs crore)	Estimates based on the study			Distortion (Rs crore)
	Recorded (Rs crore)	Unrecorded (Rs crore)	Total (Rs crore)	
1	*2*	*3*	*4 (2 + 3)*	*5 (4 − 1)*
17.59[a]	0.56	86.06	86.62	69.03

[a]Average of last 12 years from 1996–1997 to 2007–2008 (Gok 2008)

5.3.9 Extent of Distortions at the State Level

This section presents the extent of distortions in the existing system of FRA in Kerala. The distortions are expressed in terms of difference between the estimates of net benefit, which is computed on the basis of this study presented in this chapter and the revenue from forestry and logging realized by the Kerala Forest Department. It was assumed that the net benefits from forests as computed are equivalent to the revenue from the forestry sector realized by the forest department. As shown in Table 5.10, the extent of distortions at the state level was Rs 69.03 crore, and the corresponding per hectare figure is Rs 613. In this context, we would like to reiterate that the forest department does not record forest benefits such as timber, fuel wood, poles, bamboo, reeds, fodder and NTFPs collected by local people. The entire value of such benefits is treated as a distortion in the current system of FRA. The forest department can easily estimate such benefits and include them in their records for assessing the value of unrecorded contribution to the forestry sector to the regional and national economies. Thus, these distortions can be rectified.

The extrapolation has been done by assuming that the average net benefit/hectare estimated by us on the basis of three case studies in three forest divisions[4] is valid for the other divisions of the state not covered in this study. In practice, in some of the forest divisions, the actual net benefit could be higher than the estimate used by us, and in some others it could be lower than that. But it is presumed that at the state's level such deviations from the estimate of this study will cancel out each other and there will be no significant distortions in the net benefits estimated in this study and the actual revenue realized by the forest department.

[4] In Kerala, there are in all 35 forest divisions of which 3 forest divisions were selected. Thus, the size of the sample selected was approximately 9 % of the total population (35 divisions), and the selected forest divisions accounted for 8 % of the total forest area in the state. This size seems to be good enough for this exploratory study.

Table 5.11 Forest assets (forest area and growing stock) account

Particulars	Opening	Change	Closing
Physical account of forest area			
Total forest area (ha)	*1,126,500*		*1,053,481.5*
Encroachment (-ve) (ha)		7290.000[a]	
Diversion for non-forestry purposes (-ve) (ha)		65,728.50[b]	
Area subjected to forest fires in 2007–2008 (ha; no change in the area)		2381.544	
Monetary account of forest area			
Present value of forest area (Rs crore)	1,451,470.50		1,392,375.31
Loss in value due to encroachment (at Rs 8.05 lakh/ha[c]) (Rs)		5,868,450,000	
Diversion for non-forestry purposes (-ve) (Rs)		52,911,442,500	
Loss due to fires (Rs)		555,371	
Total (Rs)		*58,780,447,871*	
Growing stock			
Physical account of growing stock			
Growing stock ('000 cmt)	150,718.80		152,134.26
Logging and harvesting of timber in 2007–2008 (-ve) (cmt)		567,541.183	
Annual increment (+ve) ('000 cmt)		1983	
Monetary account of growing stock			
Growing stock (Rs crore)	904,312.80		902,768.28
Logging and harvesting of timber in 2007–2008 (-ve) (Rs)		15,445.22[d]	
Annual increment (+ve) (Rs lakh)		118,980	

[a]Total area under encroachment as on 31.3.2004
[b]Total diversion till 31st March 2008
[c]Average price, see Table 5.3
[d]Revenue received from sale of timber and other forest products

5.3.10 Forest Assets Accounts

The opening stock, change in the year and closing stock of forest assets, viz., forestland and growing stock are presented in Table 5.11.

The capital value of forest estates (forest area and growing stock) of Kerala forests was Rs 2,295,143.59 crore for the year 2007–2008.

Table 5.12 Value of recorded and unrecorded forest goods and services in Kerala in the year 2007–2008. (Source: GoK 2008)

Particulars	Recorded forest benefits	Unrecorded forest benefits	Carbon sequestration	Total (Rs lakh)
Opening value of forest benefits	153,154.26[a]	–	9,043,128.00	9,196,282.26
Change in the year 2007–2008	15,445.22	17,686.05	118,980.00	152,111.27
Closing value	168,599.48	17,686.05	9,162,108.00	9,348,393.53

In the year 2007–2008, the unrecorded forest benefits were almost equal to the recorded benefits
[a]Cumulative revenue of the 5 years from 1997–1998 to 2006–2007

Table 5.13 Recorded and unrecorded forestry sector expenditure in Kerala

Particulars	Recorded expenditure	Unrecorded expenditure	Total (Rs lakh)
Opening expenditure	14,6401.86[a]	–	146,401.86
Change in the year 2007–2008	17,287.45	9090.85	26,378.3
Closing expenditure	163,689.31	9090.85	172,780.16

[a]Cumulative expenditure of the 5 years from 1997–1998 to 2006–2007 (GoK 2008)

5.3.11 Flow Accounts: Recorded and Unrecorded Tangible Forest Goods and Services

The values of recorded and unrecorded forest benefits together form the total value of flow benefits from forests (Table 5.12).

5.3.12 Accounts of Expenditure on Forest Management and Protection

Table 5.13 presents recorded expenditure for forest management and protection and unrecorded expenditure estimated on the basis of number of days spent by the villagers for collection of forest products.

5.3.13 Consolidated Forest Resource Accounts

Based on the information about forest assets, flow of forest goods and services and expenditures that are presented in the tables in the earlier paragraphs, a consolidated forest resource account was prepared as shown in Table 5.14.

Table 5.14 A consolidated proforma forest resource account

Particulars	Forest assets account				Flow account value (Rs lakh)	Expenditure (Rs lakh)	Net value[c]
	Forestland		Growing stock				
	Area in ha	Value Rs lakh	Volume (000,cmt)	Value (Rs lakh)			
1	*2*	*3*	*4*	*5*	*6*	*7*	*[(3 + 5 + 6) − 7]*
Opening stock	1,126,500	14,514,705.00	150,718.80	9,043,128	9,196,282.26	1,464,01.86	32,607,713.40
Change in the stock in the year 2007–2008	65,728.50[a]	529,114.42[a] (-ve)	1983[b]	118,980	152,111.27	26,378.30	− 284,401.45
Closing stock	1,060,771.5	13,985,590.58	152,701.8	9,162,108	9,348,393.53	172,780.16	32,323,311.95

[a]Forest area of 65,728.50 ha diverted for non-forestry purposes, valued at Rs 529,114.42 lakh, was considered as expenditure
[b]Positive change due to increment
[c]Value of forest assets + flow value—expenditure

5.4 A Framework for Integration of FRA with SNA

Integration implies that the economic activities of the forestry sector are fully accounted for in terms of its linkages with the other sectors of the economy. However, integrating FRA into the SNA presents some difficulties, which cannot be easily eliminated. In view of this, an alternative approach, known as a satellite account for the forestry sector, is adopted. Satellite accounts are defined as being complementary to but outside the integrated SNA and hence are better referred to as 'complementary' accounts. These accounts throw light on the linkages of forestry sector with the other sectors including environment and provide guidelines for the SFM. Vincent (1999) proposes the following adjustment to be made in the conventional GDP in order to account for the linkages of the forestry sector with other sectors of the economy:

Adjusted Net Domestic Product (NDP) = Conventional GDPs + Non-market Values of Forest Benefits–Depreciation of Human Made Capital + Net Accumulation of Natural Capital

In general, what we need for integration of environmental accounts into the SNA is the estimation of a set of environmentally adjusted economic aggregates. The expansion of the asset boundary of conventional accounts for the inclusion and valuation of natural assets and asset changes permits the calculation of a range of aggregates. The aggregates can be presented as the sum total of the elements of conventional accounting identities.

A framework for integration of FRA with the SNA is shown in Fig. 5.5. The integration would facilitate an assessment of the environmental-economic performance at the sectoral and macroeconomic levels, besides reflecting the real contribution of the forestry sector at all levels and thereby providing a basis for the forest departments to claim a higher share of budgetary resources.

5.5 Conclusions

The conventional system of FRA in India suffers from several drawbacks and is not comprehensive. It does not capture the value of most of the tangible and intangible benefits and costs of forest resources including their depletion. Consequently, it underestimates the real contribution of the forestry sector to the country's GDP. The approach followed in this study is a pioneering step in the direction of strengthening the conventional system of FRA with available data, techniques/methods, infrastructure and institutional framework. The system of FRA demonstrated in this chapter is better than the conventional one in the sense that it provides for estimating, recording and valuing of all possible tangible and intangible benefits of forests. This is a significant contribution of this study in view of the fact that in India the expenditure

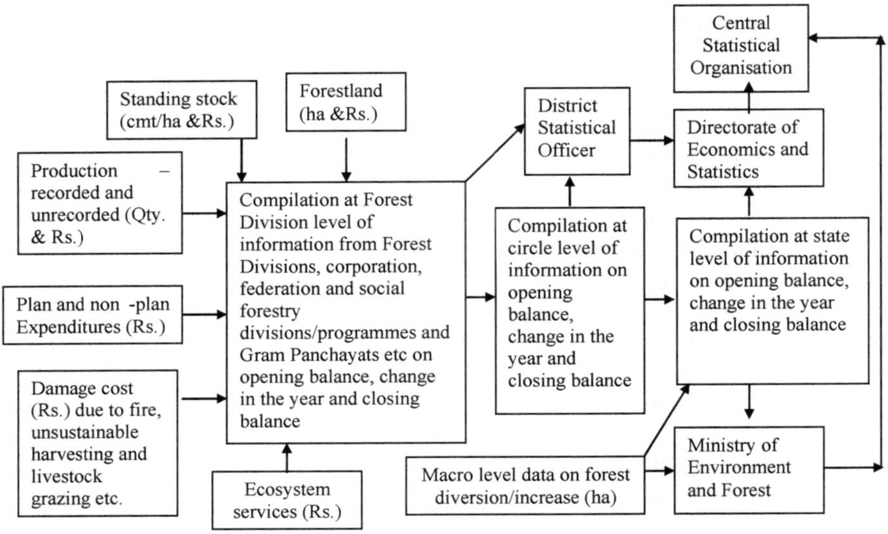

Fig. 5.5 A proposed flow diagram of the process of integration of forest resource accounts with other sectoral accounts in India

incurred on the forestry sector is almost double the revenue generated from them (CSO 2004), resulting in huge revenue deficits in most of the forest departments of states in India including Kerala. The deficits could be minimized by integrating the value of unrecorded forest benefits and associated costs through the adoption of the system of FRA suggested in this chapter.

For example, in Kerala state for the period 1999–2000 to 2006–2007, the average recorded contribution of 'Forestry and Logging' sector to the GSDP at current prices was Rs 12,932.3 million, which was 1.4 % of the GSDP. This contribution would be enhanced by 1.5 % if the value of unrecorded forest benefits of Rs 1768.60 million and carbon sequestration benefits of Rs 11,898.00 million (total Rs 13,666.60 million), as estimated in this study, were also included in the value added by the forestry sector of the state.

In case of asset accounts, 0.06 % of the total forest area has been diverted for non-forestry purposes, and 0.65 % of the total forest area was under encroachment in Kerala in the years 2004 and 2008, respectively. The positive change in the growing stock indicates that there was no depletion of forest resources in the state. Furthermore, the annual harvest does not exceed the annual increment. This indicates that the forests in the state are managed sustainably. This could be considered as a significant finding of this study and a strong rationale for institutionalizing the proposed system of FRA and integrating it with the SNA of India. This would facilitate an assessment of the environmental-economic performance at the sectoral and macroeconomic levels, besides reflecting the real contribution of the forestry sector at all levels and thereby providing a basis for appropriate policy and planning.

Acknowledgments The authors are grateful to the director of the Indian Institute of Forest Management (IIFM), Bhopal, for encouragement and providing facilities and the International Tropical Timber Organization (ITTO) for providing financial support. The authors are also thankful to Sri T. M. Manoharan, IFS, Principal Chief Conservator of Forests (PCCF), Kerala Forest Department (KFD), Thiruvananthapuram, Kerala, Sri Sri V. K. Sinha, IFS, Additional PCCF (Planning), KFD, Sri Lakhwinder Singh, IFS, Chief Conservator of Forests (CCF-Working Plan and Research), KFD, Sri Amarnatha Shetty, Managing Director, Kerala Forest Development Corporation Ltd., Kottayam, Kerala, Sri M. R. Balakrishnan, Director of Economics and Statistics, Department of Economics and Statistics, Thiruvananthapuram, Kerala, Managing Director, The Kerala State Federation of SC & ST Co-operatives Ltd., Thiruvanathapuram, Kerala, Sri N. K. Sasidharan, IFS, Conservator of Forest (SC), Eastern Circle Palakkad, Sri J. Justin Mohan, IFS, Divisional Forest Officer, Palakkad Forest Division, Sri Maheskumar, Divisional Forest Officer, Munnar Division, Sri Atasase Thongtsar, IFS, Divisional Forest Officer, Thenmala Division, Sri A. M. Babu Bonaventure, Deputy Conservator of Forests, Palakkad, Sri P. Unnikannan, District Statistical Officer, Palakkad, Secretary and staff, Ariyankavu Scheduled Tribe Service Co-operative Society Ltd, Thenmala and The Malapuzha Pattika Varga Sevana Sahakarana Sangham Ltd, Malampuzha, Palakkad, Kerala, range forest officers, block officers and staff of the Palakkad, Munnar and Thenmala Forest Divisions, presidents, secretaries and members of the selected Joint Forest Management Committees of Palakkad, Munnar and Thenmala Forest Divisions. Last but not least, we owe a deep debt of gratitude to anonymous referees of the chapter, editors Nilanjan Ghosh, Pranab Mukhopadhyay, Amita Shah and Manoj Panda and the chair and coordinators of the Indian Society for Ecological Economics (INSEE) sixth Biennial Conference on Nature, Economy and Society held during 20–22 October 2011 at the Centre for Economic and Social Studies, Begumpet, Hyderabad, India.

References

Bartelmus P (1998) The value of nature. In: Uno K, Bartelmus P (eds) Environmental accounting in theory and practice. Kluwer, London, pp 263–308

CSO (2004) 'Statistical *abstract,*' central statistical organization, ministry of statistics and programme implementation, Government of India, New Delhi. India

CSO (2007) National accounts statistics. Sources and methods 2007. Central statistical organisation, Ministry of statistics & programme implementation, Government of India, New Delhi, India

CSO (2009) National accounts statistics. Central Statistical Organization, Ministry of Statistics and Programme Implementation, Government of India, New Delhi (downloaded from website http://www.mospi.nic.in)

DES (2007) State domestic products new series from 1999–2000 to 2006–2007(modified) with base year 1999–2000. Department of Economics and Statistics, Thiruvananthapuram

FAO (2007) State of the world forests 2007. Food and Agriculture Organization of the United Nations, Rome

FSI (2009) State of forest report 2009, Forest Survey of India. Government of India, Dehra Dun, India

GoK (2008) Forest statistics 2008. Forests and Wildlife Department, Government of Kerala, Thiruvananthapuram

Lange GM (2004) Manual for environmental and economic accounts for forestry: a tool for cross-sectoral policy analysis. Working paper, Forestry Department, Rome, Italy (Source: http://www.fao.org/docrep/007/j1972e/J1972E01.htm#5536)

Maler KG (2008) Sustainable development and resilience in ecosystems. Environ Resour Econo 39(1):17–24

Manhas RK, Negi JDS, Kumar R, Chauhan PS (2006) Temporal assessment of growing stock, biomass and carbon stock of Indian forests. Clim Change 74(1–3):191–221 (Springer Netherlands)

NFC (2006) National forest commission report. Ministry of Environment and Forests, CGO complex, Government of India, New Delhi

Sachs JD, Warner AM (1995) Natural resource abundance and economic growth. Development discussion paper No. 517a, Harvard Institute for International Development, Cambridge, MA

SEEA (2003) Handbook of national accounting, integrated environmental and economic accounting 2003. United Nations, European Commission, International Monetary Fund Organisation for Economic Co-operation and Development and World Bank, New York, UN

Singh K, Mali KP (2005) Forest resource valuation and accounting: a critique of the conventional method and a framework for a new system. A report prepared for Indian Institute of Forest Management—International Tropical Timber Organization research project, Indian Institute of Forest Management, Bhopal

Singh K, Mali KP, Kotwal PC, Omprakash MD (2009) Forest resource valuation and accounting: an exploratory study in the state of Sikkim. A report prepared for Indian Institute of Forest Management-International Tropical Timber Organization research project, Indian Institute of Forest Management, Bhopal, India. ISBN 978-81-7969-071-0

Vincent JR (1999) Net accumulation of timber reserves. Rev Income Wealth 45:251–262.

Vincent JR (2003) Resource depletion and economic sustainability in Malaysia. In: Perrings C, Vincent JR (eds) Natural resource accounting and economic development: theory and practice. Edward Elgar, Cheltenham

Chapter 6
Urbanization and Water Supply: An Analysis of Unreliable Water Supply in Bangalore City, India

Krishna Raj

6.1 Introduction

Cities are increasingly facing acute water crisis in terms of imbalance between supply and demand. The demand for urban water supply and allied services is increasing rapidly as globalization accelerates economic development and brings about improvements in living standards in the country due to the dynamics of demography, i.e., the interactive effects of demographic growth and migration to cities under its push and pull effect. Provision of reliable and safe water supply services to urban habitat is an essential contribution to overall economic and welfare advancement. However, urban local bodies mandated to perform this task in India have been experiencing constant budgetary bottlenecks in regard to mobilizing resources to meet the water consumption targets of the present as well as future population. Urban water supply sector in India and particularly the study area Bangalore is facing a number of challenges and constraints in meeting one of the important components of the United Nations' Millennium Development Goals (MDGs), i.e., to supply adequate potable water to half the number of people who are currently living without access to sustainable, safe drinking water sources by 2015 (UN-HABITAT 2003) These problems and constraints include increasing scarcity of water, low pricing, high subsidy, poor cost recovery, poor maintenance, and rising unaccounted-for (UFW) and non-revenue water (NRW). Bangalore Water Supply and Sewerage Board (BWSSB) is experiencing poor cost recovery and therefore unable to generate enough revenue to meet future investment requirements of the growing water needs of the city. BWSSB is also facing serious performance gaps such as reliability, financial sustainability, environmental sustainability, and affordability due to deterioration of infrastructure.

The objective of the chapter is to assess the efficiency of water supply in Bangalore city in terms of water pricing, reliability, accessibility, etc. These include:

K. Raj (✉)
Center for Economic Studies and Policy (CESP), Institute for Social and Economic Change, Bangalore, India
e-mail: krishnaraj@isec.ac.in

N. Ghosh et al. (eds.), *Nature, Economy and Society*,
DOI 10.1007/978-81-322-2404-4_6

1. Portraying the current status of water supply
2. Examining the causes for growing inefficiency of the sector in terms of increasing unaccounted-for water and non-revenue water supply
3. Assessing economic efficiency and institutional capability to meet the water needs
4. Identifying issues and strategies for improving economic efficiency through appropriate institutional reforms
5. In conclusion, identifying major policy implications for India and other developing countries

Although this chapter deals primarily with the water supply sector of Bangalore city, the experiences and illustrations presented in this chapter represent the prevailing water supply issues of other Indian metropolitan cities as well as the cities of other developing countries. Therefore, it is hoped that the policy implications originating from this study would have relevance for other cities also. This study primarily relies on secondary sources of data from BWSSB and also various reports of World Bank and Asian Development Bank. The methodology adopted for the study includes the use of reference databases to create an overview of the water supply and demand situation in Bangalore city. The review evaluates the economic performance of water supply to illustrate the severe problems involved in pricing water, unaccounted-for water and non-revenue supply of water and lists out the myriad political economy and governance challenges in reforming the urban water supply sector.

6.2 Water Supply: Status and Coverage

Bangalore city, capital of the southern Indian state of Karnataka, popularly known as India's Silicon Valley, is characterized by high population growth and predatory geographical expansion on account of globalization causing invasion on peri-urban and rural areas. About 111 neighborhood villages have been integrated in to the city; consequently the erstwhile Bangalore Corporation is now rechristened as *Bruhat Bangalore Mahanagara Palike* (BBMP). The newly added areas to the city are in great need of infrastructure such as roads, housing, water, sanitation, and electricity. Apart from poor infrastructure, income inequality, and poverty are on the rise with successive addition of new areas and the households therein which are mostly poor. The coverage of water and sanitation services to households in unauthorized settlements is very scanty, and they generally rely on water from neighborhood taps, wells, public fountains, or from other illegal networks as securing permanent access to municipal water is just impossible without house property title. This is also one of the factors responsible for rising unaccounted-for water and non-revenue water in the city. The water subsidy meant for poor households becomes futile exercise in the absence of appropriate supply network particularly in slum areas where most of the poor live.

The main reason for the increasing demand of water is the rapid change in water-use pattern in the recent decades. Bangalore, the fifth largest metropolitan city in southern India, with a population of 9.5 million in the year 2011, has been attracting increasing number of formal and informal workers with the onset of globalization. As a consequence of the increasing population, the Greater Bangalore Metropolitan Area was formed, which resulted in the spatial growth of urban area of the city by ten times, i.e., from 69 km^2 in 1949 to 800 km^2 in 2007. This growth can be attributed to the increasing concentration of economic and commercial activities in this Silicon Valley of India as a result of the rapid growth of information technology, communication, and biotechnology sectors in recent years. Meeting water needs of the growing population together with the expansion of water supply networks to the newly added areas of Bangalore city has continued to remain a major challenge since the availability of both surface and groundwater has shrunk to threshold level. The city is now largely dependent on the river Cauvery with the commissioning of the Cauvery Water Supply Schemes (CWSS) Stages I, II, III, and IV Phase-1 for meeting the drinking water needs of its citizens. However, commissioning of the CWSS Stage IV Phase-II in 2013 has ceased the additional draft of water from the *Cauvery* River. As per the Cauvery Water Tribunal verdict, allotment of Bangalore's share of water for drinking purpose is limited to 19,000 million L^3 (TMC). The supply of water from two old reservoirs, commissioned during the British regime, viz., *Thippegondanahalli* and *Heseraghatta* reservoirs of *Arkavathi* River, a tributary of the river Cauvery, is unreliable owing to the declining flow of water in the catchments. The groundwater draft by an estimated 80,000 private borewells, well beyond the carrying capacity, has limited the future availability. The surface water of lakes and ponds situated in and around the city is highly contaminated by carcinogenic pollutants discharged by nearby industries and domestic waste. The data provided in Annual Performance Reports of BWSSB clearly show that total water being received from all surface water sources especially under normal condition was about 923 million liters daily (MLD) in 2006–2007 against the installed capacity of 959 MLD, Table 6.1 (BWSSB 2006–2007).

The present estimated supply of water varies from 100 to 110 liters per capita per day (LPCD), but in reality, the actual receipt of water by the consumers is about 75 LPCD. Presently, the availability of water supply is highly unpredictable as water is supplied on alternate days for only 4–6 h/day. The UFW, which is estimated at 48 % of total water received (fourth highest among urban water utilities in India), is evidently more than the domestic water consumption requirement (36 %). Growing UFW has become a major challenge in meeting the water requisite standard of 150–200 LPCD and 150 LPCD stipulated for metropolitan cities like Bangalore by the World Health Organization (WHO 2000) and the Central Public Health and Environmental Engineering Organization (CPHEEO), Government of India, respectively (CPHEEO, GOI). Although Bangalore city has made significant progress in extending water supply coverage from 80 to 92 % in recent years, meeting the consumption targets and minimizing disparities in water provision across different zones of the city and socioeconomic groups are the major challenges for the coming days. In this regard, the following table and figure clearly show some estimates of demand and supply gap for drinking water in Bangalore city.

Table 6.1 Water sources of Bangalore city and installed capacity

Serial. No	Sources	Year of commissioning	Distance to Bangalore (km)	Installed capacity (MLD)	Availability (MLD)
1	Heseraghatta	1986	18	36	00
2	Thippagondana Halli	1933	28	148	58
3	Cauvery Stage I	1974	100	135	145
4	Cauvery Stage II	1982	100	135	146
5	Cauvery Stage III	1993	100	270	324
6	Cauvery Stage IV Phase-I	2002	100	270	250
7	Cauvery Stage IV Phase-II[a]	2011	100	500[b]	00
	Total water availability	-	-	959	923

[a]Work has been taken up since 2007
[b]Proposed water supply

There is a visible gap between demand and supply of water in Bangalore city. The deficiency in water supply services is measured here in terms of the four main indicators: reliability, financial sustainability, environmental sustainability, and affordability. The reliability of the provision of water in the city is mainly affected by the increasing transmission and distribution (T&D) loss of water and high share of UFW, as can be seen from Table 6.2 and Fig. 6.1. The financial sustainability is critically hampered by constantly mounting NRW. According to Handbook of Statistics of BWSSB, it is estimated that out of the total 923 MLD of water that reaches Bangalore, only about 417 MLD or 45 % of the water consumed is actually billed (BWSSB 2006). Sustainability of water supply is under constant threat owing to the increasing dependence on the lone source of surface water, i.e., the river *Cauvery* and restrictions over its utilization. The financial ability of BWSSB is worsened as treated water has to be pumped from 100 km distance and to a height of 540 m for distribution in Bangalore. The process includes pumping, treatment, transmission, maintenance, etc., and involves huge costs. BWSSB spends about 65 % of the total revenue towards power charges alone, which constitutes a huge part of the total production and operational costs. Further, diminishing water resources and high cost of provision of water due to increasing energy costs significantly affect financial viability of the board. The estimated cost of production and distribution of water to Bangalore has increased from ₹12 to 23/kl in the last 5 years, which is now the highest among urban water utilities of cities like Mumbai (₹2.17) and Chennai (₹5.73). The average cost of water supply has increased from ₹3.60 to 13.20/m^3, while the average revenue has increased only from ₹2.60 to 8.67/m^3. The undue increase in the cost of water supply and the sluggish increase in revenue earnings are found to be

Table 6.2 Demand and supply gap for drinking water (in MLD) in Bangalore

Year	Population (Lakhs)	Water potential available	Net water supply	UFW	Water needs/demand		Shortage	
					At 150 LPCD	At 200 LPCD	At 150 LPCD	At 200 LPCD
2000	57	705	433	218 (33)	862	1150	−157	−445
2001	62	705	458	220 (32)	934	1245	−229	−540
2002	64	995	462	250 (35)	966	1288	29	−293
2003	67	995	484	269 (36)	999	1333	−4	−338
2004	64	959	512	331 (39)	956	1275	3	−316
2005	65	959	542	340 (39)	981	1309	−22	−350
2006	67	959	531	372 (41)	1006	1342	−46	−383
2007	70	959	550	509 (48)	1032	1376	−73	−417

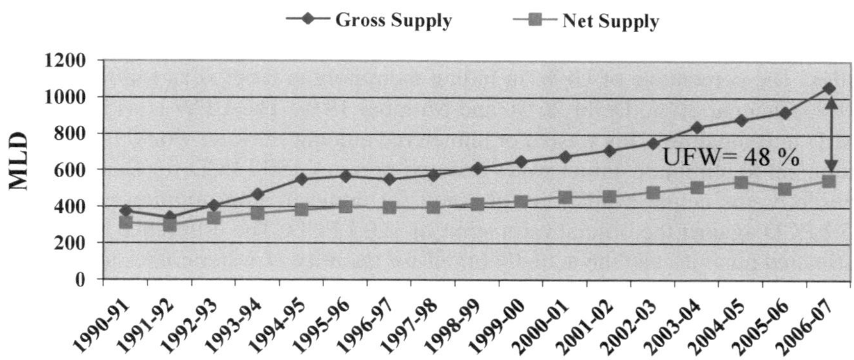

Fig. 6.1 Gross, net supply, and unaccounted-for water in Bangalore city. (Source: Nanjundaiah 2008)

due to inefficient management. While the demand for water increased by 16 % during the period 1991–2001, the increase was by 20 % during the period 2001–2010 (Saleth and Sastry 2004). The increase in the production cost of water, particularly when the demand for water is growing, is highly distressing, given the lowering groundwater table and the limited surface water available.

6.3 Causes and Effects of Inefficiency in Urban Water Supply

Increasing share of total water loss, UFW (the difference between water produced minus water consumed and water in store) in total water production has significantly reduced the water availability for domestic and nondomestic/productive use. About

Table 6.3 Type of leakages in the pilot project area (Source: BWSSB Annual Performance Report 2005–2006)

Serial No.	Type of leakages	%
1	Main	38.1
2	Service pipe	32.8
3	Stand post	17.6
4	Main valve	6.6
5	Meter joint	2.0
6	Stop valve	1.6
7	Ferrule	0.7
8	Air valve	0.1
9	Others	0.5
	Total	100

48 % of water currently distributed is UFW, which is very high as compared to the level of UFW in well-managed urban water supply systems (15–20 %) across world cities. The percentage of UFW in Indian metropolitan cities is as follows: Kolkata 50 %, Chennai 20 %, Delhi 26 %, and Mumbai 18 %. The UFW is as high as 509 MLD in Bangalore. This wasted or unmetered amount of water would be more than sufficient to eliminate unmet water demand norm of 150 LPCD for Bangalore city. Presently, the actual availability of water for domestic consumption is just about 75 LPCD against the official estimation of 120 LPCD. The difference between the estimated quantity and the actually billed-for quantity of water comes to 45 LPCD. This huge loss is directly attributed to the water seepage at various stages of supply. Water leakages largely take place at distribution mains, service pipes, and stand posts, and together account for 88.5 % of water spillover, the rest being low leakages at main valve, meter joint stop valve, ferrule, air valve, and others (Table 6.3 and Fig. 6.2).

The augmented water supply through various stages of Cauvery river water supply projects has been offset by the predominance of UFW; conversely, enhancement of per capita water availability to 150 LPCD remains unfulfilled as several unauthorized water connections siphon off half the water produced. Despite the offer of BWSSB to regularize unauthorized connections if applied within a stipulated time and the threat of imposing a penalty of ₹10,000 apart from 6 months imprisonment for unauthorized water pilferage continues unabated in the city.

Poor recovery of operational, maintenance, and capital costs from user charges is attributed to the inefficiency of water supply management in Bangalore city. BWSSB charges water tariff at ₹5 per 1000 L of water. However, supplying of 1000 L of water costs ₹40 and as a result of which, the board incurs an annual revenue loss of ₹27 crores owing to the prevalence of high UFW and poor tariff collection. Thus, the revenue generated from water supply services is insufficient even to meet the O&M costs of BWSSB; therefore, the board is perennially dependent on the state government for funding. Financial support of government has become

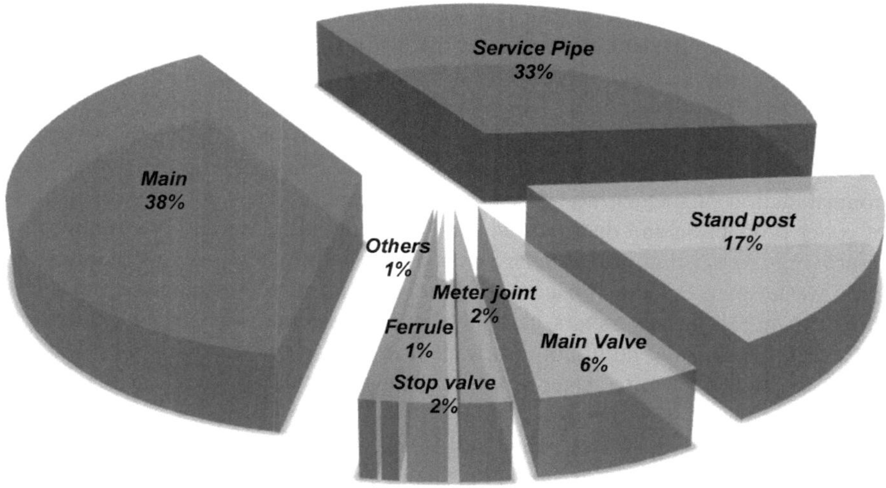

Fig. 6.2 Type of unaccounted-for water leakages

inevitable in the present context of rising electricity charges, poor O&M, deterioration of water supply service assets, and unwillingness of decision makers to increase water tariffs. These factors have further aggravated the poor health of water supply services. Research evidence shows that in Bangalore politicians are unwilling to raise water tariffs despite evidence of people's willingness to pay more for good-quality water supply services. The poor quality of water supply services in the city is attributed to the unrevised domestic tariff rates coupled with poor metering, increase in non-revenue water, and deteriorating efficiency in O&M of infrastructure. Water quality has deteriorated in the city due to inadequate sewage, effluent discharge, and wastewater treatment facilities. This in turn has resulted in the occurrence of water-borne diseases in several localities, and contributed to productivity losses and mounting health costs. Poor water delivery is also imposing indirect coping costs on households in terms of time spent in fetching water, which is increasingly depriving education to many girls, particularly those living in the slums of the city.

6.4 Growing Scarcity of Drinking Water in Bangalore: Issues

Bangalore city has prospered not only due to industrialization but also due to rapid urbanization and population growth. Consequently, demand for water has been witnessing an increasing trend over the years. Projections of water demand in Bangalore under various future scenarios have shown that there will be growing mismatch between demand and supply of water. Plugging this demand–supply gap and ensuring sustainability of resources and their efficient utilization are major policy challenges particularly in the context of existing demand–supply mismatch. Thus,

BWSSB now faces the herculean task of addressing both physical and economic scarcity of water. Physical scarcity arises mainly owing to increasing dependence on Cauvery river for drinking water needs and also due to appalling fall of groundwater table, death of lakes and pollution of water bodies, and improper sewage treatment in the city. Economic scarcity denotes the adverse impacts of mounting unaccounted-for water and non-revenue water mainly caused by leakages in pipes, water theft, noninstallation of meters, non-revision of water tariffs, etc., on the sustainability of water production and distribution. This has resulted in scarcity of resources for improving the existing infrastructure as well as creating new ones including recycling wastewater, reducing unaccounted-for water through replacing old and rusted pipelines, installing new meters, etc. The above shortcomings in the city's water sector reflect the abysmal failure of government in putting the water supply system on even keel. In other words, water problem in the city is due to governmental apathy and nothing else.

A comparison of BWSSB water utility services with water utilities in major cities, based on international norms or benchmarks of performance such as water supply coverage, per capita consumption, annual production of water, UFW, average tariff, unit production cost, operating ratio, revenue collection efficiency, accounts receivable, and ratio of staff per 100 connections, reveals that the performance of BWSSB is much below the scores of water utilities in all developed countries as well as those in top 25 % of developing countries. However, when compared to water utility services of other Indian metropolitan cities, performance indicators of BWSB, especially in regard to water supply coverage, installation of meters, operating ratio, and revenue collection, give a positive picture. However, performance indicators, such as UFW, water availability, per capita consumption, and average tariffs, are negative for BWSSB. UFW is estimated at 48 % of water supplied in 2006–2007, which is fourth highest among Indian cities. The high prevalence of NRW has pushed up the cost of supply and constitutes a substantial part of the operating subsidies contributed by Government of Karnataka. Although the coverage of water supply now stands at about 93 % and water supply duration is 4.5 h on alternate days, it still is much below acceptable standards. Further, inadequate pressure during water supply leads to insufficient availability of water. The water availability for consumption is only 74 LPCD, which is far below CPHEEO normative standard of 150 LPCD, whereas the score of Ahmadabad and Mumbai conforms to this standard (Table 6.4). BWSSB's performance indicator in regard to installation of meter is quite satisfactory at 95.5 % of total connections in the city. Although operating ratio of BWSSB is good at 0.8, accounts receivable at 7.1 months is the fifth highest. Average tariff of ₹20.55 per 1000 L is the highest among other metros. BWSSB operates on no-loss no-profit basis, which means high UFW keeps the total revenue just enough to cover the costs. Still, BWSSB needs to increase the staff strength per 1000 connections, which is currently about 5.2. Reduction of NRW, improving water availability tariff collection, and metering of connections are other areas where the board should pay special attention.n

Further, the problems of water supply consists of both quality and quantity issues. The water delivery is restricted to 2–4 h on alternate days and quantity supplied

Table 6.4 Performance indicators for water utility in Bangalore in comparison with water utilities in major metropolitan cities of India. (Source: MoUD and ADB (2007)

Serial No	Description	Bangalore	Chennai	Ahmadabad	Kolkata	Mumbai
1	Water supply coverage (%)	92.9	89.3	74.5	79.0	100.0
2	Water availability (h)	4.5	5.0	2.0	8.3	4.0
3	Per capita consumption (LPCD)	74	87	171	130	191
4	Production/ population (m^3/d/c)	0.185	0.131	0.168	0.246	0.246
5	Unaccounted-for water (%)	45	39	ND	35	13
6	Connection metered (%)	95.5	3.5	3.0	0.1	75.0
7	Operating ratio	0.80	0.44	1.43	4.73	0.49
8	Accounts receivable (Months)	7.1	1.1	8	2.4	11.8
9	Revenue collection efficiency (%)	112	152	67	100	189
10	Average tariff (₹/m^3)	20.55	10.87	1.39	1.13	4.60
11	New connection fees	1740	1930	100	1000	660
12	Capital expenditure/connections (₹)	787	10,080	427	2247	3790
13	Staff/1000 connections ratio	5.2	13.3	2.2	14.7	17.2

stands reduced from 120 to 75 LPCD. Water quality has continuously deteriorated due to poor maintenance of infrastructure, particularly distribution lines and consequent mixing of sewage water with potable water. The investment on water filters and boiling has become people's common strategies to avoid health risks. The questionable potability of the water supplied is forcing sizeable number of even economically weaker citizens to buy drinking water from private water markets. The exponential growth of bottled water supply in Bangalore is a sad reflection of the poor quality of water delivered at the consumer's faucet.

These views are also supported and endorsed by international funding agencies like World Bank and Asian Development Bank. However, few countries like Singapore have succeeded in their effort to manage water efficiently in an integrated way by managing catchment areas, ensuring adequate quality and quantity of water, recycling wastewater, conserving water as well as water bodies, and ensuring public participation. The Phnom Penh, the capital city of Vietnam, has indeed shown the way to other cities in terms of efficient management of urban water by significantly reducing its unaccounted-for water from 75 % in 1993 to 9 % in 2006 and thereby providing uninterrupted 24 h drinking water supply, which is by international standards, much better than in many European cities. It is a paradox that Bangalore, despite having better management and technical skills, and financial resources, has failed to prevent mounting volume of unaccounted-for water. Therefore, the board needs to review its policies in the light of the positive experiences and successful models of water utilities in other countries and identify their best practices which could be adopted with suitable changes to meet local requirements and overcome challenges. It is understandable that managing water and wastewater is not an easy task given Bangalore's uncontrolled urbanization and population growth; but with clear policy vision and enlightened leadership the existing and foreseeable water problems can be solved.

6.5 Drying Sources of Water: Now or Never

Water scarcity of Bangalore city has worsened with the unprecedented drying up of most water sources in the Catchments of Cauvery, Kabini, and Hemavathi rivers. This may be attributed to various reasons including climatic change, reduced rainy days and rainfall, deforestation in Western Ghats, high deposit of silt at the dams, overdraft of groundwater for various needs like agriculture, industry, and human consumption. Consequently, the city relies completely on Cauvery water which is pumped from a distance of 100 km to a height of 920 m above mean sea level. BWSSB has the obligation to supply water to 7.22 lakhs connection serving 9.5 million people in Bangalore city by pumping 1125 MLD water everyday. The pumping point, i.e., Shiva Balancing Reservoir faces shortage of water during almost every summer. The water shortage also impacts hydroelectricity production at the four hydel plants as water needs to be diverted as and when there is demand for more water from *Cauvery Neeravari Nigam Ltd.* The allocation of scarce water for various needs is already under stress with the recent CWDT award, and the increasing apportioning of water by Bangalore city has created rift within the *Cauvery* basin of Karnataka state. The drying up of *Hesaraghatta* Lake and *T G Halli* reservoir and the prospect of relying on Cauvery river for the total water needs of the city hangs like the proverbial Damocles' sword over the heads of Bangalore's citizenry.

Further, the rapid urbanization and population growth have already affected the water supply efficiency adversely in Bangalore city as evidenced by the falling groundwater table and limited surface water availability. Bangalore city is identified as a "groundwater quality hotspot" due to exploitation of groundwater beyond rechargeable limits. This has resulted in increasing number of semi-critical, critical, overexploited, and over developed regions emerging in many watersheds of Bangalore urban and rural districts. The groundwater quality deteriorated with high presence of iron, fluoride, nitrate, and conductivity with drastic decline in water table. The volume of groundwater extraction in Bangalore urban district is alarmingly high as 207,000 ML/year. The volume of groundwater draft for domestic and nondomestic uses in Bangalore Metropolitan Region (BMR) has been estimated at 237,000 ML/year and 207,000 ML/year, respectively. The annual recharge of groundwater in BMR from all sources together is estimated at about 81,100 ML/year. The recharge in BBMP area is approximately 71,000 ML/year as against the groundwater utilization of 103,000 ML/year. The report on Groundwater Resources of Karnataka in 2004 prepared by the Department of Mines and Geology, Government of Karnataka, reveals that there is alarmingly increasing gap between groundwater exploitation and its recharge in Bangalore city (Fig. 6.3).

The dwindling yield and sources of water in catchment areas of the city call for a shift in policy focus in regard to resource use efficiency if we are to mitigate water scarcity and improve water availability especially under the fast decline in water availability. Water resource use efficiency has received less attention in policy circle in developing countries despite having reached scarcity era. Sadly, the mindset of institutional framework of water supply agencies is still attuned to the bygone surplus era. This mindset is reflected in the policy of treating water as a free good and not as a commodity that requires enormous investment to supply at the consumers' end. Undoubtedly, the stumbling block is the reluctance to attune policies to the emerging water supply management issues. Therefore, new institutions armed with new polices only can help to shift strategy towards management of demand through economic instruments, i.e., water pricing and rationing. Further, such novel strategies only can assure proper allocation of scarce water among various economic needs by taking into account economic, social, and environmental gains for the society. In this direction, it may be noted that many cities have already switched over to demand-driven water management based on consumers' choice and willingness to pay. This policy shift will help to attract investment on urban water infrastructure which would hopefully result in augmentation of water availability, improving the quality of water supplied and recycling of wastewater.

6.6 Economics of Urban Water Supply

Finding solution to water scarcity and unreliable water supply from the perspective of economic efficiency is very crucial for ensuring sustainable urban habitat. Proper valuation of multiple benefits of water provides its actual utility construed as value-in-use, as asserted by Adam Smith (Batten 2007). Zerah (2000) observes that many instances of water shortages stem from the widespread failure to value water's true worth. Growing scarcity of drinking water in metropolitan cities is reflected in the increasing economic value of water resources. However, urban water utilities have largely failed to factor in the importance of economic aspects of water supply in major policy frameworks. Demand for water is estimated based on projections and normative requirements of water for the growing population,

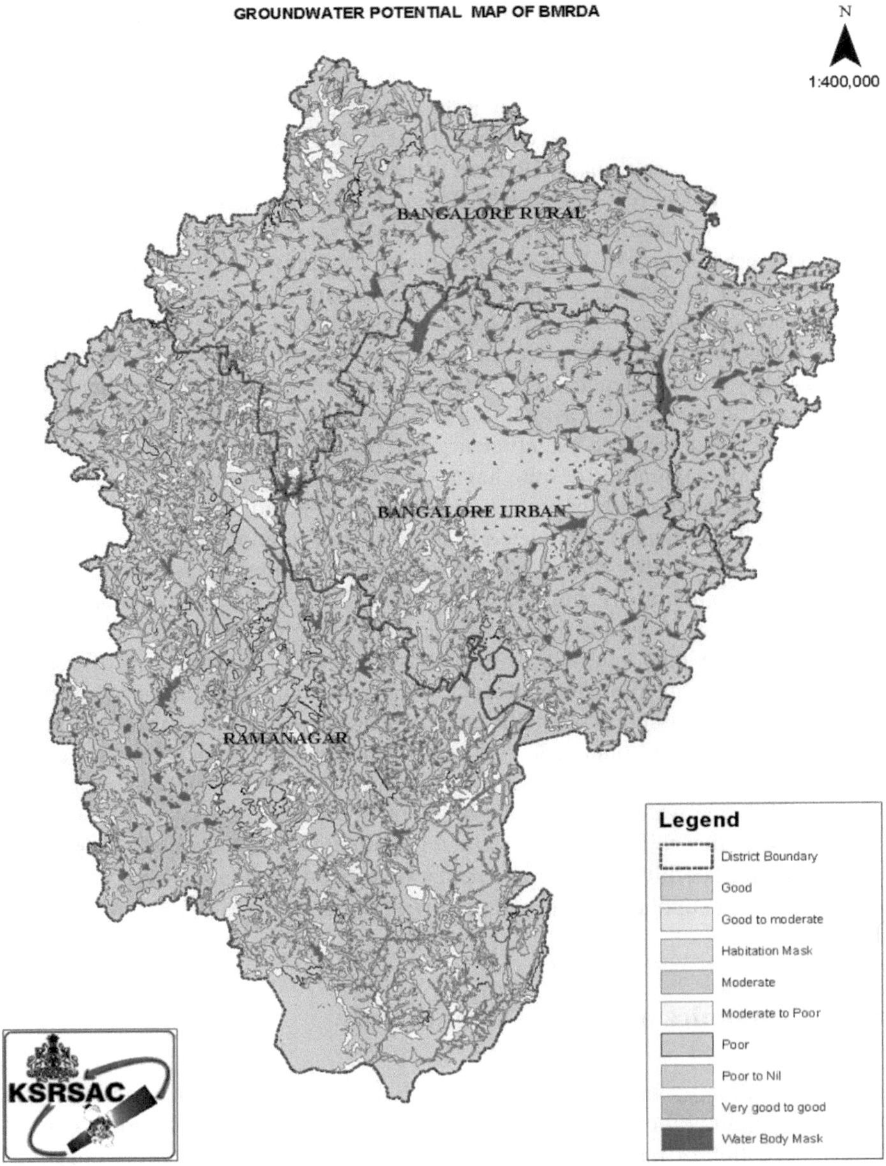

Fig. 6.3 Groundwater potential for Bangalore

but the effect of price on the demand for water is seldom recognized. The quantity of water demanded is significantly affected by the price of water and other socioeconomic factors. Demand for water is inversely related to its price, and positively to individual income; however, a certain minimum quantity of water will always be in demand regardless of price. Therefore, augmentation of water supply is given higher priority than demand management in the policy circles, and this leads to the emergence of "low-level-equilibrium-trap" (Reddy 1999). Whenever the demand for water exceeds supply, urban water utilities quickly design water supply strategies, giving little importance to demand control or management. Failure of water supply authorities to incorporate demand-side factors in their policies leads to "system-collapse" or "institutional failure." For instance, in Bangalore city, without first addressing the towering level of UFW (48 %), the BWSSB has tried to overdraw from already strained water sources at huge public investment without offering "economic justification" for taking up new water supply projects. But the demand-side management principles advocate operating within the limits of current supplies. Even today, supply augmentation strategies get precedence in urban water resource planning in developing countries over demand management opportunities, despite the fact that most developed countries have overturned their dominance in water policies. Under these circumstances it is difficult to achieve economic efficiency in water utilization unless demand management options get priority over supply enhancement options (Biswas 2008). Even today, it is noneconomic factors such as political considerations that get priority in urban water management decision making over economic or market instruments; therefore, there is a need to factor in both economic and welfare aspects of water supply to ensure the long-term sustainability of scarce water resources. Presently, most developing countries lack the capacity/will to evolve proper market-based instruments for ensuring allocation of water efficiently and equitably. Efficiency of public water supply system in developing countries is often constrained by a combination of factors like financial stringency, poor maintenance system, etc., as a result the percentage of UFW goes up, which is often two to three times higher than that in developed countries. The cost of inefficiency in terms of unreliable water imposes high cost on the households, forcing adoption of compensatory strategies. The cost of unreliable supply of water in Delhi has been estimated at ₹3000 million annually, which works out to an average cost of ₹2182 per household per year. The greater efficiency of water use is found positively associated with fixing the price of water in line with the real cost of supplying it. Better pricing—levying and collecting appropriate user charges—solves water scarcity by promoting allocation of this scarce resource more efficiently and equitably.

6.7 Economic Efficiency and Institutional Capability

Research studies in urban water management policy have clearly identified the reasons for inefficiency as the lack of recognition of water as a scarce or economic commodity, and the failure of market or pricing policies to account for the full economic costs of its production and supply. It is essential to include opportunity costs related to alternate uses of water, user costs, and cost of externalities such as groundwater depletion, pollution of water bodies, etc. If cost recovery is based merely on water production and supply, it may lead to underpricing of water and result in inefficiency in operation and also inadequate investment on new projects and even in operation and maintenance. It is seen from experience that economic instruments such as higher water pricing or tariffs have remarkably reduced household water consumption levels and also promoted awareness about water conservation. Market signals the scarcity value and also opportunity cost of water, which helps in efficient allocation of water among various competing sectors. Realization of an ideal state of economic efficiency in urban water supply system is not likely in the near future, as already a number of cities of the world have been afflicted with water distress and mounting water production and supply costs as also political compulsions in this matter. Niemczynowicz (1996) observes that "Water-related problems in [the] cities are already enormous, and further degradation is expected. Water shortage is a growing problem and delivery of safe drinking water cannot be assured." However, improvement in economic efficiency depends on short-term and long-term institutional capability to address the main water supply indicators such as reliability, financial sustainability, environmental sustainability, and affordability. Economic efficiency should also reflect the goal of supplying clean and quality water for 24 h by providing universal access to water for all. Improvement of economic efficiency depends on evolving an effective water pricing mechanism taking into account the economic status and ability of households to pay. Economic status influences the peoples' willingness to pay for improvement in access and quality of water supply. Policy of water pricing, therefore, needs to achieve two objectives simultaneously: recovery of the full cost of production from supply of water and provision of efficient and reliable water supply service to the consumers. Without any provision for better water supply services, mere increase in water tariffs are doomed to fail in this objective as established by many research studies.

6.8 The Role of Economic Instruments in Water-Use Efficiency

Economic instruments such as increasing water tariffs have acted very positively in rationalizing the behavior of water users thereby reducing substantial water consumption (unnecessary wastage of water) in cities (Oliver 2010). This has positive impacts on the water supply agency as well as households to rationalize their thinking. On the one hand, reduction in water consumption helps in conserving water and meeting water supply needs in all areas, and reduces water uncertainty and accessibility; on the other hand, increased revenue collection helps to build new infrastructure and also maintain existing water networks. Satisfactory revenue collection also helps in cross-subsidy from rich to poor. Charging rich for their higher consumption of water for other than drinking and household essential needs such as car washing, gardening, etc., will help to conserve water while revenue augmentation will help to effect cross-subsidization in favor of poor consumers. However, fixing water tariffs should have the objective of ensuring affordability, equity, and economic efficiency. Therefore, urban water sector reforms based on economic principles will continue to shape the behavior of water supply agency and consumers. It will also achieve the long-term objective of sustainability, resource conservation, and economic efficiency in urban water supply and demand management. The drive for economic efficiency should include appropriate tariff structure to send correct signals to both water suppliers and consumers to maximize benefits from the deal. In other words, both suppliers and consumers should take into account the total economic value of the water that consists of use and nonuse values (economic and environmental costs of water use). Fixed water tariffs or charges are used in Bangalore city even though water scarcity is high; therefore, there should be economic incentives to limit consumption and choose water conservation. The optimal price fixation reflects operation and maintenance cost recovery, and marginal cost pricing reflects efficiency of water supply. In the case of Bangalore city, all users enjoy subsidies though it seldom reaches the targeted categories of low-income consumers. This is because large chunks of these categories stand excluded from the BWSSB water supply network through which consumption subsidy is channeled, and consequently the poorer households end up in water networks of richer consumers and public fountains and fail to benefit from subsidy benefits. To overcome inefficiency in water supply and also to include the poorest in supply networks and consumption subsidies, increasing block tariffs (IBT) method is devised, which helps to achieve three main objectives, viz.: (1) affordability and fairness (this is highly subsidized or subsistence block or lifeline rate), (2) resource conservation (high water consumption attracts higher water tariffs), (3) economic efficiency (upper block covers the short-run marginal cost of using water). However, meeting the first objective of highly subsidized tariff for poor households is not an easy task, unless poor households sharing the water connection live in separate areas with dedicated water lines. However, in reality, the potential role played by the economic instruments for greater cost efficiency is ignored in this model. From the above analysis, it is construed that considering the issue of price as a policy instrument for efficient urban water resource management is unlikely to be debated in policy circles due to lack of political will. Therefore, drinking water quantity and quality requirements continue to suffer with respect to WHO standards and regulations.

6.9 Accounting for Unaccounted Water

Bangalore city faces growing gap between demand and supply of water due to limited availability of surface water. The projected demand for water supply by the year 2025 and increasing gap between supply and demand are portrayed in Table 6.5 and Fig. 6.4. By 2025, the average water consumption demand would have progressively increased to 154 LPCD from the present consumption rate of 101 LPCD (including 48 % of UFW). UFW is the major stumbling block in achieving the target of supplying 150 LPCD, which is currently 48 % of the total water produced, and

Table 6.5 Water supply and demand in Bangalore City

Serial No.	Description	Year				
		1991	2001	2007	2011	2025
1.	Population (Lakhs)	40.8	58.0	65.26	73.4	97.0
2.	Available water (MLD)[a]	372	647	923	1459	1800
3.	Water demand (MLD)[b]	924	1433	1464	1575	2314
4	Supply and demand gap	552	786	541	116	514

[a]Projections are subject to availability of water from River Arkavathi
[b]Subject to 20 % of UFW up to 2010 and 15 % from 2011 onwards

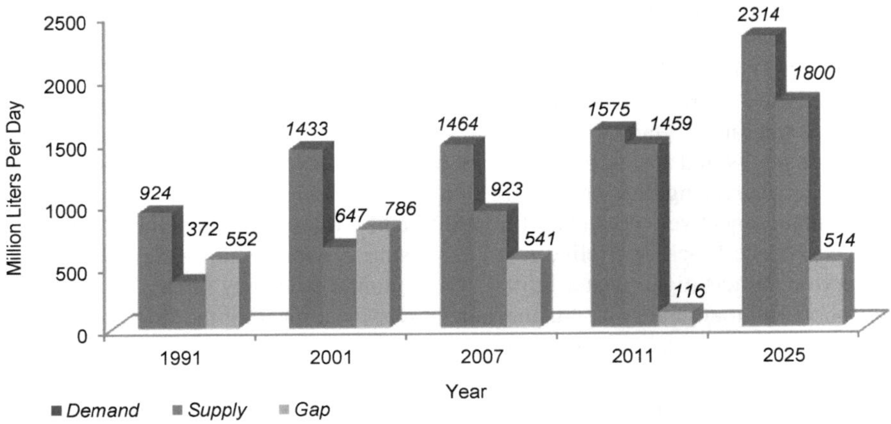

Fig. 6.4 Demand and supply gap in water supply in Bangalore city

alarmingly, higher than even the total domestic consumption (35.67 %). The projections for 2011 and 2025 are based on the assumption that there would be reduction in UFW to 20 % by 2010 and 15 % from 2011 onwards and also subject to availability of water from River *Arkavathi*. It is projected that the supply and demand gap will increase to 514 MLD by 2025, and the water availability then will be 1800 MLD only against the demand of 2314 MLD. Again the projected availability of water is on the assumption that there would be 25 MLD of rainwater harvesting, 198 MLD of effluent reuse, and 239 MLD of groundwater utilization. To overcome this shortage, a long-term and actionable blueprint needs to be implemented on priority basis.

The reduction of non-revenue water to the international best standard level of 10 % from present level of 48 % is a real challenge before BWSSB. The poor human and physical infrastructure has hindered the work of water leak detection, replacing corroded and old pipes, and installing water meters. The deteriorating financial performance of BWSSB, therefore, continues to impact the efficiency of water supply on account of rising unaccounted-for water, non-revenue water, increased electricity charges, and high operation and maintenance cost. The mounting fiscal crisis has made BWSSB to heavily depend on state government for meeting O&M costs without considering ways and means of improving cost recovery and the performance of water supply system. Therefore, the water utility in the city is unlikely to improve if effective steps are not taken to cap UFW in water distribution system, especially heavy seepage due to corrosion of main pipes, eliminate thefts and overflows, improper metering, and illegal connections. Metering of public fountains will effectively reduce UFW. Demand-side management of urban water supply necessarily requires reduction of UFW from the existing level of 48 %. Scientific assessment of water leakages, whether due to real loss or apparent loss, has to be prioritized by the BWSSB. Real loss of water is mainly due to technical snags, i.e., unscientific recording of total water supplied on account of leakages, theft, and overflow prior to reaching end users. Apparent losses are those stems from improper recording of total water utilized by the consumers owing to faulty meter reading and administrative inefficiency. It can be inferred from the water leakages estimation that the real loss of water is indeed higher than the apparent loss. Extensive replacement of corrosive pipes is required to reduce UFW in Bangalore city. The high UFW is mainly due to two types of distributional losses: first, distributional loss due to damages and leakages in water supply system and, second, unauthorized water connections, which are not billed at all. The growing scarcity of water sources and increasing cost of production (water treatment) and distribution essentially calls for treating water as of economic good, rather than just a renewable resource. Water should be essentially treated as an "economic good" or an "economic and social good" to emphasize the urgent need for its rational use by human beings. Further, instead of defining water as a "renewable resource," water should be redefined and reclassified as "conditionally renewable resources" as water recharge or natural replenishment has been much below the exploited quantity in recent years. Water conservation is the immediate need for improving groundwater table to ensure availability of water for productive use. Augmentation can be done by means of storage of water in surface bodies such as lakes, reservoirs, tanks, ponds, and wells and groundwater bodies such as soil. Government should also prioritize rejuvenation of degraded traditional surface water bodies such as lakes and ponds in Bangalore urban and rural districts. Good efforts towards water conservation inevitably demand some sort of economic incentives in the form of payment and rewards for the restoration of water ecosystems. Therefore, devising promotional economic instruments like subsidies, payments, or compensations are the need of the hour. Rainwater or roof water harvesting is considered to be the best alternative and cost-effective option, but this has to be implemented strictly by making new laws with adequate economic incentives. Enhancing wastewater treatment capacity can reduce the quantum on fresh water requirement in the future. Judicious and economic use of both tap and groundwater has to be encouraged by imparting awareness about growing concerns in respect of fresh water scarcity. Prevention of contamination of water bodies from industrial and domestic pollutants at point source and nonpoint source needs to be given priority in the policy.

6.10 Strategies for Improving Economic Efficiency

BWSSB, after realizing its failure to reduce UFW and non-revenue water, is planning to bring appropriate institutional reforms in water supply. Its major priorities and challenges are to achieve economic self-sufficiency by effectively reducing/eliminating subsidized supply of water on the lines of other market-based institutions. BWSSB also needs to educate the masses about the urgency inevitability and economic justification of increasing water tariffs. It also needs to switchover to efficient management practices such as computerizing water supply, achieving 100 % metering, charging appropriate user charges, overcoming NRW through reduction in UFW, and water losses. There is also a need for a long-term plan to augment water availability from present 75 to 150 LPCD, apart from extending coverage to all sections of the society for 24/7, by 2011–2012. BWSSB has given the highest priority to plugging water leakages in the distribution system, the volume of which is currently equivalent to 7-day's water supply, and improve revenue collection by 20–30 %. The cost that involved ₹4000 million is being met with financial assistance from the Japan International Cooperation Agency (JICA). The Government of Karnataka has shown its interest in reducing water scarcity through its Agenda for Bangalore Infrastructure Development, 2008. What is envisaged is to address the water shortage of 349 MLD or 40 % of drinking water needs of the city, by harvesting water from lakes in and around the city. The state government has also come up with a major plan costing ₹1700 million to replace all old and corroded water pipelines and install dual pipelines in new layouts of the city within two years. However, adequate attention has not been given to wastewater treatment. Bangalore city generates about 1000 MLD of wastewater; out of which just about 721 MLD is treated, and the rest goes without any economic use. Only 70 MLD of wastewater under tertiary treatment is currently used by the Bangalore International Airport, *Bharath* Electronics Limited, Rail Wheel Factory, and other industries for non-potable purpose. Enhancement of tertiary treatment capacity to achieve at least 10 % of reuse of treated wastewater will reduce the current heavy dependence on surface and groundwater. Bangalore city receives uninterrupted and well-distributed rainfall with an average rainfall of 970 mms annually. Rainwater harvesting involves low capital and transaction costs; besides it helps to reduce water stress of the city by over 20 % in the future. Therefore, an integrated water resource management approach, combined with water supply augmentation and demand management strategies, is certain to improve economic efficiency of the city's water supply system (Table 6.6).

Poor governance and political interference have generally acted against induction of knowledge, experience, and technology in water management in the city. The

Table 6.6 Water supply enhancement and demand management strategies for Bangalore

Supply enhancement strategies	Demand management strategies
Harness rainwater	Reduce non-revenue water
Wastewater treatment and reuse	Metering of taps
Reduction of unaccounted-for water	Raise water rates
Lakes rejuvenation	Water rights and conservation option
Tapping of leakages	Educate water users

governance structure needs urgent reforms to bring about greater accountability and coordination as also to affect devolution of powers to user groups on the basis of economic rewards. There is also urgent need to bring about cohesive coordination among political, social, economic, educational, and environmental institutions in order to achieve common objectives in water sector. Incentives need to be evolved on the basis of market-based instruments for bringing improvements in environmental governance, i.e., effective management of water resources. BWSSB can adopt several strategies to improve resource generation like reducing unaccounted-for water from the current level of 48–15 %, minimizing non-revenue water by fixing meters, plugging leakages, and preventing pilferage or theft by households, industries, and commercial establishments. Investment on water infrastructure is very important for improving water delivery and water quality. Though wastewater treatment and reuse need huge investment, its benefits are certain to justify the investment. The harvesting rooftop rainwater has been made mandatory in building bylaws to save and conserve water with low investment by households; however, it is still in inception stage, and how it translates into reduction in urban water scarcity that needs to be watched. In conclusion, the water crisis of Bangalore city needs to be addressed with better policy options of both supply and demand-side management perspectives in the long term.

References

Bangalore Water Supply and Sewerage Board (BWSSB) (2006) Handbook of statistics

Bangalore Water Supply and Sewerage Board (BWSSB) (2006–2007) Annual Performance Report.

Batten DF (2007) Can economists value water's multiple benefits? Water Policy 9:345–362

Biswas AK (2008) Water management in India in 2020 and Beyond, Shastri Indo-Canadian Institute, New Delhi

Central Public Health & Environmental Engineering Organization, Ministry of Urban Development, Government of India. http://cpheeo.nic.in. Accessed 10 June 2010

Nanjundaiah C (2010) Water supply and sanitation, State of Environment Report for Bangalore, published by Department of ecology and environment, Government of Karnataka. http://parisaramahiti.kar.nic.in/soerbangalore2008.html. Accessed 10 June 2010

Niemczynowicz J (1996) Megacities from a water perspective. Water Inter 21:198–205

Oliver A (2010) Water tariff and consumption drop: an evaluation of households' response to a water tariff increase in Manaus, Brazil. Water Policy 12:564–588

Reddy RV (1999) Quenching the thirst: the cost of water in fragile environments. Dev Change 30(1):89–124

Saleth RM, Sastry GS (2004) Water supply and sanitation sector of Karnataka, India: status, performance and change. Water Policy 6:161–183

United Nations Human Settlements Program (UN-HABITAT) (2003) Water and sanitation in the world's cities: local actions for global goals. Earthscan Publications, London

World Health Organization and United Nations Children's Fund (2000) Global Water Supply and Sanitation Assessment Report. WHO

Zerah MH (2000) Water: unreliable supply in Delhi. Manohar Publishers, Delhi

Chapter 7
Energetics of Irrigation Under Surplus Rainfall Conditions

Nandan Nawn

7.1 Introduction

This chapter estimates the energy value (use-value) of irrigation under surplus rainfall conditions. This represents the value of the unsaved rainwater, an ecosystem function, to the farmer-cultivator, for meeting the water requirements of the planted crops, an ecosystem service.

Rainwater, in terms of function, supports many interconnected ecosystem services to human beings and other biotic and abiotic species, be it in the form of broader supporting, provisioning, regulating or cultural, or their sub-divisions, following the classification by the Millennium Ecosystem Assessment (2005). Admittedly, the presence of linkages among the services makes the valuation of the function difficult, while presence of feedback effects adds to the complexity. Nevertheless, for any specific service (say, irrigation), for a particular set of beneficiaries (say, farmers), resources (i.e. labour, machine, and material) spent on the alternatives represent its opportunity cost, and can be measured in quantitative terms. Such costs of irrigation are the value of the rainwater to the farmer, available but unsaved due to the absence of adequate storage infrastructure.

The author wishes to thank Debobroto Ghosh for the navigational assistance in traversing through the dataset and the advices towards adopting reasonable and realistic assumptions and also to the members of the Indian Society for Ecological Economics (INSEE) community, especially Kanchan Chopra, Purnamita Dasgupta, and Seema Purushothaman, for their comments on the draft and the participants of the session on 'Towards Green and Sustainable Socio-Economic Systems: Including Ecological Agriculture, Energy Efficient Industry and Sustainable Urban Habitats' in 'Nature, Economy and Society: Understanding the Linkages', Sixth Biennial Conference of INSEE, October 20–22, at the Centre for Economic and Social Studies (CESS), Hyderabad. The chapter has been partially sponsored by the project titled Socio-economic Study of Rainfed Agriculture and Low Input Animal Husbandry in India conducted by CSSP-Economic Research Unit at the Centre for Studies in Science Policy, Jawaharlal Nehru University, New Delhi.

N. Nawn (✉)
Department of Policy Studies, TERI University, New Delhi, India
e-mail: nnletter@gmail.com

N. Ghosh et al. (eds.), *Nature, Economy and Society*,
DOI 10.1007/978-81-322-2404-4_7

There are many locations across the world, where the seasonal effective rainfall is more than the consumptive water use (CU) of the planted crop; the state of West Bengal located in eastern part of India is one such. Admittedly, in other locations, the rainfall can only partially meet the crop water requirements (CWRs) and hence its value to the farmers will be proportionately less. Valuations of the kind undertaken in this chapter can contribute in resolving important and pressing policy choices, of various kinds.

Started in 2005, following the enactment of the Mahatma Gandhi National Rural Employment Guarantee Act, the Government of India has already made quite a large allocation as well as expenditure towards the creation of durable assets, through National Rural Employment Guarantee Scheme (NREGS) in which (re)construction of new and rejuvenation of defunct water structures have received a rather high priority.[1] While the presence of such public assets can crowd-in private investments and thwart the ongoing agrarian distress considerably, more often than not, the supposed financial constraint is posed as a hurdle to extend and expand this programme. The quantitative estimates offered by the case study in this chapter using primary data collected by a government agency from 380 households belonging to 38 selected blocks spread across 5 agro-climatic zones across West Bengal, for the year 2004–2005[2] can provide the policymaker a kind of 'cost-benefit analysis' that was hitherto unavailable. This assumes some significance following the changes in the 'principles, approaches and strategies of water management' as envisaged in the Twelfth five year plan (Shah 2013).

The subject and the approach in this chapter necessitated combining the labour theory of value, energy theory of value (ETV), and aspects of agronomy. Further, while there have been reasonable attempts to value clean water in monetary terms, especially in urban locations, the same is not the case for water in farming.[3]

Part II contains the analytical framework for classifying various types of irrigation, a brief discussion on the method in broad terms, and a few qualifying remarks on the ETV. Part III describes the field for the case-study and the data sources. Part IV elucidates the method for arriving at the energy value under each type of irrigation. Method of identifying 'surplus' and 'deficit' rainfall *tehsil*/blocks are discussed in part V, while the results are included in part VI, along with a discussion. Part VII contains a few concluding remarks.

---

[1] As per the District Management Unit (DMU) report for 2010–2011 on physical assets, water conservation and water harvesting had received the highest (20.27 %) committed expenditure.

[2] This is the last agriculturally normal year for which the dataset was available for academic use in 2011, the year of the conference where the paper was presented.

[3] H T Odum had worked tirelessly in establishing energy values of various aspects of the biosphere, including biotic and abiotic resources and ecosystem services, including rainfall (Odum 1984); our approach differs from Odum's considerably. For a critical analysis, see, Jones (1989, p. 345).

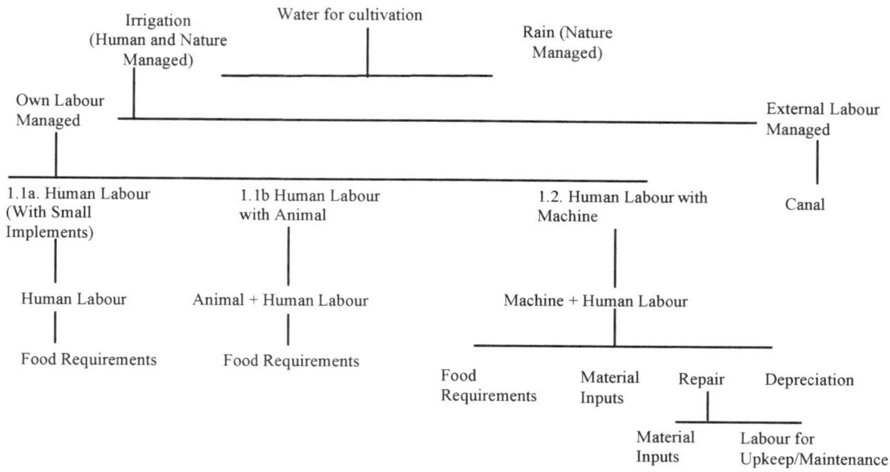

Fig. 7.1 Typologies of sources of water during cultivation

7.2 Typologies of 'Irrigation' and Their Valuation

7.2.1 Typologies of Irrigation

Irrigation literally means 'to bring a supply of water to a dry area, especially in order to help crops to grow', and 'to supply (land) with water by artificial means, as by diverting streams, flooding, or spraying'. Its etymological root lay in Latin *irrigationem* meaning 'a watering' that shares the same parentage as old English 'regn' or rain.

Water is supplied to the plants either only by nature (rain) or with additional contributions from labour (irrigation). Arguably, the retention capacity of the soil can also play a significant role in the duration and extent of availability of water for the plants, and there can be indirect contributions from labour here; we have kept it outside the scope, however. Nonetheless, agro-climatic zone is an identifier in our analysis, which is based on soil characteristics, apart from rainfall and other parameters.

Irrigation can be of five mutually exclusive types: (a) nature managed (rain), (b) external labour managed (canal), (c) own labour managed with small implements (case 1.1a), (d) own labour managed with animal labour (case 1.1b), and (e) own labour managed with machine (case 1.2) (Fig. 7.1).

Usually, the purview of irrigation precludes both case 1.1a and 1.1b that involves the 'refurbished natural depressions or dug-out ponds and checkdams for harvesting and storing surplus run-off' (Sharma 2011), the mainstay of farmers in rain-fed areas for irrigation. The implicit assumption has been that of conjunctive use of equipments or structures or machines in irrigation. An example is the instructions in the official manual for data collection against 'Net Area Irrigated by Source':

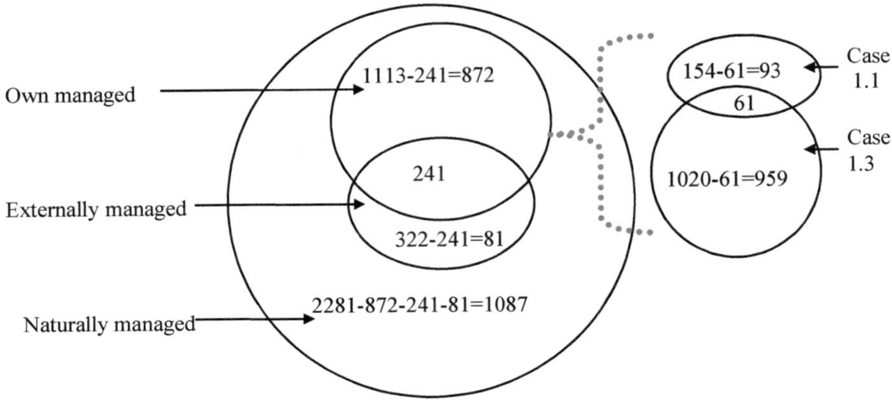

Fig. 7.2 Distribution of types of water availability across plots

> Provision has been made to record separately area irrigated by four sources of irrigation. A parcel operated by a holder would normally be irrigated by one of these four sources [Canals, Wells (with pump-sets, both electric and diesel and without pump-sets), Tube-wells (electric and diesel), and Tanks]. In exceptional cases, where any other source is utilized, the primary worker can record the area under 'others' column specifying the source of irrigation. (15. Instructions for Filling Schedules, S. 15.7 Schedule H: (Detailed data on Holding), Block F: Net Area Irrigated by Source (in ha.) (DAC 2010, p. 35, 53)

Such 'otherness' is also evident in the dataset provided by AQUASTAT (Information System on Water and Agriculture, FAO 2007), where only those 'equipped' are covered, leaving aside even the irrigation from receding flood water.

In this chapter, however, irrigation includes the possibilities of both with and without the use of structures, machines, and the likes. In fact, it is identified with the use of labour for 'irrigation' as a type of 'crop operations'. In particular, the case(s) 1.1 (a and b) and 1.2 have been considered for making quantitative estimates (Sect. 7.3).

The case of irrigation managed by external labour had been ignored, for reasons of difficulty in making calculations and existence of only 10 % of plots having access to such a facility during the period of study (Fig. 7.2). The data on canal irrigation has been made available only on the basis of a lump sum figure in the dataset used, presumably being the product of per hectare irrigation rates and the area under cultivation, without any corresponding volumetric information.[4] This leads to a slight underestimation of the value of rainfall. In addition, the issue of heterogeneous water quality has been assumed away, across the sources.

[4] The data difficulties had been noted by the Committee on Pricing of Irrigation Water (Planning Commission1992) as well. Even with efforts at the highest official level it had to depend mainly on the information obtained during the team's visits. Even then, the quality and the extent of data coverage were not to its satisfaction (1992, p. 10).

7.2.2 Method of Energy Analysis

The method employed in this chapter was pioneered by the workshop conducted on Energy Analysis and Economics by International Federation of Institutes of Advanced Study (IFIAS) at Lidingö, Sweden, 22–27 June 1975 which had prepared a summary of recommendations including nomenclature, method, and the definition of various terms (IFIAS 1978). This method had received widespread attention since the 1973 oil crisis, with its applications mostly in agriculture under different names: energy accounting, energy balancing, energy budgeting, and energy costing (Pimentel et al. 1973; Stanhill 1974; Mitchell 1979; Pimentel 1980; Stanhill 1984; Fluck 1992, among others).

However, there still exists considerable diversity within the adopted methodologies, expected of any intellectual endeavour of a recent vintage. Conflicting assumptions across methods, over aggregation, quantification, handling of inputs, especially labour (Norum 1983) as well as unsettled questions over its nature, scope, boundary of analysis, research questions, accounting nomenclature, treatment of inputs, interpretation of results, and limitations (Jones 1989, p. 340) have remained. However, each of the difficulties can be addressed with careful handling.

There are well recognised limitations of this method at the theoretical plain as well: of translation and reverse translation from (largely intra-substitutable) energy equivalent to (mostly non-substitutable) matter in practice, of the yet to be resolved issue of handling capital through straight-line method of depreciation or a gradual depreciation of energy efficiency, of the relationship between energy quality, technological change and monetary value, of energy quality, among others. Yet, such effects are ignored in the homogeneity assumption of much empirical work (Judson 1989), owing to a number of advantages, not offered by any other method.

At the practical level, the main constraint is the considerable resources of data and time needed to calculate these values. This chapter uses fairly detailed plot level data on all aspects of crop production to address it. Additional problems like treatment of joint production/consumption, etc., are also addressed through appropriate assumptions.

Jones (1989, pp. 352–353) argued that once the 'not-inconsiderable' costs of obtaining the information is surpassed, most beneficial will be the economists who often face energy related situations 'where existing economic approaches cannot provide all the answers'. Apart from its use in the longer-term national level planning rather than as a guide for day-to-day decision making, this technique had been recommended for making general comparisons, rather than for analyzing specific concerns.

This method extends a consistent treatment across all the irrigation types under consideration. Further, it is more advantageous over the monetary evaluation, for the market failures/distortions with the latter, especially in the presence of household labour and the subsidy on a variety of energy sources, from diesel to electricity.

7.2.3 Qualifications Over the 'ETV'

There are a number of approaches with diverse ideologies and objectives, all termed under 'ETV' and just like the phrase, 'classicalist approaches to value', there lay crucial differences among them, which often results in unnecessary and avoidable confusion. The commonality ends with the consideration of embodied ETV which is essentially a cost of production theory and where all costs are carried back to the primary input, the only 'scarce' factor of production, the solar energy required to produce them, with labour, manufactured capital, and natural capital, all being considered as 'intermediate' inputs. Energy is seen here as the physical measure of environmental and social impacts, as well as of material, capital, and human resource requirements, and of reserve quantities. The point of debate is over the purpose of this exercise.

On the one hand, there are a few who argue that the energy use intensity and money use intensity could be compared, which was just a short step before treating energy values as the 'values in exchange' (price); this approach has been claimed to be a close parallel of the neo-Ricardian labour-embodied theory of value with energy replacing labour as the primary costs of production (Farber et al. 2002, for example). Such primacy of energy as an input has been criticised for the presumption about the reproducibility of the other factors in terms of pure energetics (Burkett 2003, p. 139). Certainly, energy can serve as an ordinary standard of measure, but not as a 'standard commodity' as its composition 'cannot plausibly be expressed in pure energy terms, given the material (physical, chemical and biological) differentiation of the production system' (Burkett 2003, fn 5).

The other strand following this second approach consisting mainly of ecologists, thermodynamists, biologists, and energy analysts had placed energy analysis as a tool for supplementing economic analysis in policymaking. They declare,

> [...] we do *not* subscribe to any 'energy theory of value'. Our approach while resting on same scientific principles [...] aims to provide *description,* not *evaluation* [...]. Logically there is no way that descriptions of what *is* can be used to deduce what *ought* to be done. The decision about what to do involves a value judgment and, to be clear, this should be explicitly separated from the description of what is. (Chapman and Roberts 1983; emphasis as in citation)[5]

This chapter follows the latter position, of treating energy value as the use-value, and not as an exchange value. The descriptions can provide useful information for comparing the alternatives consisting of different types of energy sources. The value judgment exemplified by the prioritisation over them through the policy can follow, thereafter.

---

[5] P F Chapman and F Roberts, 1983, *Metal Resources and Energy*, Butterworths, cited in Common (1995, p. 210).

7.3 Description of Study Area and Data Sources

7.3.1 Rainfall Patterns in the Study Area

National Agricultural Research Project (NARP) of Indian Council of Agricultural Research (ICAR), New Delhi in its 1991 report (which still remains the latest) had identified 120 agro-climatic zones across the country (Ghosh 1991). West Bengal was divided into 6 zones: (I) Hills, (II) Terai, (III) Old Alluvial, (IV) New Alluvial, (V) Red and Laterite Soil, and (VI) Coastal Saline. On the other hand, Agro-climatic Regional Planning Unit (ARPU), Planning Commission, New Delhi that started in 1988, had divided the country on the basis of 15 broad agro-climatic zones with further sub-zones: West Bengal covered seven such sub-zones. Finally, in the classification of the National Bureau of Soil Survey & Land Use Planning (NBBS&LUP), Nagpur of ICAR, West Bengal falls under three agro-ecological zones (Sehgal et al. 1992). All the three classifications could be approximately tallied with each other (Table 7.1). This was necessary for combining the data obtained from different government agencies that do not follow identical zonal classifications.

Notwithstanding the terminological divergences across classifications, for each specific location, a few characteristics including rainfall converged for obvious reasons. Indeed, of all the climatic parameters in West Bengal, rainfall is the single most dominating variable which influences agriculture (Chatterjee 1991, p. 12). From the rainfall data it appears that about 80 % of the rainfall is concentrated in the kharif season (June–October) throughout the state, which corresponds to the cultivation period of *aman* rice, the main rain-fed crop of the state. Later, some precipitation during the *boro* season takes place. Barring a few disturbances in the north, the *rabi* season remains comparatively dry. Hills and foothills of the north enjoy the maximum annual rainfall, followed by the coastal tracts of the south. Even the south central region is blessed with moderately high rainfall while the north central part and western plateau enjoy lower yet not negligible precipitation (Chatterjee 1991).

The pertinent point here is that substantial portions of the state do receive adequate rainfall which remains unutilised due to inadequate storage capacity. On the other hand, due to seasonal, uncertain, and variable nature of the rainfall pattern, in a state with average annual rainfall over 2000 mm, rainfall has appeared as a critical constraint (Chatterjee 1991, p. 12). Indeed, such discontinuities in rainfall are not unusual. The Committee on Pricing of Irrigation Water (Planning Commission 1992, p. 389) had recommended an assumption of one or two waterings of 75 mm 'even though on monthly basis the rainfall may be sufficient to meet the crop water requirement'.

Even the Planning Commission had accepted that for the fall in crop productivity in Regions III & IV (Lower and Middle Gangetic Plains) covering 15 of 19 districts of the state, flood/water logging, improper drainage, salinity/alkalinity, and arsenic contamination are responsible (Planning Commission 2008, p. 9). This region indeed had been typified with 'rich water and soil resource', and a sequential rice based multi cropping system throughout the year, indicating the richness of the

Table 7.1 Correspondence between alternative zonal classifications for selected tehsils, West Bengal. (Source: Ghosh (1991a, 1996), Tehsil list for the Crop Complex 2002–2005 (CCS WB 2002–2005))

Planning commission			ICAR (NARP)		CCS (DES)		
ACZ	ACSZ	Districts	ACZ	ACSZ	Zone no.	List of selected tehsils (number)	Districts corresponding to selected tehsils
II. Eastern Himalayas Region	II.1. Himalayan Hills	Darjeeling	II. Eastern Himalayan Region	WB-1. Hill Zone	1	N/A	N/A
II. Eastern Himalayas Region	II.5. Upper Brahmputra valley, Barak valley and part of WB	Jalpaiguri, Cooch Behar	II. Eastern Himalayan Region	WB-2. Terai Zone	2	Dinhata-I (1), Coochbihar-I (3), Sitalkuchi (4), Moynaguri (5), Falakata (6), Karandighi (9)	Cooch Behar, Jalpaiguri, Uttar Dinajpur
III. Lower Gangetic Plains Region	III.1. Barind Plains	West Dinajpur, Malda	III. Lower Gangetic Plains Region	WB-3. Old Alluvial Zone	3	Raiganj (10), Harishchandrapur-II (11), Chanchal-II (12), Murshidabad-jiaganj (13), Raninagar-II (14), Jalangi (15), Tehatta-II (16), Kaliganj (17), Ranaghat-I (18), Habra (20), Singur (22), Polba-dadpur (23)	Uttar Dinajpur, Malda, Murshidabad, Nadia, 24 Parganas (S), Howrah, Hoogly
III Lower Gangetic Plains Region	III.2. Central Alluvial Plains	Howrah, Hoogli, Bardhaman, Midnapur, Nadia, Murshidabad	III. Lower Gangetic Plains Region	WB-4. New Alluvial Zone	4	Burdwan (29), Memari-I (30), Ketugram-II (31), Mangalkot (32), Kalna-II (33), Bhagawanpur-II (35), Sabang (37), Pingla (39)	Hoogly, Bardhaman, Bankura, Midnapur (E), Midnapur (W)
III Lower Gangetic Plains Region	III.3. Alluvial Coastal Saline Plains	24 Parganas, Kolkata	III. Lower Gangetic Plains Region	WB-5. Laterite and Red Soil Zone	5	Murarai-I (50), Dubrajpur (51), Keshpur (57), Binpur-I (60)	Bardhaman, Birbhum, Bankura, Midnapur (E), Midnapur (W)

Table 7.1 (continued)

Planning commission			ICAR (NARP)		CCS (DES)		
III Lower Gangetic Plains Region	III.4. Rarh Plains	Bankura, Birbhum	III. Lower Gangetic Plains Region	WB-6. Coastal Saline Zone	6	Bishnupur-II (43), Shyampur-I (45), Nandigram-I (46), Ramnagar-II (47)	24 Parganas (S), Howrah, Midnapur (E)
VII Eastern Plateau and Hills Region	VII.4. Eastern Plateau	Purulia	III. Lower Gangetic Plains Region	WB-5. Laterite and Red Soil Zone	5	Raghunathpur-II (53), Manbazar-I (54), Jhalda-II (55), Para (56),	Purulia

ACZ agro-climatic zone, *ACSZ* agro-climatic sub-zone, *CCS* cost of cultivation survey, *WB* West Bengal

natural biotic and abiotic infrastructure. Further, the report on Agro-climatic region specific priorities for the Eleventh five year plan against Lower Gangetic Plains clearly specified that the inadequacy of water is not a problem but it is the drainage system. Even the subregion specific priorities were in similar lines (Singh 2006, pp. 27–28, 35, annexure V):

> (III.1) Barind plains (1587 mm rainfall p.a.): Water harvesting and recycling;
>
> (III.2) Central alluvial Plains (1449 mm) & (III.3) Alluvial coastal saline Plains (1607 mm): Reclamation of soil salinity in coastal areas and soil acidity in lateretic soils, conjunctive use of canal and ground water, and crop management in flood prone areas; [...]
>
> (III.4) Rarh Plains (1302 mm): strengthening of tanks for effective water harvesting.

Elsewhere in the state, priorities included crop management in flood prone areas, harnessing ground water potential in the valley, and in situ water harvesting (II.5. Upper Brahmputra, 2809 mm), alongwith delineation and development of watershed at macro and micro levels, and construction of check dams for conserving rainwater and its recycling (VII.4. Eastern plateau, 1369 mm) (Singh 2006).

The argument for sufficiency of rainfall gains further strength from an evaluation of various drought characteristics of the state (Bandyopadhyay 1987); in fact a significant increasing trend has been observed in the annual rainfall for the state over 1901–2003 (Guhathakurta and Rajeevan 2006, p. 9). While both types of meteorological droughts, that is, seasonal and contingent, as per 'Thornthwaite's classification', can be found in West Bengal (Chatterjee 1991), none of them are chronic, as per Indian Meteorological Department's (IMD) classification. Further, the other IMD criterion of 'coefficient of rainfall variability' was found to be less than the cut-off of 30 %, for rainfall to be considered erratic or the district to be classified as drought prone (Chatterjee 1991). Even the index of seasonal water balance, on weekly or monthly basis, calculated by subtracting the PET from normal rainfall did not show prevalence of any agricultural drought, the more important indicator for cultivation (Chatterjee 1991).

7.3.2 Data Sources

Farm level primary data on 600 households cultivating 2317 parcels together resulting in 3479 unique plot-season-crop combinations (PSC) were collected from the Bidhan Chandra Krishi Vishwavidyalaya, Kalyani University, West Bengal, under the 'Comprehensive Scheme for Studying Cost of Cultivation/production of Principal Crops (CCS)', Directorate of Economics & Statistics, Department of Agriculture and Cooperation, Ministry of Agriculture, Government of India.[6] Monthly block

[6] For a critical evaluation of the data source, see Nawn (2013). The directly used human labour data (under all irrigation types) was taken from data-fields (DFs) 12–19, 22–25 from record type (RT) 710 with irrigation in DF 8 (Operations), with or without the use of associated machine (DF 30–33)

level rainfall data in West Bengal for June 2004–May 2005 were collected from the office of Agricultural Meteorologist, for those meteorological stations which were situated either within one of the tehsils selected by the CCS or were in reasonably close vicinity to them. Thirty-eight such blocks were identified by the Agricultural Meteorologist himself fulfilling such a criterion. They were spread across all the five agro-climatic zones covered by CCS with the following distribution: 6 (II), 12 (III), 8 (IV), 8 (V), and 4 (VI). Three hundred and eighty households within these tehsils had cultivated 1455 parcels, with an average area of 0.397 ha. This resulted in selection of 2281 PSCs, with 1205 in season I, 796 in season II, and 280 in season III. Average area for each plot was 0.367 ha, while the range was 0.02–1.57 ha.

7.4 Measuring Energy Value of Irrigation

The energy value of irrigation per acre is expressed either as megajoule (MJ)/ha or barrel oil or tonne coal equivalents (BOE/ha or TCE/ha). The latter can be converted to the monetary equivalent, and will be varying with the market price for oil; the greater is the scarcity, the energy value of irrigation increases, and so will be the value of rainwater to the farmer under surplus rainfall locations.

Calculation of energy values involved consideration of the energy requirements of labour in direct engagement (both human and animal) and also in maintenance of machines, embodied energy in material flow inputs, and material for repair/maintenance, alongwith the depreciation of machines in energy terms (Fig. 7.1). Following the accepted norms, the embodied energy in naturally available resources, namely, rain, wind (including atmospheric moisture), heat, and soil (nutrients available) had been ignored.

7.4.1 Case 1.1a (Irrigation with Human Labour)

Within this practice, human labourer uses small implements to carry rainwater accumulated in natural or human assisted depressions near the plots under cultivation. Calorie values for the agricultural labourer—considered as a moderate activity (Gopalan et al. 1989, p. 10)—for both men and women were taken from ICMR (1990). In the absence of supporting data, reference weight of an adult male labourer had been taken as 60 kg and of an adult female as 50 kg (ICMR 1990, p. 70). Calorie requirements for children were taken as the average of adolescents, calculated from

---

or animal (DF 26–29), for each PSC. RT 440 provided information on the types of owned machines (DF 9–12) like capacity, age, and remaining life. RT 742 furnished data on the material consumed directly or in the maintenance for owned machines, while human labour data for maintenance was taken from RT 740. RT 743 provided the data on hired-out machines.

Table 4.11 of ICMR (1990, p. 25), given that irrigation work warrant involvement of children only beyond a certain age.

In particular, 2800 and 2250 Cal per day respectively for adult men and women had been assumed. For hired children, they were taken as 2500 and 2000 Cal. These norms for both hired and household labourer had been used for direct labour input in cases 1.1a, 1.1b, and for repair/maintenance of machines in case 1.2.

Labour-hour(s) had been converted to labour-day(s) by using a working-day norm of 6 h. Though the official position is to consider 8 h as the per day working time, 6 h physical work takes place on average. The collected data is based on actual physical work after all, which excludes the time for taking the lunch and a two hour rest period in between.

The other input in this type of irrigation is the various small implements. Many such tools are made of a number of naturally available resources like branches of tree, bamboo, logs, etc. Keeping in mind the convention of energy analysis, they were not considered. In any case, the amount would have been miniscule. At times household utensils like drums are used simultaneously; for such items the depreciation will also be very small.

7.4.2 Case 1.1b (Irrigation with Human and Animal Labour)

There had been just two entries against animal based irrigation, an extremely rare practice in West Bengal. However, they have been taken into consideration for the sake of completeness. Dietary energy requirement of bullocks of 300 kg, the most common variety of cattle in the state for power requirements, was taken as 35,249 Cal per pair day following Rao (1984). In terms of results, this case has been clubbed with 1.1a, to form case 1.1.

7.4.3 Case 1.2 (Irrigation with Human Labour and Machine)

In this case, labour is involved in two ways: directly, and also for upkeep/maintenance of machines. All the labour hours across age, gender, and source (men/women/children, household head/family/servant/exchange/casual) used in conjunction with a machine used for irrigation purposes (tubewell, pumping set, pipes, oil engine, electric motor, or submersion pump) were considered as a sedentary activity, as the power level in these activities is much less than in either of the other cases. For adult men and women, 2400 and 1900 Cal per day respectively have been assumed and 1900 Cal per day for adult men and women, respectively, have been assumed.

Embodied energy in the most standard variety of various types of machines used in the state had been computed using standard coefficients for various materials following Batty and Keller (1980), keeping in mind the individual material requirements under each type of machine (see, annexure).[7] Accordingly, annual depreciation for own machines were calculated following the straight line method, in consideration with the various characteristics of machines.[8] Similar calculations were also done for the various materials consumed, directly or for upkeep/maintenance, following Cervinka (1980). Annual consumption data on material inputs against these machines at the household level were apportioned monthwise for each PSC on the basis of machine hour data.[9] All other apportionments were carried out on an identical basis. Depreciation of the unused machines and equipments, however, had not been taken into account, following the norms of energy analysis.

For hired machines, consumption of material was derived from the average of machines used only in tehsils 30 and 60, due to the relatively superior quality of data. The computation of depreciated embodied energy and energy expenditure on maintenance also followed a similar averaging, as for the owned machines.

7.5 CWRs in West Bengal

CWR equals the sum of water required for its survival, growth, and development. Variables that influence CWR includes root water uptake, soil evaporation, runoff, hydraulic conductivity, soil water flux, rooting pattern, growth period, stages of growth, etc. (Allen et al. 1998). From the requirement of planning, the relevant variables are related as follows:

> Consumptive Use (transpiration loss through leaves + evaporation losses through soil surface within cropped area + crop metabolic activities) + application loss + water for operations = Effective Rainfall + supplementary irrigation + contribution from ground water table.

The first term in the RHS represents naturally managed water. Micro studies on water management take specific values for all the terms in the LHS, in consideration

---

[7] This information was obtained from the fieldwork, through the discussions with the enumerators of CCS.

[8] Let the value in construction/purchase of the asset be V_0, and at year t, it is V_t, with t taking the value from 1 to n. Considering uniform rate of depreciation r, $V_t = (1-r)^t \cdot V_0$. As n becomes large, $(1-r)^t$ approaches zero, and so does V_t. Typically, use of the machine ceases much before V_t equals or even be close to zero. Doering III (1980: 11) assumed reliable life to be 82 % for farm machinery and buildings. Typically, with $r = 0.08$, at $t = 20$, $(1-r)^t$ takes the value of 19 %. Further, as the average life span of pumping sets (code: 6502) was 19 years, this particular value of r was assumed. Depreciation or the change in capital stock due to wear and tear for the ith year was taken as $V_{t-1} - V_t = r \cdot (1-r)^{t-1} \cdot V_o$ t was obtained from RT 440 (see, fn 6 above). Calculations of V_0 for different machines along with the coefficients for conversion are available from the author on request.

[9] Annual consumption data against each machine were used instead of the monthly data in order to minimise the possibility of errors in reporting as it is usually conducted through visual observations instead of actual measurement.

of the agro-ecological properties of a particular location at the macro level and soil and crop properties at the micro level. They are reflected through various indices, like evapotranspiration (ET) or consumptive use, potential evaporation (PE), daily soil water balance, etc. Undertaking such an analysis at the plot level demands considerable supplementary data (Vaidyanathan and Sivasubramaniyan 2004, p. 2990). Overviews are available at best, and that too for specific indices, crops, and locations (for example, Tomar and O'toole (1980) captured ET values across rice fields in selected locations in Asia).

Understandably, the chapter had to adopt certain simplifying assumptions, each of which can be relaxed subsequently, however, with the availability of required data. Notwithstanding these limitations, the method of valuation remains valid. Adoption of such simplified assumptions is not unprecedented; they remain often unspecified though. Recommended crop specific consumptive use values (CU) at the state level, for instance, do make assumptions about 'normal' growth, growth periods for the most common variety, average agro-ecological conditions, etc.

7.5.1 Consumptive Use, ET, and Crop Coefficients

Crop specific CU added with site specific evaporation (E) results in ET. For any region, with a given E, CU varies widely between crops depending on the nature, season, length of growing period, and stages of growth and so does the corresponding ET. On the other hand, E varies across agro-climatic/agro-ecological regions (AER). ARPU had documented monthly potential evapotranspiration (PET) and rainfall for agro-climatic subregions (as per its own classification) of 14 states including West Bengal (Planning Commission 1992, p. 400). This information was used to find PE and rainfall values for each agro-climatic subregion classification followed by CCS (DES) (Table 7.2).

Table 7.2 also compares tehsil specific rainfall data in contrast to the agro-climatic/agro-ecological sub-zone based ones, followed by state- or country- wise studies. Differences between the observed monthly data and the average rainfall assumed by Planning Commission (1992) or Vaidyanathan and Sivasubramaniyan (2004a, 2004) indicates to the potentialities of each *tehsil* in terms of availability of rainwater over and above PET more clearly. This provides a justification for holding the *tehsils* as the more appropriate unit for agro-climatic planning which is similar to the earlier proposal by the National Advisory Council for conceiving planning for the natural resource management at the 'natural village or hamlet' level instead of the usual, that is, revenue village (Indian Express 2011).

Table 7.2 Season-wise values of potential evaportanspiration (PET) and rainfall (RF) for selected blocks, West Bengal, 2004–2005. (Source: Planning Commission (1992) for columns (1)–(6) and the Office of the Agricultural Meteorologist, Government of West Bengal for columns (7)–(9))

Zone	Tehsil name	Tehsil no	Kharif PET	Kharif RF	Winter PET	Winter RF	Summer PET	Summer RF	Met station	Kharif RF	Winter RF	Summer RF	Gap Kharif	Gap Winter	Gap Winter
			(1)	(2)	(3)	(4)	(5)	(6)		(7)	(8)	(9)	(10) = (7) − (2)	(11) = (8) − (4)	(12) = (9) − (6)
II	Dinhata-I	1	554	2318	245	76	378	627	Dinhata-I	2344	14	401	26	− 62	− 226
II	Coochbehar-I	3							Coochbehar-I	3021	36	648	703	− 40	21
II	Sitalkuchi	4							Sitalkuchi	2462	13	419	144	− 63	− 208
II	Moynaguri	5							Moynaguri	2862	28	443	544	− 48	− 184
II	Falakata	6							Falakata	3188	31	614	870	− 45	− 13
II	Karandighi	9							Karandighi	930	11	79	− 1388	− 65	− 548
III	Raiganj	10	576	1187	285	39	501	175	Raiganj	953	19	131	− 235	− 21	− 44
III	Harishchan-drapur-II	11							Harishchan-drapur-I	2298	39	173	1111	0	− 2
III	Chanchal-II	12							Chanchal-I	1668	22	180	481	− 18	5
III	Murshidabad-Jiaganj	13							Jiaganj	863	4	93	− 324	− 35	− 82
III	Raninagar-II	14							Raninagar	1295	30	118	108	− 9	− 57
III	Jalangi	15							Jalangi	1182	15	191	− 5	− 24	16
III	Tehatta-II	16							Tehatta-I	1346	8	11	159	− 31	− 164
III	Kaliganj	17							Kaliganj.	953	2	65	− 234	− 37	− 110
III	Ranaghat-I	18							Ranaghat-I	1417	61	201	230	22	26
III	Habra	20							Habra	1270	5	155	83	− 34	− 20
III	Singur	22							Singur	1268	17	291	81	− 23	116
III	Polba-dadpur	23							Polba-dadpur	1208	6	163	21	− 33	− 12

Table 7.2 (continued)

Zone	Tehsil name	Tehsil no	Kharif PET	Kharif RF	Winter PET	Winter RF	Summer PET	Summer RF	Met station	Kharif RF	Winter RF	Summer RF	Gap Kharif	Gap Winter	Gap Winter
			(1)	(2)	(3)	(4)	(5)	(6)		(7)	(8)	(9)	(10) = (7) − (2)	(11) = (8) − (4)	(12) = (9) − (6)
IV	Burdwan	29	591	1144	324	51	536	155	Burdwan	1007	35	169	− 137	− 16	14
IV	Memari-I	30							Memari-I	873	39	236	− 271	− 12	81
IV	Ketugram-II	31							Ketugram-II	949	21	96	− 195	− 30	− 60
IV	Mangalkot	32							Mongolkote	1278	29	133	134	− 22	− 22
IV	Kalna-II	33							Kalna-I	837	40	165	− 307	− 11	10
IV	Bhagawanpur-II	35							Bhagaban-pur-II	1511	18	407	367	− 33	252
IV	Sabang	37							Sabong	1378	17	200	234	− 34	45
IV	Pingla	39							Pingla	1123	20	295	− 21	− 31	140
VI	Bishnupur-II	43	500[a]	1133	300[a]	51	400[a]	133	Bishnupur-II	1135	38	278	2	− 13	145
VI	Shyampur-I	45							Shyampur –II	1294	6	298	161	− 45	165
VI	Nandigram-I	46							Nandigram-I	1280	25	389	147	− 26	256
VI	Ramnagar-II	47							Digha	1406	21	184	273	− 30	51
V	Murarai-I	50	566	1323	329	56	434	176	Muraroi-II	1097	28	163	− 226	− 28	− 13
V	Dubrajpur	51							Dubrajpur	1010	56	112	− 313	0	− 64
V	Raghunathpur-II	53	565	1159	302	67	500	88	Hatwara-Purulia	1020	40	163	− 139	− 27	75
V	Manbazar-I	54							Manbazar-I	1372	48	109	213	− 19	21
V	Jhalda-II	55							Jhalda-I	1002	48	68	− 157	− 19	− 20
V	Para	56							Para	901	14	25	− 259	− 53	− 63
V	Keshpur	57	566	1323	329	56	434	176	Keshpur	949	7	308	− 374	− 49	132
V	Binpur-I	60							Binpur-II	1578	53	167	255	− 3	− 9

[a] Assumption

7.5.1.1 Measuring Consumptive Use

While it is difficult to measure CU directly for every location, it is possible to follow a roundabout way. The upper limit of ET for most field crops equals PE from a free, open surface of water. Thus, with the reasonable assumption that (CWR)s must have been met during a normal year, CU could be estimated, with a few additional considerations. The fact of 2004–2005 being a normal year was supported by the yield data.[10] Indeed, the presence of irrigation justified such an assumption (see Vaidyanathan and Sivasubramanian 2004b, p. 2990).

a. As actual ET is less than PE during the nongrowth stages (for instance, early stages), '[i]n any given location [. . .], for most crops, actual ET over the growing season is expected to be less than PET' (Vaidyanathan and Sivasubramaniyan 2004, p. 2990). On the other hand, actual evaporation can fall below PET in situations with soil moisture stock less than the capacity. Former makes the estimate of CU downward, while the latter contributes in an upward bias. It is virtually impossible to separate the two, other than under carefully conducted field trials. Vaidyanathan and Sivasubramaniyan 2004a, p. 32) found it reasonable to 'assume that the estimates of consumptive use are nearer, if not at, the upper limit'. This chapter holds the same.
b. While official statistics designate three main crop seasons as kharif (June–October or (1) as per CCS), *rabi* (October–March or (2)), and summer (March–June, or (3)), there exists widespread variations in length of growing period, land preparation, and harvesting dates between irrigated and unirrigated areas and even for a given crop in a given *tehsil*. While the CCS dataset permits such fine-tuning possibilities, due to the absence of agro-climatic zone specific ET data against different stages of growth, we had to adopt a simplifying assumption: here, season 1 starts with 1st day of month 6, season 2 with month 11, and season 3 with month 3 for all crops and all plots, in contrast to the duration of seasons in Vaidyanathan and Sivasubramaniyan (2004a).
c. The ratio of consumptive use to PET varies between crops, as captured by the crop coefficient, recommended by the FAO. Under Indian conditions for most seasonal crops this ratio ranges between 0.8 and 0.11, with the exception of Rice (1.2), as it requires standing water (Vaidyanathan and Sivasubramaniyan 2004a, p. 31). Given the predominance of paddy among the PSC combinations (69 %); two types of calculations had been made. In one CU was taken equal to PET values while the other had considered the crop coefficient of paddy.

---

[10] In 2004–2005, West Bengal stood 6th in the yield rank for paddy at 2574 kg/ha, accounting for the highest share in total area under paddy in the country (13.79 %) and production (17.9 %; Table 4.6 (b) in DES 2007). In the CCS data, only 231 parcels reported eight types of problems together, including animal damage (53), drought (47), disease (46), flood (22) and seepage from irrigation (22). Further, the harvested percentage was at least 95 % for 2145 PSCs out of 2281, and 2050 of them recorded 100 %.

d. The effective rainfall reaching the crop was assumed as 80 % in totally rain-fed conditions by Vaidyanathan and Sivasubramaniyan (2004a). In other words, '[u]nirrigated consumptive use [was] taken as PET*crop factor or 80 % of rainfall whichever is less' (Vaidyanathan and Sivasubramaniyan 2004a, p. 31). Arguably, adoption of such a norm holds true only under certain agro-ecological conditions but the same had been followed for all locations in this chapter. For instance, substantial number of plots under consideration was located in soils which could 'carry forward' soil moisture. This injected a downward bias to the estimates of consumptive use.

7.5.1.2 CU Values

Kar and Verma (2005, p. 31) had computed CWR for the dominant rice crop in sub regions of AER 12.0 (eastern Chhotanagpur plateau), using climatic water balance in three seasons (450–550 mm in kharif, 600–720 in winter, and 775–875 in summer). Water requirement of rice without stress was taken as 50 mm/week in eastern India holding both the daily ET and percolation losses around 3–4 mm/day by Singh and Singh (2000, p. 200). In contrast, stage-wise recommendation of Government of West Bengal was followed by Chatterjee (1991) for kharif rice. Vaidyanathan and Sivasubramaniyan (2004a) had computed consumptive values for irrigated crops using crop coefficient (of FAO) and PET values across agro-climatic sub-zone-wise PET values while for rain-fed crops the determinants were effective rainfall, crop coefficients, and PET.

Strictly speaking, CU shall be PET $\times$ crop coefficient, with the condition that it cannot exceed 0.8 * rainfall (Vaidyanathan and Sivasubramaniyan 2004a, p. 36). Against each agro-climatic zone such an evaluation was made so as to locate the possibilities of adequacy of rainfall. Seasonal calculations had also been done with holding blocks of months together for both PET and rainfall, with the assumption that entire surplus water from one particular month/season is fully carried forward to the next month/season.

7.6 Results and Discussion

Table 7.3 shows two kinds of calculations: (a) CU equals the PET value and (b) CU is taken as the crop coefficient for paddy. Each was done for five different cases: (i) only kharif, (ii) only winter, (iii) only summer, (iv) all three together (or annual), and (v) kharif and winter together. Highest number of 'surplus' tehsils were found in (i) under (a) for obvious reasons, followed by (v) under (a). Corresponding number of 'surplus' tehsils under (b) were much less; indeed, in (i), two tehsils (30, 33) had showed a negative surplus and quite a few with less than 100 mm of surplus. Clearly, rainwater alone was inadequate for cultivation of paddy in these areas.

Table 7.3 'Deficit' and 'surplus' rainwater tehsils (selected), West Bengal, 2004–2005

Zone	Tehsil name	District	Tehsil no	(a) Effective rainfall—PET					(b) Effective rainfall—PET × crop coefficient of paddy (1.2)				
				(i) Kharif	(ii) Winter	(iii) Summer	(iv) Annual	(v) Kharif + Winter	(i) Kharif	(ii) Winter	(iii) Summer	(iv) Annual	(v) Kharif + Winter
II	Dinhata-I	Coochbehar	1	*1320.88*	− 234.12	− 56.88	*1029.88*	*1086.76*	*1210.08*	− 283.12	− 132.48	*794.48*	*926.96*
II	Coochbehar-I	Coochbehar	3	*1862.96*	− 215.88	*140.24*	*1787.32*	*1647.08*	*1752.16*	− 264.88	*64.64*	*1551.92*	*1487.28*
II	Sitalkuchi	Coochbehar	4	*1415.44*	− 234.28	− 42.96	*1138.20*	*1181.16*	*1304.64*	− 283.28	− 118.56	*902.80*	*1021.36*
II	Moynaguri	Jalpaiguri	5	*1735.76*	− 222.92	− 23.76	*1489.08*	*1512.84*	*1624.96*	− 271.92	− 99.36	*1253.68*	*1353.04*
II	Falakata	Jalpaiguri	6	*1996.24*	− 220.36	*113.20*	*1889.08*	*1775.88*	*1885.44*	− 269.36	*37.60*	*1653.68*	*1616.08*
II	Karandighi	Uttar Dinajpur	9	*190.08*	− 236.36	− 314.72	− 361.00	− 46.28	*79.28*	− 285.36	− 390.32	− 596.40	− 206.08
III	Raiganj	Uttar Dinajpur	10	*186.00*	− 270.20	− 396.20	− 480.40	− 84.20	*70.80*	− 327.20	− 496.40	− 752.80	− 256.40
III	Harishchan-drapur-II	Malda	11	*1262.32*	− 253.48	− 362.92	*645.92*	*1008.84*	*1147.12*	− 310.48	− 463.12	*373.52*	*836.64*
III	Chanchal-II	Malda	12	*758.00*	− 267.8	− 357.16	*133.04*	*490.20*	*642.80*	− 324.80	− 457.36	− 139.36	*318.00*
III	Murshidabad-Jiaganj	Murshidabad	13	*114.64*	− 281.8	− 426.92	− 594.08	− 167.16	− 0.56	− 338.80	− 527.12	− 866.48	− 339.36
III	Raninagar-II	Murshidabad	14	*460.00*	− 261.00	− 406.44	− 207.44	*199.00*	*344.80*	− 318.00	− 506.64	− 479.84	*26.80*
III	Jalangi	Murshidabad	15	*369.68*	− 272.68	− 347.88	− 250.88	*97.00*	*254.48*	− 329.68	− 448.08	− 523.28	− 75.20
III	Tehatta-II	Nadia	16	*501.12*	− 278.60	− 492.20	− 269.68	*222.52*	*385.92*	− 335.60	− 592.40	− 542.08	*50.32*
III	Kaliganj	Nadia	17	*186.64*	− 283.40	− 448.84	− 545.60	− 96.76	*71.44*	− 340.40	− 549.04	− 818.00	− 268.96
III	Ranaghat-I	Nadia	18	*557.92*	− 236.36	− 339.88	− 18.32	*321.56*	*442.72*	− 293.36	− 440.08	− 290.72	*149.36*
III	Habra	24 Par-ganas (S)	20	*440.32*	− 280.76	− 376.68	− 217.12	*159.56*	*325.12*	− 337.76	− 476.88	− 489.52	− 12.64

Table 7.3 (continued)

Zone	Tehsil name	District	Tehsil no	(a) Effective rainfall—PET					(b) Effective rainfall—PET × crop coefficient of paddy (1.2)				
				(i) Kharif	(ii) Winter	(iii) Summer	(iv) Annual	(v) Kharif + Winter	(i) Kharif	(ii) Winter	(iii) Summer	(iv) Annual	(v) Kharif + Winter
III	Singur	Hooghly	22	*438.16*	– 271.80	– 268.04	– 101.68	*166.36*	*322.96*	– 328.80	– 368.24	– 374.08	– 5.84
III	Polba-dadpur	Hooghly	23	*390.40*	– 280.20	– 370.76	– 260.56	*110.20*	*275.20*	– 337.20	– 470.96	– 532.96	– 62.00
IV	Burdwan	Burdwan	29	*214.52*	– 295.84	– 400.64	– 481.96	– 81.32	*96.32*	– 360.64	– 507.84	– 772.16	– 264.32
IV	Memari-I	Burdwan	30	*107.56*	– 292.48	– 346.88	– 531.80	– 184.92	– 10.64	– 357.28	– 454.08	– 822.00	– 367.92
IV	Ketugram-II	Burdwan	31	*167.88*	– 307.36	– 459.60	– 599.08	– 139.48	*49.68*	– 372.16	– 566.80	– 889.28	– 322.48
IV	Mangalkot	Burdwan	32	*431.00*	– 300.48	– 429.52	– 299.00	*130.52*	*312.80*	– 365.28	– 536.72	– 589.20	– 52.48
IV	Kalna-II	Burdwan	33	*78.28*	– 291.84	– 403.76	– 617.32	– 213.56	– 39.92	– 356.64	– 510.96	– 907.52	– 396.56
IV	Bhagawanpur-II	Midnapur (E)	35	*618.12*	– 309.44	– 210.72	*97.96*	*308.68*	*499.92*	– 374.24	– 317.92	– 192.24	*125.68*
IV	Sabang	Midnapur (W)	37	*511.56*	– 310.56	– 376.32	– 175.32	*201.00*	*393.36*	– 375.36	– 483.52	– 465.52	*18.00*
IV	Pingla	Midnapur (W)	39	*307.40*	– 308.24	– 299.76	– 300.60	– 0.84	*189.20*	– 373.04	– 406.96	– 590.80	– 183.84
VI	Bishnupur-II	24 Parganas (S)	43	*408.00*	– 269.76	– 227.92	– 89.68	*138.24*	*308.00*	– 329.76	– 317.92	– 339.68	– 21.76
VI	Shyampur-I	Howrah	45	*535.36*	– 295.20	– 211.60	*28.56*	*240.16*	*435.36*	– 355.20	– 301.60	– 221.44	*80.16*
VI	Nandigram-I	Midnapur (E)	46	*524.08*	– 280.08	– 138.56	*105.44*	*244.00*	*424.08*	– 340.08	– 228.56	– 144.56	*84.00*
VI	Ramnagar-II	Midnapur (E)	47	*624.56*	– 283.20	– 303.04	*38.32*	*341.36*	*524.56*	– 343.20	– 393.04	– 211.68	*181.36*

Table 7.3 (continued)

Zone	Tehsil name	District	Tehsil no	(a) Effective rainfall—PET					(b) Effective rainfall—PET × crop coefficient of paddy (1.2)				
				(i) Kharif	(ii) Winter	(iii) Summer	(iv) Annual	(v) Kharif + Winter	(i) Kharif	(ii) Winter	(iii) Summer	(iv) Annual	(v) Kharif + Winter
V	Murarai-I	Birbhum	50	*311.68*	−306.60	−303.60	−298.52	*5.08*	*198.48*	−372.40	−390.40	−564.32	−173.92
V	Dubrajpur	Birbhum	51	*241.68*	−284.28	−344.56	−387.16	−42.60	*128.48*	−350.08	−431.36	−652.96	−221.60
V	Raghunathpur-II	Purulia	53	*251.00*	−269.68	−369.28	−387.96	−18.68	*138.00*	−330.08	−469.28	−661.36	−192.08
V	Manbazar-I	Purulia	54	*532.92*	−263.44	−413.12	−143.64	*269.48*	*419.92*	−323.84	−513.12	−417.04	*96.08*
V	Jhalda-II	Purulia	55	*236.68*	−263.60	−445.60	−472.52	−26.92	*123.68*	−324.00	−545.60	−745.92	−200.32
V	Para	Purulia	56	*155.40*	−290.80	−480.32	−615.72	−135.40	*42.40*	−351.20	−580.32	−889.12	−308.80
V	Keshpur	Midnapur (W)	57	*193.36*	−323.40	−187.76	−317.80	−130.04	*80.16*	−389.20	−274.56	−583.60	−309.04
V	Binpur-I	Midnapur (W)	60	*696.40*	−286.92	−300.08	*109.40*	*409.48*	*583.20*	−352.72	−386.88	−156.40	*230.48*

Effective rainfall = 80 % of actual rainfall figures taken from column (7)–(9) in Table 7.2, PET figures taken from column (1)–(3) in Table 7.2

Overall, besides (CCS) zone II lying at the northernmost part of the state, there were many *tehsils* belonging to other zones with 'surplus' rainfall under (a) and (b) both. More prominent was the zone VI belonging to the southern part of the state. In (v) under (a), all the four *tehsils* recorded a surplus rainfall, while under (b), three of them did so. Figure 7.3 shows the location of deficit and surplus tehsils in (v) under (a).

Three typical examples follow, from calculations under (a), for illustrating the irrigation possibilities and their limitations. Harischandrapur-II (tehsil 11, zone III) had a surplus of 645.92 mm in (iv) and 1008.84 mm in (v). Even then, plots belonging to all the CCS size-classes (based on gross cropped area; 4 in number, with smallest corresponding to 1) had to be provided with machine-aided irrigation in winter season, with considerable labour and material inputs. In summer, while plots belonging to the largest two size-classes had been cultivated with machine-aided irrigation, no cultivation could be carried out in plots belonging to the lowest size-class. In contrast, farmers in Binpur-I (tehsil 60, zone VI) had to irrigate under case 1.1 during (i), to take care of the irregular but sufficient rainfall surplus of 696.4 mm (net of PET). Polba-dadpur (tehsil 23, zone III), endowed with a surplus rainfall of 390.4 mm in (i) and 110.2 mm in (v), depicts another possibility. While in kharif most of the farms belonging to the largest two size-classes had to spend some labour and material under machine-aided irrigation, in winter, non-machine aided irrigation was practiced by farmers of all size-classes.[11]

All such 17 surplus tehsils for paddy cultivation only with rainwater in kharif and winter together were identified, from the last column of Table 7.3. For those *tehsils,* irrigation costs (in BOE/ha) were calculated under both case 1.1 and 1.2, with respect to CCS size-classes and three seasons (Table 7.4). Given that 1 BOE equals 1 barrel of crude oil, priced on average at US$100, under case 1.2, average per hectare irrigation cost was 0.697 BOE for kharif, 4.305 BOE for winter, and 2.96 BOE for summer.[12] Each of these estimates was 100 times higher on average than the corresponding seasonal value under case 1.1. Assuming US$1 = ₹ 40, corresponding monetary costs are ₹ 2788, ₹ 17,200, and ₹ 11,840, respectively. Size-class based season-wise irrigation costs reveals that for all *tehsils* together, average per hectare cost reduces with an increase in size-class, for all seasons (Fig. 7.4). Further, for each size-class, per hectare energy cost remained highest for winter, followed by summer and then kharif, which is an expected result.

[11] Detailed per hectare energy values (in MJ/ha) of different irrigation practices for selected blocks (CU = PET) across seasons and CCS size-classes in West Bengal for 2004–2005, are available from the author on request.

[12] Leaving aside the very large energy costs corresponding to tehsils 20, 32, and 47, which requires further inquiry for explanation, average irrigation costs are 0.697 BOE for kharif, 2.145 BOE for winter and 1.29 BOE for summer.

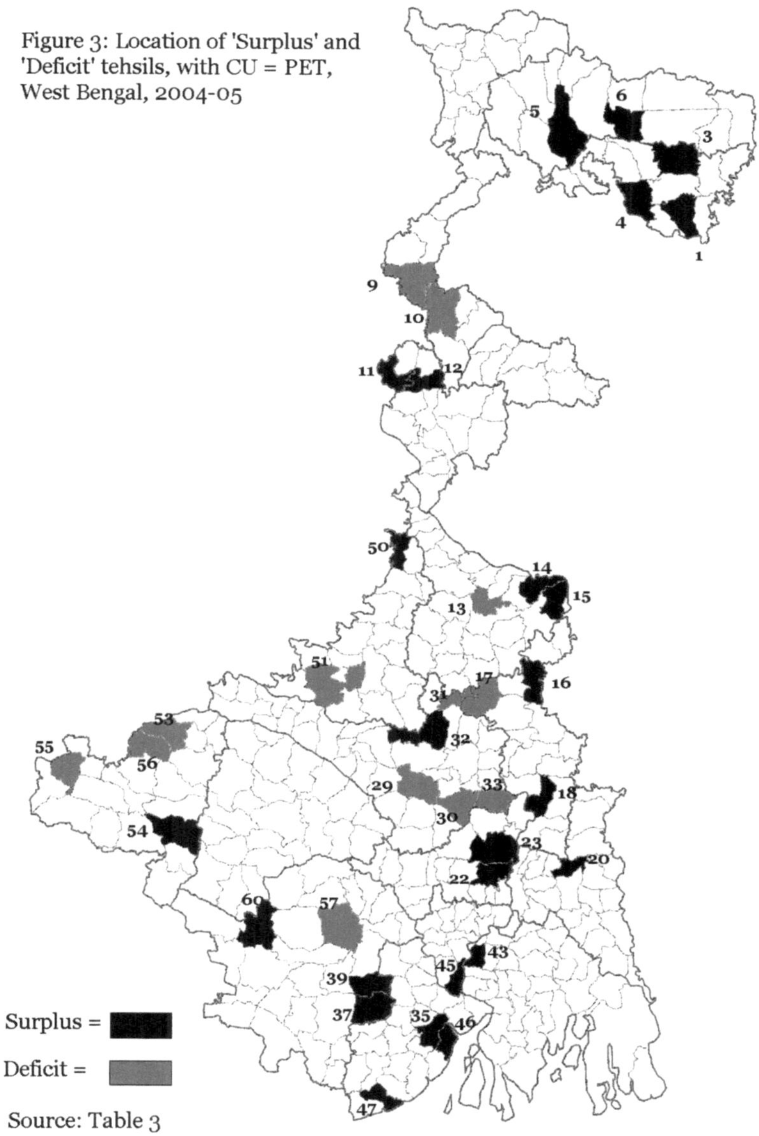

Fig. 7.3 Location of 'surplus' and 'deficit' tehsils, with CU = PET, West Bengal, 2004–2005. (Source: Table 7.3)

Table 7.4 Per hectare value of rainfall to the farmer (in barrels of oil equivalent/hectare) in surplus tehsils for paddy cultivation in kharif and winter seasons

Irrigation type		1.1	1.2	1.1	1.2	1.1	1.2
Tehsil no	Size-group	Season 1		Season 2		Season 3	
1	2				1.63		
	3				2.11		
3	2				1.80		
	3				2.26		0.78
6	1				0.58		
	3				1.50		
11	1				3.30		
	2				3.26		0.53
	3				2.58		0.41
12	3				0.74		
14	1				1.76		0.44
	2		2.36		4.32		0.93
	3		1.20		2.32		0.80
15	2		0.86		1.06		
	3		0.63		0.82		
16	1		0.88		2.15		0.50
	2	0.03	0.16	0.12	1.80	0.02	0.69
	3		0.83		1.38		0.45
18	1				6.31		1.91
	2				1.08		0.74
	3			0.29	1.78		1.43
20	1				26.93		19.03
	2				20.32		16.87
	3				16.28		7.31
22	1		0.62		2.32		
	2		0.37		1.20		
	3		0.45		1.81		
23	1			0.08			0.47
	2		0.22	0.05	2.27		0.74
	3		0.07	0.04	0.02		0.35
32	1				16.66		
	2		0.43		2.85		
	3				1.79		
35	2				0.63		
37	1				4.38		2.19
	2			0.11	2.96		1.43

Table 7.4 (continued)

Irrigation type		1.1	1.2	1.1	1.2	1.1	1.2
Tehsil no	Size-group	Season 1		Season 2		Season 3	
46	3			0.02			
47	2				19.08		
	3				8.13		
54	1				1.37		
	2				0.97		
60	1	0.01					
	2	0.01			0.30		
	3	0.01			0.54		

Selection of *tehsils* from last column in Table 7.3 ((v) under (b)) Blank signifies absence of irrigation expenditure

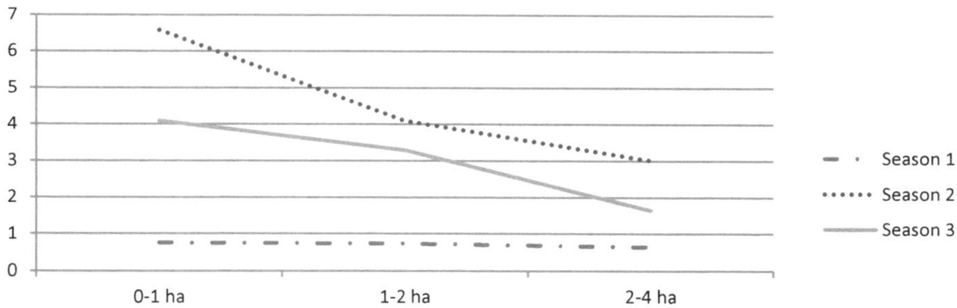

Fig. 7.4 Season-wise per hectare energy cost of machine-aided irrigation (in BOE/ha) across size-calsses of selected 19 tehsils, West Bengal, 2004–2005. List of tehsils from last column of Table 7.3; data reflects average in Table 7.4 under case 1.2

Arguably, 'structures' for holding rainwater also involve costs, primarily of human labour, both intellectual and physical. Most of these are one-time expenditure, and even if it may involve higher initial monetary costs than the machine aided alternatives, savings on operation and maintenance costs far outweigh them. Further, the custom of exchange labour[13] widely prevalent in rural West Bengal and the possibility of mobilisation of collectivised labour can reduce the monetary expenditure substantially. Even the additional space required for the farm ponds may not also involve a very large opportunity cost, given the possibility of growing fish simultaneously, which is quite common in this part of the country.

Importance of rainwater to the farmer, and hence agriculture, is well established. In India even after exhausting the full potential of machine-aided irrigation, nearly

[13] It is the hired-in labour against which corresponding hired-out labour takes place, between the owners of such labour.

50 % of the net cultivated area will remain as rain fed (Sharma 2011; Wani et al. 2009),[14] though officially it has been informed that of 139.89 million ha of 'Ultimate irrigation potential', 109.77 million ha have been created till 2010–2011.[15]

Out of 4000 billion cubic metres (BCM) of rainwater that is received by India annually, 1600 BCM falls on agricultural land. About 240 BCM equivalent is available for harvesting in small scale storages. Areas with annual rainfall upto 1000 mm produce a substantial run-off, to the extent of 114 BCM. Of this quantum, only about 19 % (or 28 BCM) can provide a supplemental irrigation of 100 mm depth at the growing stage of rain-fed crops in an area of 27.5 Mha, that can result into a production of roughly 10 million t of foodgrains (Sharma et al. 2010).

Different methods of rainwater harvesting were developed to suit the geographical and meteorological conditions across the country (Ramakrishna et al. 2006). In regions with agro-climatic characteristics of sub-humid/humid high rainfall regions, like most of West Bengal, many researchers had explored the possibilities of 'run-off recycling' so as to provide adequate water supply to the crop (rice in particular). A positive conclusion about the feasibility of such systems with a favorable benefit cost ratio is rather common (Srivastava 2001, pp. 38–39), be it tanks or embankments around the rain-fed rice field (Mishra and Mohanty 2004, p. 120; Mishra et al. 1998). The question however remains on how to incentivise, mobilise, and organise farmer-cultivators to exploit such a blessing.

7.7 Concluding Remarks

While Shah (2013) had reported about the 'paradigm shifts' in the water management as envisaged in the near future by the government, there are a few disturbing signs. For example, Bhattacharjee (2012) had reported after surveying various parts of the state of West Bengal that due to faulty administrative planning and bureaucratic red-tapism under NREGS, that includes rationing of work across as many villages as possible, seldom the digging or desiltation works are completed so as to yield a productive asset.[16]

---

[14] Of the 140.86 million hectares (Mha) net area sown (Gross Cropped Area—area sown more than once), (net) irrigated area is 62.29 Mha (44.22 %). Of the Gross Cropped Area of 195.83 Mha, gross irrigated area is 87.26 Mha (44.56 %) in India. (Computed from Table 14.1: Agricultural land by use in India, DES 2010). Planning commission (2008, p. 25) had pegged the share at 60 %.

[15] Unstarred question no 3590, answered on April 26, 2012 in the Parliament, by Minister of State in the Ministry of Water Resources.

[16] A survey of more than 100 ponds under the NREGS, conducted just after the monsoon season (October–December) in 2010, had found that 88 % did not have any water. During the previous 4–5 years, it was these ponds on which 74–89 % of the total expenditure under NREGS was spent (Bhattacharya 2012).

At the same time, there are signs of hope as well. In perennially water-starved regions of Maharashtra, the state government has decided to de-prioritise the construction of large dams in favour of check-dams and other such small structures, thanks to the ever-escalating cost due to the long delays and problems related to land acquisition (Iyer 2012). From the same state, as per the study conducted by Mahatma Phule Agriculture University, Rahuri, over a lakh of farm ponds, each with about 1200 m^3 of water have been constructed over the last five years (Rashid 2012).

Given the expected changes in the rainfall pattern, of the increase in frequency and magnitude of high rainfall events and the fall in the number of rainy days, for the majority of rain-fed farmers across the country, storing of water remains the one and only sustaining solution, in ecological and hence in economic terms.

References

Allen RG, Pereira, L S, Raes D, Smith M (1998) Crop evapotranspiration—guidelines for computing crop water requirements, FAO Irrigation and drainage paper 56. FAO, Rome

Bandyopadhyay J (1987) Political ecology of drought and water scarcity: need for an ecological water resources policy. Econ Polit Wkly 22(50):2159–2161, 2163–2169

Batty JC, Keller J (1980) Energy requirements for irrigation. In: Pimentel D (ed) Handbook of energy utilization in agriculture. CRC Press, Boca Raton, pp 35–44

Bhattacharjee S (2012) Pukur Kete Ki Hawbe, Jodi Tatey Jol Na Thake (Why dig, if water does not remain in the pond. Anandabazar Patrika, Kolkata edition, 24 July

Burkett P (2003) The value problem in ecological economics: lessons from the Physiocrats and Marx. Organ Environ 16:137–167

Cervinka V (1980) Fuel and energy efficiency. In: Pimentel D (ed) Handbook of energy utilization in agriculture. CRC Press, Boca Raton, pp 15–21

Chatterjee N (1991) Agro-climatic perspective on irrigation in West Bengal, India: a district level analysis. Singap J Trop Geogr 12(1):12–30

Common M (1995) Sustainability and policy limits to economics. Cambridge University Press, Cambridge

DAC (2010) Agricultural census 2005–06: manual of schedules and instructions for data collection (non-land record states). Agricultural Census Division, Department of Agriculture & Cooperation, Ministry of Agriculture, Government of India, New Delhi

DES (2007) Agricultural statistics at a glance 2006. Directorate of Economics & Statistics, Department of Agriculture & Cooperation, Ministry of Agriculture, Government of India, New Delhi

DES (2010) Agricultural statistics at a glance 2010. Directorate of Economics & Statistics, Department of Agriculture & Cooperation, Ministry of Agriculture, Government of India, New Delhi

Doering OC III (1980) Accounting for energy in farm machinery. In: Pimentel D (ed) Handbook of energy utilization in agriculture. CRC Press, Boca Raton, pp 9–14

FAO (2007) Glossary, global map of irrigation areas version 4.0.1. AQUASTAT Programme. FAO, Rome

Farber SC, Costanza R, Wilson MA (2002) Economic and ecological concepts for valuing ecosystem services. Ecolog Econ 41:375–392

Fluck RC (ed) (1992) Energy in farm production. Energy in world agriculture 6. Elsevier, Amsterdam

Ghosh SP (1991a) Concepts of agro-climatic zones for research and development. In: Ghosh SP (ed) Agro-climatic zone specific research. Indian Council of Agricultural Research, New Delhi, pp 438–462
Ghosh SP (ed) (1991b) Agro-climatic zone specific research. Indian Council of Agricultural Research, New Delhi
Ghosh SP (1996) Regionalization experiences in Indian agriculture. In Basu DN, Guha GS (ed) Agro-climatic regional planning in India, 2 vols, vol 1. Concept Publishing Co., New Delhi
Gopalan C, Shastri BVR, Balasubramanian SC (1989) Nutritive value of Indian foods, revised edition, 1995 reprint. National Institute of Nutrition, ICMR, Hyderabad
Guhathakurta P, Rajeevan M (2006) Trends in the rainfall pattern over India. Research report No. 2/2006. National Climate Centre, India Meteorological Department, Pune
ICMR (1990) Nutrient requirements and recommended dietary allowances for Indians: a report of the expert group of the ICMR, 2000 reprint. Indian Council of Medical Research, New Delhi
IFIAS (1978) Energy analysis and economics: IFIAS workshop report. Res Energy 1:151–204
Indian Express (2011) Plan NREGS works at hamlet level: NAC. 29 July, New Delhi edition
Iyer K (2012) Drought of new ideas in old crisis. Indian Express, 16 May. New Delhi edition
Jones MR (1989) Analysis of the use of energy in agriculture approaches and problems. Agric Syst 29:339–355
Judson DH (1989) The convergence of neo-ricardian and embodied energy theories of value and price. Ecol Econ 1:261–281
Kar G, Verma HN (2005) Climatic water balance, probable rainfall, rice crop water requirements and cold periods in AER 12.0 in India. Agric Water Manage 72:15–32
Millennium Ecosystem Assessment (2005) Ecosystems and human well-being: synthesis. Island Press, Washington, DC
Mishra A, Mohanty RK (2004) Productivity enhancement through rice-fish farming using a two-stage rainwater conservation technique. Agric Water Manage 67:119–131
Mishra A, Ghorai AK, Singh SR (1998) Rainwater, soil and nutrient conservation in rainfed rice lands in Eastern India. Agric Water Manage 38:45–57
Mitchell R (1979) The analysis of Indian agro-ecosystems. Interprint, New Delhi
Nawn N (2013) Using cost of cultivation survey data: changing challenges for researchers. Econ Polit Wkly, 58(26, 27):139–147
Norum L (1983) Problem formulation and quantification in energy analysis. Energy Agric 2:1–10
Odum HT (1984) Energy analysis of the environmental role in agriculture. In Stanhill G (ed) Energy and agriculture. Springer, Berlin, pp 24–51
Pimentel D, Hurd LE, Bellotti AC, Forster MJ, Oka IN, Sholes OD, Whitman RJ (1973) Food production and the energy crisis. Science 182(4111):443–449
Planning Commission (1992) Report of the committee on pricing of irrigation water. Planning Commission, Government of India, New Delhi
Planning Commission (2008) Eleventh five year plan 2007–2012, 3 volumes, volume 3: agriculture, rural development, industry, services and physical infrastructure. Planning Commission, Government of India, New Delhi
Ramakrishna YS, Rao GGSN, Rao VUM, Rao AVMS, Rao KV (2006) Water management water use in rainfed regions of India. In Mannava V, Sivakumar K, Motha RP (ed) Managing weather and climate risks in agriculture. Springer, New York
Rao AR (1984) Bioenergetics of bullock power. Energy 9(6):541–543
Rashid A (2012) Challenging the drought crisis, with farm ponds. Indian Express, New Delhi edition, 10 May
Sehgal JL, Mandal, D K, Mandal C, Vadivelu S (1992) Agro-ecological regions of India, NBSS Publication no. 24, second edition. National Bureau of Soil Survey & Land Use Planning, Indian Council of Agricultural Research and Oxford & IBH Publishing Co, New Delhi
Shah M (2013) Water: towards a paradigm shift in the twelfth plan. Econ Polit Wkly 58(3):40–52
Sharma KD (2011) Rainfed agriculture could meet the challenges of food security in India. Curr Sci 100(11):1615–1616

Sharma BR, Rao KV, Vittal KPR, Ramakrishna YSR, Amarasinghe U (2010) Estimating the potential of rainfed agriculture in India: Prospects for water productivity improvements. Agric Water Manage 97(1):23–30

Singh P (2006) Agro-climatic zonal planning including agriculture development in north-eastern India, Final report of the working group on agriculture, XI five year plan (2007-2012), two volumes, volume II: Interim report of the sub-groups. Planning Commission, Government of India, New Delhi

Singh VP, Singh RK (ed) (2000) Rainfed rice: a sourcebook of best practices and strategies in eastern India. International Rice Research Institute, Los Baños

Srivastava RC (2001) Methodology for design of water harvesting system for high rainfall areas. Agric Water Manage 47:37–53

Stanhill G (1974) Energy and agriculture: a national case study. Agro Ecosyst 1:205–217

Stanhill G (ed) (1984) Energy in agriculture. Springer, Berlin

Tomar VS, O'toole JC (1980) Water use in lowland rice cultivation in Asia: a review of evapotranspiration. Agric Water Manage 3:83–106

Vaidyanathan A, Sivasubramaniyan K (2004a) Efficiency of water use in Indian agriculture. Working paper no. 183. Madras Institute of Development Studies, Chennai

Vaidyanathan A, Sivasubramaniyan K (2004b) Efficiency of water use in agriculture. Econ Polit Wkly 39(27):2989–2996

Wani SP, Sreedevi TK, Rockström J, Ramakrishna YS (2009) Rainfed agriculture-past trends and future prospects. In Wani SP, Rockström J, Oweis T (ed) Rainfed agriculture: unlocking the potential. CAB International, ICRISAT and IWMI, Oxfordshire

Chapter 8
Governing Firms' Corporate Environmental Response: Success or Failure? Evidence from a Panel Data Based Analysis on Solid Waste Management in Agri-Food Processing Sector of Sri Lanka

Menuka Udugama and U. K. Jayasinghe-Mudalige

8.1 Introduction

Waste management in general and solid waste management in particular is a key environmental challenge, especially in industrialized urban areas throughout the world where Sri Lanka is no exception. Without an effective and efficient program to manage solid waste, the waste generated through various human activities, both industrial and domestic, can result in numerous health hazards and have a negative impact on the environment. We may see that concerns for an effective and economical program to manage solid waste is ever increasing and governments and private sectors in both developed and developing countries are therefore called to play a significant role in this connection.

Where Sri Lanka is concerned, over the past decade, the problem of generation and accumulation of waste is seen to be exacerbated mainly by the absence of proper management systems at the level of firm and by the existence of various types of agri-food processing industries in large numbers[1].

Economists, bringing out examples from both the developed and the developing country context, elaborated that market-based actions are, in general, more effective than government-oriented "first best" solutions to deal with the problems associated with public goods. In the context of environmental policy, however, it is rather difficult to formulate a set of appropriate policies that can be put into practice at the level of firm mainly due to the limited knowledge of the level and nature of

[1] Being the largest manufacturing sector in Sri Lanka with more than 80 % of firms operating in the provinces of very high population density (i.e., Western, Central, North Western, and Southern with more than 500 people per km^2).

M. Udugama (✉) · U. K. Jayasinghe-Mudalige
Department of Agribusiness Management, Faculty of Agriculture and Plantation Management, Wayamba University of Sri Lanka, Gonawila (NWP), Sri Lanka
e-mail: menukaudugama@gmail.com

N. Ghosh et al. (eds.), *Nature, Economy and Society*,
DOI 10.1007/978-81-322-2404-4_8

economic incentives available, in both market and nonmarket context, for a firm to comply with and/or adopt such controls at the firm level (Khanna 2001; Segerson and Miceli 1998).

The literature on environmental economics and management points out four socioeconomic processes, namely: (1) Efficiency, (2) Market, (3) Regulatory, and (4) Ethical motives that can influence firms in implementing environmental management controls. These studies specifically shed light on various factors governing environmental responsiveness. For example, the study by Hettige et al. (1996) focuses on: (1) different types of government regulation that affect incentives for abatement and the associated costs, (2) informal regulations by citizens and market characteristics that can lead firms to improve environmental performance, and (3) the voluntary environmental protection, which is to be carried out by a firm on its own willingness [private].

Based on the evidence drawn from a cross section of firms operating in Bangladesh, India, Indonesia, and Thailand, Hettige et al. (1996) also test the importance of plant characteristics, economic considerations, and types of external pressures in determining the environmental performance of a firm. The results suggest that, in general, pollution intensity is negatively associated with scale, productive efficiency, and the use of new process technology, while it is strongly and positively associated with public ownership. Amongst the external sources of pressure, the presence or absence of community action (informal regulation) emerges as a clear source of interplant difference. Using the data from 243 firms operating in Indonesia and belonging to different sectors, Pargal and Wheeler (1996) examined the impact of informal regulation on industrial pollution.

Blackman and Bannister (1997) carried out an econometric analysis to determine the role of community pressure and clean technology among traditional brick makers in Mexico. The outcomes of this empirical investigation suggests that it is possible to successfully promote the adoption of a clean technology by intensely competitive informal firms even when the new technology significantly raises variable cost, and community pressure applied by competing firms and private-sector local organizations can generate incentives for adoption. Dasgupta et al. (2000) studied the effects of regulation, plant-level management policies and several other factors on the environmental compliance of Mexican manufacturers and found that many firms in Mexico avoid complying with regulations because of sporadic monitoring and enforcement, while others over-comply with the regulations because their abatement decisions are strongly affected by extra legal factors. Blackman 2008) used plant-level data from more than 60,000 facilities to identify the drivers of participation in Mexico's Clean Industry Program. The results show that the threat of regulatory sanctions can drive firms to participate in this particular program and it could attract, in particular, relatively dirty firms into the program.

For the purpose of identifying and managing the potential impacts that those commercial and industrial operations can have on the environment, various types of environmental management systems/practices were developed and/or adopted by the private sector. The legal framework required for management of solid waste in Sri Lanka is provided under the Local Government Act. The local authorities are charged in terms of the Act with the responsibility of collection and disposal of solid waste at the municipal, urban, and *Pradeshiya Sabha* (i.e., local government) level. Despite all the formal regulations in place, many stakeholders in the food processing sector claim that the implementation and effective enforcement of formal regulations aiming at the management of solid and liquid waste generated in this sector is very poor, since regulations themselves vary significantly across local authorities as well as at the level of the provincial governments.

For the purpose of this study, we used the special case of adoption of enhanced environmental controls, which was formulated by the Ministry of Environment (and Natural Recourses then) (MENR), by firms operating in the agri–food processing sector in Sri Lanka. Being the largest manufacturing sector in Sri Lanka with more than 80 present firms operating in the provinces of very high population density (i.e., more than 500 people per km^2), the generation and unhygienic accumulation of waste through these agri-food processing firms has become a growing problem in Sri Lanka.

As a solution to this problem, the Ministry of Environment has already formulated the *National Strategy for Solid Waste Management* and also a number of specific procedures that firms in the food processing sector should adopt to manage the solid waste generated in them are introduced. These include: (1) *Sorting of waste based on 3R system*—establishment of necessary infrastructure facilities in appropriate places and allocating labor for the purpose; (2) *Composting*—the conversion of solid waste materials into compost, in which the heavy metals composition should be maintained below the recommended standards; (3) *Biogas Technology*—establishing unit in accordance with the guidelines provide by the MENR; (4) *Biodegradable packing materials*—using materials such as a paper, glass, cloth, etc. instead of polythene and other non biodegradable plastic; (5) *Sanitary land filling*—the maintenance of a site for which the firm should obtain clearance based on the guidelines provided by the Central Environment Authority (CEA) in Sri Lanka; (6) a set of *Good Manufacturing Practices* (GMPs); (7) Regular *Waste Auditing* system; and (8) ISO 14000 *Environmental Management System.* An individual business can select either one or a combination of these practices or any other appropriate mechanism that they deem to be effective in rectifying the problem associated with the generation of waste in their premises.

Under the Local Government Act for governing solid waste in Sri Lanka and the municipal, urban, and *Pradeshiya Sabha* (local government) are endowed with the responsibility of commissioning the entire process. Despite all the formal regulations in place, firms' corporate environmental response in this respect—reflected by the level of adoption of government recommended solid waste management practices (SWMPs) in the firm—claimed to be poor; thus, generation and accumulation of solid waste from various industries has surfaced as a major concern of environmental management.

Jayasinghe-Mudalige and Udugama (2011), empirically explored the impact of regulatory incentives (REG), market forces (MRF), internal efficiency (IEF), and environmental altruism (EAL) for the agri-food processing firms to adopt SWMPs in their firms[2]. However, the changes that occurred in the perceptions of firms about relative effectiveness of different incentives prevailing at the level of firm have not been explored empirically to date, especially in the context of the agri-food processing sector in Sri Lanka. In light of this, the specific objective of this study was to identify the impact of various incentives and firm level characteristics governing the adoption of recommended SWMPs by the firms in the agri-food processing sector in Sri Lanka with a special focus on the perceptual changes over time.

8.2 Methodology

8.2.1 Conceptual Framework

We can conceptualize that there are four social processes, namely: (1) market, (2) political, (3) judicial, and (4) ethical governing the firm's decision to implement environmental management controls. Let us assume that the environmental policy of a firm that works to create a "waste-free non-polluted environment" is characterized by the utility function $U_i = u\,[v\,(D_i \mid I_{ji}, F_{ki})]$ of the decision maker/management of the firm i (where, $i = 1, 2, 3 \ldots n$) and u(v) is concave in its arguments. The management of the firm is responsible for complying with the regulatory requirements of the government. At the same time, the firm may decide to adopt various types of strategies voluntarily to manage the waste generated in the firm. Consequently, the term v in the above equation represents the overall gains to the firm through its responsible behavior towards the quality of environment where it operates. The responsiveness of a firm towards the environment (D) is reflected by the different environmental management practices ($SWMP_i$) adopted by the firm, which depend on the individual incentives faced by the decision maker/management (I_{ji}), where j = types of incentives ($j = 1, 2, 3 \ldots m$) and the characteristics of the firm (F_{ki}), where k = size or type of the firm, etc. Following Nakamura et al. 2001 from the maximization of the utility function, we derive the following empirical expression of the determinants ith firm's environmental management practices (where ε_i is an error term):

$$SWMP_i = \alpha_i + \beta_j I_{ji} + \gamma_k F_{ki} + \varepsilon_I \tag{8.1}$$

Based on the conceptual framework presented above and findings from past studies, four individual incentives are identified to represent market, regulatory, and liability

[2] Visit: http://www.sandeeonline.org/uploads/documents/publication/937_PUB_WP_60_Udith_Jaysinghe.pdf or contact the Corresponding Author at: *menukaudugama@gmail.com* to have a copy of the Working Paper published by the South Asian Network for Development & Environmental Economics (SANDEE), which explains the entire procedure associated with identifying these incentives.

incentives. These include: (1) REG, (2) MRF, (3) IEF in the firm, and (4) EAL (Caswell et al. 1998; Jayasinghe-Mudalige and Henson 2006a, 2006b; Khanna and Anton 2002; Segerson 1999; Bansal and Roth 2000).

8.2.2 *Econometric Specification of the Model*

We can extend Eq. (8.1) expressed above to specify the following econometric model:

$$SWMP = \sigma_0 + \beta_1 {*}REG_i + \beta_2 {*}MRF_i + \beta_3 {*}IEF_i + \beta_4 {*}EAL_i + \gamma_1 {*}FT_i + \gamma_2 {*}FS_i + \gamma_3 {*}VT_i$$

where: $SWMP_i$ denotes the dependent variable (i.e., SWMPs adopted by a firm). The right hand side variables include: σ_0 = intercept, β_j = coefficients of four individual incentives ($j = 1, 2, 3, 4$) considered in the analysis and γ_k = coefficients of characteristics of a firm (F_{ki}) denoted by dummy variables such that *FT* = firm type [five types based on the major products processing (see Sect. 5.1)]; *FS* = firm size [five categories such as: large, medium, small based on annual returns (see Sect. 6.1)]; *VT* = Vintage ($1 = \geq 10$ years; $0 = < 10$ years).

8.2.2.1 Specification of the Dependent Variable (SWMPi)

For the purpose of this analysis, it was presumed that the number of SWMPs adopted by a firm reflects its degree of responsiveness towards environmental quality at each stage, since the Ministry of Environment does not suggest any recommended order in which to adopt the practices listed above in a food processing firm and none of which is endowed with a higher value over the others. Under these circumstances, there is a possibility that certain firms may decide to adopt a single or two practices at a time, whereas others may even go beyond (i.e., four or five) depending on the gains to the firm by doing so. On the other hand, there may be firms that do not adopt a single practice out of the list given above. In such case, an analyst may come up with a series of zeros as she works on a scale of: Adoption = 1; Non-adoption = 0 to report the status of adoption of these practices in the firm on an individual basis. At times, the analyst may therefore experience excess zeros. Principles of Count Data Regression models, thus, can be employed to estimate the coefficients of the econometric model specified above, which uses the number of SWMPs adopted by a firm as the estimable dependent variable (SWMPi; Chowdhury and Imran 2010).

8.2.2.2 Specification of Explanatory Variables (Iji)

Independent variables include the four incentives, firm characteristics, firm size, and vintage. It is unable to include the four individual incentives listed above directly into the econometric model specified as its explanatory variables, mainly due to: (a) Mutual exclusivity and endogeneity—the prevalence of an individual incentive as an element of the system (Nakamura et al. 2001; Shavell 1987); (b) Subjectivity—the management of the firm perceives these incentives in terms of potential benefits and costs to the firm (Buchanan 1969); and (c) Unobservability—the management cannot directly observe the nature of the incentives prevailing at the firm level (Hair et al. 2006). To overcome these difficulties, the confirmatory factor analysis (CFA) techniques [i.e., a multivariate data analysis technique that comes under structural equation modeling (SEM)] were employed to develop estimable variables for the individual incentives (see, Jayasinghe-Mudalige and Udugama 2011).

8.2.3 Data Collection

The database used in Jayasinghe-Mudalige and Udugama (2011), which includes primary data gathered on various aspects related to a firm's performance on environmental quality management from 325 firms in 2008/2009 (Stage I), was considered as the sampling framework for select firms to collect panel data in Stage II. We have identified that certain firms selected into the sample were not in operation (i.e., plant-exit) or under-operation (i.e., partial-exit) and/or the ownership/management has changed from Stage I. The data collected from each firm in Stage II were coded in a database, which was derived from the original South Asian Network for Development and Environmental Economics (SANDEE) database were matched with the corresponding data from Stage I to facilitate count data analysis. Given the time and budgetary constraints, it was decided to carry out Stage II of this study with 50 % of firms which participated to Stage I of the study (i.e., about 150–160 firms). To select the best representative sample of firms to collect panel data, a systematic procedure was followed.

First, we have contacted each firm participated in Stage I over the phone and were informed about the intention of study, and more specifically, about their contribution to the Stage I and the importance of participation to the Stage II in which each firm's progress on this issue is investigated. For the firms who have "given their consent to participate" in this phone conversation (i.e., about 200 firms), a letter of request for an appointment together with the summary of the outcome of Stage I was sent. A *face-to-face* interview supported by the structured questionnaire[3], which is slightly deviated from the format used in Stage I to accommodate panel data, followed by an inspection of the site for the cases where permission was granted, was carried out with the same person that responded to the Stage I, or in cases where the same person was not in position (i.e., resign, busy, changed position, etc.), the top-most executive who possesses executive powers to make decisions with respect to environmental quality at the point of contact to collect data.

[3] See, Jayasinghe-Mudalige and Udugama (2011).

8.2.4 Data Analysis

8.2.4.1 Count Data Model

The firms' responsiveness towards environmental quality (dependent variable) is measured by the number of SWMPs adopted by firms. Observations on the dependent variable are represented by non-negative integer quantities, and failure to account for the integer nature of the data can bias results (Hellerstein 1991, 1992; Hellerstein and Mendolsohn 1993; Haab and McConnell 1996). Furthermore, any resulting policy measure based on continuous demand models when the variable is a non-negative integer is inaccurate and misleading and it is well known that count data models are deemed appropriate when the dependent variable takes on integer values and hence is useful in dealing with problems of this nature.

Count data analyses are advantageous when there are many practices adopted as in this case and when their use intensities are the center of interest. Count data analyses are quite handy, especially when other econometric techniques such as Multinomial Logit or Multivariate Probit models present computational difficulties. These difficulties may arise when the number of practices adopted becomes greater than two, in the case of Multinomial Logit, or four or five, in the case of Multivariate Probit (Rahelizatovo and Gillespie 2004). Count data techniques are inconsequently capable of delivering valuable understandings drawn through the identification and analysis of the most intensive adopters, which are then used to model policies targeting less intensive adopters. We explicitly consider the possibility of *excess zeros* problem and *over-dispersion*, therein, estimate zero-inflated count data models that are deemed appropriate when both use intensity and probability of zero adoption are of concern.

Considering all above, we have estimated the coefficients of the econometric model specified earlier using count data analysis techniques for the data from both Stage I and Stage II separately.

8.2.4.2 Derivation of Index

The panel data collected were then used to develop indices to reflect each incentive that can be used to explore the extent to which a manager of a firm perceived the importance of each incentive considered above, individually and collectively, in the firm's attempt to adopt the SWMPs recommended; thus, it can be used as a yardstick to signal the changes that took place in perceptions of these incentives overtime (Oppenheim 1992).

The scores provided by respondents on a five-point Likert Scale on each statement reflecting individual market based and regulatory incentive were subject to

CFA and principle component analysis (PCA)[4] to test for their unidimensionality (Hair et al. 2006) and the steps used in this connection were analogous to that used in Jayasinghe-Mudalige and Udugama (2011) to derive respective indices ranging from 0–1, where a value of 1 denotes that a firm perceives the said incentive to be "absolutely important" for the firm to look into adoption of these SWMPs, while a 0, on the other end, denotes "not at all."[5]

8.3 Results and Discussions

8.3.1 Level of Adoption of Recommended SWMPs

The sample comprised 32 % "Large," 33 % "Medium," and 35 % "Small" scale firms. With regard to the type of SWMPs adopted by firms in Stage II, it was observed that "Good Manufacturing Practices (GMPs)" has become the most popular practice over time followed by the "3R system" and "Composting" (Fig. 8.1). "3R System" which was the most popular in Stage I had lost its popularity to "GMPs" by Stage II. It was evident that firms show a keen interest in adopting these practices over time. "Bio gas" as a practice showed the least increase in the adoption rate. This may be due to the fact that, in comparison to other two practices, this requires relatively higher effort, space, and human resources during implementation.

Figure 8.2 shows that the level of adoption of SWMPs has improved overtime, i.e., (mean, standard deviation) (1.2, 1.60) in *Stage I* to (1.8, 1.33) in *Stage II*, suggesting the sector, as a whole, becoming "greener."

As shown by the left-skewed distribution curve in Fig. 8.2a, during Stage I, about 52 % of the firms did not possess any of the recommended practices (i.e., non-adopters). However, Fig. 8.2a clearly depicts that this percentage has dropped to 15 % in Stage II. Further, only 30, 11, and 4 % of firms adopted one, two, or three out of the eight practices recommended, respectively, in Stage I. These percentages, respectively, have been changed to 41, 29, and 15 % in Stage II creating a shift in the distribution curve to the right. However, it was evident that in Stage II, the firms that adopted more than four practices have chosen to operate with lesser number of practices by shifting to the most technically and cost effective and sustainable practices rather than maintaining several different practices in place.

Interestingly, the number of SWMPs adopted by a firm varied to a greater extent vis-à-vis the type of the firm and its size. With regard to firm size, large firms, not surprisingly, tended to adopt a higher number of SWMPs. For example, nearly

---

[4] CFA and PCA are technique stated under the Multivariate Data Analysis techniques that is used commonly to define the underlying structure among a set of variables of an analysis objectively, was employed to test this condition.

[5] The steps are not presented here due to page limitations; thus, the interested readers are directed to follow Jayasinghe-Mudalige and Udugama (2011) for further elaboration of methods used.

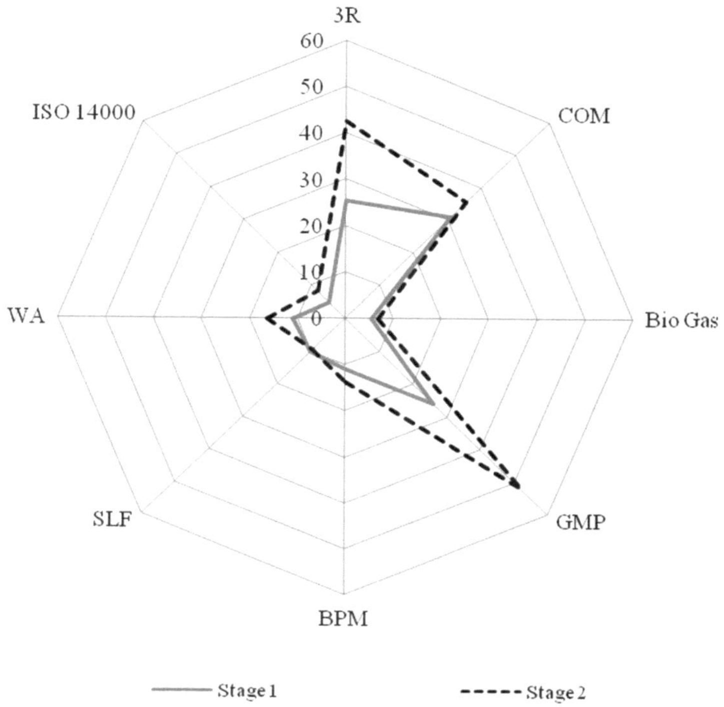

Fig. 8.1 Type of SWMP adopted

20 % of large firms adopted more than four such practices in plant, compared to 67 % of small firms who did not adopt a single practice in Stage I. By Stage II, an evident increase of 47 % was shown by the large firms, while the small firms without a single practice in place (i.e., non-adopters) have decreased to 49 % indicating a move towards higher environmental responsiveness.

8.3.2 Outcome of the Count Data Analysis

The first step towards a count data analysis was to examine the excess zeros and over-dispersion of the data. The results show that it was distributed with a mean (standard deviation) of 1.153 (± 1.559) (i.e., Variance = ± 2.430). This shows that there is an over-dispersion. Therefore, we decided to estimate a model other than the Poisson model in which the two are constrained to be equal. Figure 8.2 depicts that the response variable obtained shows that the number of zeros is excessive. These suggest that it is best to estimate the econometric model with other option/s available, including zero-inflated Poisson (ZIP) and zero-inflated negative binomial

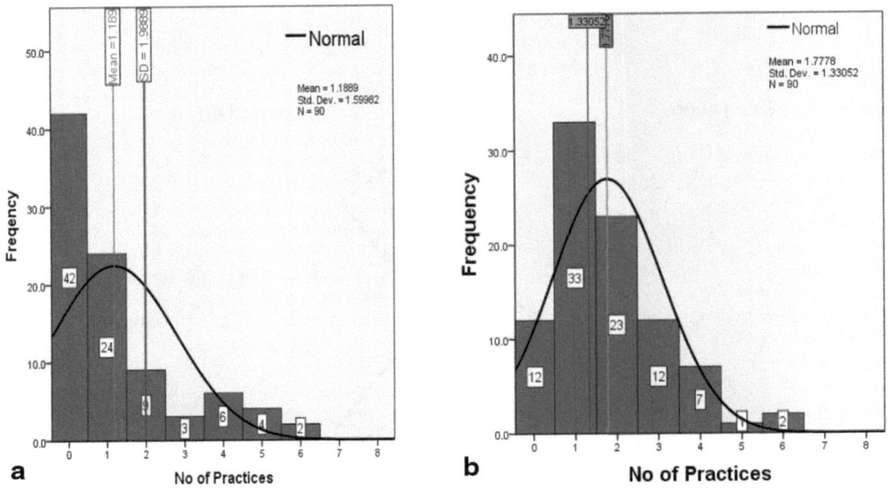

Fig. 8.2 Number of SWMPs adopted by firms. **a** No. of SWMPs in Stage 1, **b** No. of SWMPs in Stage II

(ZINB) models that could account for this over-dispersion. We report the statistical outcome of ZIP and ZINB models in Table 8.1.

The Vuong statistic (V =3.46 in Stage I and 8.02 in Stage II) compares the ZIP with ordinary Poisson Regression (PR) models (Table 8.1). Since it is significant, we prefer ZIP to the PR model. Where Negative Binomial Model (NBM) is considered, the Vuong *t*-test ($V = 3.83$) result further suggests that the ZINB outperforms its parent specification, the negative binomial model (NB). The likelihood ratio (LR) test indicates that the ZINB model reduces to the ZIP model ($P > 0.05$ in Stage I and $P > 0.05$ in Stage II). Thus, the results from this test demonstrate that the LR test statistic, favors the ZINB model over the ZIP model.

The ZINB model results for both stages reveal the impact of the variables of concern. It is evident that, regulation posed a significant impact on the rate of adoption irrespective of time where as market-based incentives and IEF was only perceived to be significantly important in Stage I. However, interestingly, altruism although not perceived to be significantly important in Stage I was perceived important overtime. As regulation continued to pose a significant impact on the rate of adoption irrespective of time could be due to the fact that firms do perceive REG, existing regulation of the country, anticipated regulation, and liability laws important in both stages and they were of the view that with the recent regulatory emphasis on the environment, the regulations in the future will pose a greater impact on their responsiveness on the environment.

Although unexpected, the firms did not perceive MRF and IEF to be as important as it was by Stage II. This may have been due to the fact that the firms who already had the required practices in place or who intended to adopt them in the future did not perceive market incentives such as sales and revenue, reputation of the firm, commercial pressure as well as IEF in the form of technical and human

Table 8.1 Outcome of count analysis

Covariate	Zero inflated poison (ZIP)				Zero inflated negative binomial (ZINB)			
	Stage 01		Stage 02		Stage 01		Stage 02	
	Coe	Prob	Coe	Prob	Coe	Prob	Coe	Prob
Incentives								
REG	0.448**	0.01	1.819**	0.01	0.448**	0.01	1.819**	0.01
MRF	0.008**	0.02	1.601	0.42	0.008**	0.02	1.601	0.42
IEF	0.370**	0.03	2.028	0.51	0.370**	0.03	2.028	0.51
EAL	0.258	0.15	1.128**	0.00	0.258	0.15	1.128**	0.00
Constant	0.028	0.053	0.031	0.08	0.028	0.053	0.031	0.08
Firm characters								
COP	0.313	0.94	1.456**	0.03	0.313	0.94	1.456**	0.03
ESO	– 0.173**	0.03	1.016**	0.02	0.173**	0.03	1.016**	0.02
NAB	2.046	0.20	1.251**	0.00	2.046	0.20	1.251**	0.00
OPP	0.393	0.58	0.012	0.10	0.393	0.58	0.012	0.10
Large	0.745**	0.01	0.099**	0.04	0.745**	0.01	0.099**	0.04
Medium	0.184	0.67	0.108	0.12	0.184	0.67	0.108	0.12
Vintage	0.510**	0.01	0.014	0.465	0.510**	0.01	0.014	0.465
Log like hood	– 80.08		– 238.08		– 80.08		– 238.08	
LR chi2	34.26		31.11		21.35		38.17	
Model	Probit		Probit		Logit		Logit	
Vuong test	3.36**		7.12**		8.42		1.55	
Likelihood ratio test					1.000**		1.044**	
Prob > chi2	0.1691		0.0649		0.000		0.000	

*** Significant at prob = 0.01;** Significant at prob = 0.05;* Significant at prob = 0.10

resource efficiency to have a significant impact as they may have not realized a marginal benefit over time. EAL did not have a significant impact on adoption in Stage I. However, by Stage II it is shown to have a significant impact. This may be due to the fact that over time, firms' decision makers have realized the importance of environmental responsibility and adoption of these practices and its impact on the overall performance of the firms which may have led to this attitudinal and behavioural change towards environmental response.

This may have been due to the fact that, with firms paying more attention on their revenue, the focus on environmental responsiveness seems to decrease over time (by almost 75 % by Stage II). Interestingly, while firms continued to consider anticipated regulation to be more important over time, they additionally perceived more on the importance of liability laws. Therefore, it is evident that regulatory framework is critical in governing a firm's action although they do perceive certain failures in the regulatory framework. Nevertheless, they have been inclined to deem that a financial liability (i.e., may be in the forms of fines, compensation, closure of firms, etc.) or rather a liability law would have a greater impact on their behavior as it will make them feel more responsible and liable towards their actions. Simultaneously, certain other firms in the sample felt that a liability law would also have implications on their reputation as well as the profitability of the firm which would hold them responsible for adoption.

Firm level characteristics also seemed to have a varying impact on adoption. Where the type of firm/producer was of concern, while in Stage I, only essential oil producers had a notable impact on adoption, with time, it was evident that there was a tendency for coconut product producers and nonalcoholic beverage producers to move towards compliance. As expected, only large scale firms significantly continued as adopters while others showed insignificant adoption levels stating several hindering factors such financial constraints, lack of awareness, etc. This is further proven by the significant impact of cost on their compliance. Furthermore, vintage—experience of the firm or the years it has been in existence—did not continue to be significantly important through Stage II. This may be due to the fact that the period of 3 years considered in the study did not make a significant difference towards motivating adoption.

8.4 Conclusion and Policy Implications

The outcome of analysis that looks into the role of regulation, MRF, IEF, and EAL in governing the adoption of SWMPs in the agri-food processing sector in Sri Lanka over the years suggests that firms under study are mainly governed by the regulatory framework of the country and anticipate much stricter regulations in the near future. It is clear that this perception on government regulation, together with their altruism can effectively regulate these firms.

The results further imply that firms, futuristically show a positive affinity in adopting SWMPs overtime. However, it is evident that they have laid comparatively lower importance on the internal and external market based and efficiency incentives while more importance was placed on REG and EAL in driving adoption. Policy makers could use the "carrot and stick" approach suggested by Segerson (1999), which imposes voluntary and mandatory public environmental controls at the firm level as the basis for designing appropriate regulatory measures for environmental controls, to create an incentive-based regulatory system for all firms. An alternative would be a "bottom-to-top" approach (i.e., firm-to-regulator approach) that would reflect the individual incentives faced by firms that takes into account the characteristics of firms and of the market in which they operate. As a "stick," the government could develop a more stringent "command and control" type environmental quality management program for all firms and apply it irrespective considering the characteristics of the firm such as its type, size, or whether it functions in the domestic or international markets. Hence, it is important to bring current public regulatory regimes in developing countries like Sri Lanka towards co-regulation, which is practiced by developed countries to facilitate businesses to come up with their own solutions for environmental quality.

Thus, the outcome emphasize that local governing institutions must work further to design an incentive-based facilitative and regulatory regime to incentivize firms while promoting self regulation within firms into a scheme of co-regulation so that their efforts in governing private actions of firms towards environmental quality becomes a sustainable mechanism.

Acknowledgements Authors acknowledge the financial assistance provided by the South Asian Network for Development and Environmental Economics (SANDEE) (Stage I) and the Wayamba University of Sri Lanka for the Competitive Research Grant Scheme (SRHDC/RP/01/10-01) (Stage II) to carry out this study.

References

Bansal P, Roth K (2000) Why companies go green, a model of ecological responsiveness. Acad Manage J 43(4):717–737

Blackman A (2008) Can voluntary environmental regulation work in developing countries: lessons from case studies. Policy Stud J 36(1):119–141

Blackman A, Bannister G (1997) Community pressure and clean technology in the informal sector: An econometric analysis of the adoption of propane by traditional Mexican brick makers. Journal of Environmental Economics and Management 35(1):1–21

Buchanan JM (1969) Cost and choice: an inquiry in economic theory. Markham, Chicago

Caswell JA, Bredahl ME, Hooker NH (1998) How quality management meta-systems are affecting food industry. Review of Agricultural Economics 20:547–557

Chowdhury T, Imran M (2010) Morbidity costs of vehicular air pollution: examining Dhaka city in Bangladesh. Working paper 47-10. South Asian Network for development and environment economics, Kathmandu, Nepal

Dasgupta S, Hettige H, Wheeler D (2000) What improves environmental compliance? Evidence from Mexican industry. J Environ Econ Manage 39:39–66

Haab TC, McConnell KE (1996) Count data models and the problem of zeros in recreation demand analysis. American Journal of Agricultural Economics 78(1):89–102

Haab C, Mcconnell K (2002) Valuing environmental and natural resources: the econometrics of non-market valuation, new horizons in environmental economics series. Edward Elgar, Cheltenham

Hair JF, Anderson RE, Tatham RL, Black WC (2006) Multivariate data analysis with readings. Prentice Hall, Englewood Cliffs

Hellerstein DM (1991) Using count data models in travel cost analysis with aggregate data. Am J Agric Econ 73:860–866

Hellerstein DM, Mendelsohn RA (1993) Theoretical foundation for count data models. Am J Agric Econ 75:604–611

Hettige H, Huq M, Pargal S, Wheeler D (1996) Determinants of pollution abatement in developing countries: evidence from South and Southeast Asia. World Dev 24:1891–1904

Jayasinghe-Mudalige UK, Udugama JMM (2011) Motives for firms to adopt solid waste management controls: the case of food processing sector in Sri Lanka, Working Paper 60-11. South Asian Network for development and environmental economics, Kathmandu, Nepal

Jayasinghe-Mudalige UK, Henson S (2006a) Economic incentives for firms to implement enhanced food safety controls: Case of the Canadian red meat and poultry processing sector. Review of Agricultural Economics 28(4):494–514

Jayasinghe-Mudalige UK, Henson S (2006b) Use of confirmatory factor analysis techniques to overcome the problems of subjectivity & unobservability of incentives. Sri Lankan Journal of Applied Statistics 7:71–89

Khanna M, Anton WRQ (2002) Corporate environmental management: regulatory and market-based pressures. Land Economics 78:539–558

Khanna M (2001) Economic analysis of non-mandatory approaches to environmental protection. Nat Res 4:42–46

Nakamura M, Takahashi T, Vertinsky I (2001) Why Japanese firms choose to certify: a study of managerial responses to environmental issues. J Environ Econ Manage 42:23–52

Oppenheim AN (1992) Questionnaire design, interviewing and attitude measurement. Pinter, London

Pargal S, Wheeler D (1996) Informal regulation of industrial pollution in developing countries: evidence from Indonesia. J Polit Econ 104(6):1314–1327

Rahelizatovo NC, Gillespie JM (2004) the adoption of best-management practices by Louisiana dairy producers. Journal of Agricultural and Applied Economics 36(1):229–40

Segerson K (1999) Mandatory versus voluntary approaches to food safety. Agribusiness 15(1): 53–70

Segerson K, Miceli TJ (1998) Voluntary environmental agreements: good or bad news for environmental protection? J Environ Econ Manage 36:109–130.

Shavell S (1987) Economic analysis of accident law. Harvard University, Cambridge

Chapter 9
'Livestock for Development' in Resource-Constrained Environment: Would Induction of External Buffalo Breeds Help?

Bhagya Laxmi S, Ravindra Adusumilli and A. Vijay Mohan Rao

9.1 Introduction

Diverse, integrated and adaptive production systems have evolved over time in the dry land areas[1] of India to balance the adversities of scarce and variable rainfall, low crop-growing periods, poor soils and low levels of irrigation development. Livestock production in the dry lands evolved as a system with high dependency on crop residues and common pool natural resources as against intensive fodder systems managed mainly with residues and cultivated fodder in private property regimes in the irrigated areas. Availability of leftover crop residues, common lands, private fallows and forests for grazing so common in the dry land areas has enabled an entry for the poor with little or no ownership on land and water into the livestock production systems.[2] About 50–60 % of the total feed for livestock comes from crop residues in semiarid India. It is estimated that about 91 % of households in Andhra

The authors are grateful to Shahin Begum Giyas for sharing primary data on characterizing the local buffalo population collected as a part of her larger interest to promote recognition of the local breeds. The rest of the data, analysis and views expressed in the chapter are entirely of the authors. The authors acknowledge support from the Ford Foundation, India, as a part of the Revitalizing Rainfed Agriculture programme.

[1] The chapter mainly focuses on arid and semiarid regions of the India—broadly termed as dry lands.

[2] For example, it is estimated in study of arid and semiarid areas that the grazing hours of livestock vary from 4.65 to 6.05 in Tumkur and Anantapur districts, respectively (Mishra et al. 2010).

B. Laxmi S (✉) · R. Adusumilli
Watershed Support Services and Activity Network (WASSAN), Secunderabad, India
e-mail: bhagyawn@gmail.com

R. Adusumilli
raviwn@gmail.com

A. V. M. Rao
Department of Animal Husbandry, Hyderabad, Andhra Pradesh, India
e-mail: avmohanrao@gmail.com

N. Ghosh et al. (eds.), *Nature, Economy and Society*,
DOI 10.1007/978-81-322-2404-4_9

Pradesh are dependent on open grazing on common lands for an average supply of 35 % of total forage. A study reported that about one-third of the dry matter comes from grazing during monsoon (Mishra et al. 2010).

One of the major adaptations of livestock systems in dry lands is to sustain livestock over both seasonal and cyclical periods of scarcity of biomass and water; an objective that weighs heavily in the choice of breeds in resource-constrained environments. Livestock development programmes related to cattle and buffaloes in India, to the contrary, have an explicit 'productivity' criterion in the choice of animals even for the ecologically distressed dry land areas. Productivity traits are linked to high 'milk yield' in response to good management which often decides the choice of breeds in government programmes.[3] These preferences also permeate livestock components of the poverty alleviation programmes, as most often the technical recommendations on livestock purchases are given by the Animal Husbandry Department.[4] The overarching livestock breeding policies that determine the scope of government interventions in livestock do not have special criteria for resource-constrained areas in the choice of breeds.

Performance of the externally introduced breeds in the resource-constrained harsh environments stressed for water and biomass with low-input production systems has remained a serious issue in livestock development and poverty programmes. The potential of the so-called nondescript livestock and the comparative performance of the 'high-yielding' exotic animals (crossbreeds in cows and external breeds in buffaloes) when introduced into the harsher environments of arid and semiarid regions are underexplored.[5]

Located in the above context, the study pursues two central questions; what happens when external high-yielding milch animal breeds are introduced into harsh dry land regions with typical seasonal and cyclical scarcity in biomass and water? And, how is their performance in the real-time situation of farmers compared with the existing local 'nondescript' breeds?

The study is based on an analysis of Pashu Kranthi Programme (PKP) of Government of Andhra Pradesh with a focus on the distribution of Murrah breed of buffaloes brought from Haryana under the programme in the semiarid Medak district. The study is based on a primary census survey of all the beneficiaries receiving these buffaloes in two Mandals—Regode and Alladurgin Medak district under PKP. For the comparison, the study draws upon the extensive unpublished documentation of the 'nondescript' buffalo breeds by a local veterinarian of the Department

[3] See Guidelines of the scheme 'National Project on Cattle and Buffalo breeding', Department of Animal Husbandry, Government of India (http://dahd.nic.in/dahd/schemes/animal-husbandary.aspx).

[4] A study based in Andhra Pradesh by International Livestock Research Institute concludes that 'farm with livestock' is the best route to escape poverty. The study (Shaheen Akter 2008) provides various dimensions of livestock contribution to escaping poverty.

[5] For an overview on the subject, see Köhler-Rollefson (2008).

of Animal Husbandry.[6] This is an exploratory study to provide a primary analysis of the issues based on a substantial field data collected from 2008 to 2011. While we have good reasons to believe in the general validity of the results reported here across several semiarid and arid agro-ecological regions[7], an in-depth multidisciplinary analysis in diverse agro-ecological regions may be needed for generalization of the reported results, and to get deeper insights into these questions. The study results raise some serious policy issues in livestock development in the ecologically stressed dry land areas.

9.1.1 Buffalo Breeds in India and Andhra Pradesh

India is one of the richest centres for buffalo germplasm accounting for about 56 % of the world population of buffaloes. Though the indigenous breeds such as Murrah, Nili-Ravi, Surti, Mehsana, Marathwada, Jaffarabadi, Bhadawari, Nagpuri, Pandharpuri, Toda and Bunny are officially recognized in India, there is almost exclusive focus on two or three breeds, Murrah being the dominant one. These breed types constitute about 40 % of the buffalo population in the country while rest 60 % are referred to as 'nondescript'. Performance information of many and even the recognized breed types in their breeding tracts and under farmer management conditions has not been adequately documented (Das 2008).

The livestock development programmes, particularly in Andhra Pradesh, primarily deal with Murrah breed for its 'high productivity'. Upgradation of the local buffalo population through artificial insemination is a popular programme. The combined population of Murrah and Graded Murrah in the state now constitutes about 29 % of the total buffalo population. In spite of these substantial breed upgradation programmes, 69 % of the buffalo population in the state remains classified as 'nondescript' (Table 9.1).

Murrah and Graded Murrah populations are highly concentrated (70.95 %) in seven well-endowed irrigated coastal districts of Andhra Pradesh indicating a much higher density of 'nondescript' buffaloes in the rest of the 16 mainly dry land districts of Telangana, Rayalaseema regions and rainfed districts of northeast Andhra as per the 18th Livestock Census (2007). The potential of improving productivity of 69 % of the buffalo population spread mainly in the rainfed areas of the state has not received any attention.

---

[6] The primary data on the productivity traits and characterization of local breeds were shared by Shahin Begum Giyas; the authors acknowledge her support in this regard. The analysis and views are of the authors.

[7] See Sainath (2011) of similar outcome in Vidharbha with crossbred cattle, Köhler-Rollefson (2009) narrates an experience in Rajasthan.

Table 9.1 Distribution of cattle and buffaloes across breeds in Andhra Pradesh. (Source: Department of Animal Husbandry 2010)

Cattle			Buffaloes		
Cattle breed	*Population*	*%*	*Breed*	*Population*	*%*
Jersey CB cattle	1,574,411	14.03	Murrah	159,178	1.20
H.F. CB cattle	32,318	0.29	Graded Murrah	3,709,402	27.95
Total crossbred cattle	1,897,592	16.91	Jaffarabadi	20,276	0.15
Ongole cattle	257,661	2.30	Nagpuri	12,883	0.10
Deoni	23,928	0.21	Godavari	193,998	1.46
Hallikar	190,798	1.70	Nondescript	9,175,977	69.14
Gir	657	0.01	Total	13,271,714	100.00
Punganur	733	0.01			
Nondescript	8,851,675	78.87			
Total cattle	11,223,044	100.00			

9.2 Study Area and Pashu Kranthi Programme

The study was undertaken in Alladurg and Ragode Mandals of Medak district in the northwestern part of Andhra Pradesh falling under semiarid central Telangana agro-climatic zone with the mean annual rainfall of 890 mm. The district received deficit rainfall in 30 years out of 51 years (58 %) from 1958 to 2009, indicating high levels of drought vulnerability.[8] With a population density of 274 per sq. km, the district has a net sown area of 45 % of its geographical area and area sown more than once at 11.5 %. High levels of current fallows at 16 % along with permanent pastures (3 %), other fallows (10 %), cultivable waste (2 %), barren and uncultivated land (6) and forests (9 %) constitute the major grazing resources for livestock. Only about 22 % of the gross cropped area is irrigated by tanks, borewells and other sources. The district has about 17.5 and 18.7 % of buffalo and cattle population, respectively.

The two study Mandals—Alladurg and Regode—resemble the larger district profile with 23 and 19 % of net sown area under irrigation, respectively; the rest being rainfed. Buffaloes are about 13 and 11 % of the total livestock population in the two Mandals (Table 9.2).

Government provided subsidy of ₹ 15,000 per milch animal and the rest was arranged as loan by the local bank. Each beneficiary was provided with one milch animal initially, and the second was provided after 6 months of induction of the first animal. Support for rearing the female calves of inducted animals, feed and fodder, health care and breeding was also provided to an extent of ₹ 5300 per animal as

[8] http://agri.ap.nic.in/agroclimatezon.htm.

Table 9.2 Livestock population in the study Mandals

	Alladurg	Regode
Cattle	8207 (23.46)	5840 (18.84)
Buffaloes	4376 (12.51)	3490 (11.26)
Sheep	12,216 (34.91)	18,243 (58.84)
Goat	9905 (28.31)	3129 (10.09)
Pigs	285 (0.81)	304 (0.98)
Total	34,989	31,006

Figures in parenthesis are percentages of total

Table 9.3 Details of milch animals inducted

Year	Medak	Andhra Pradesh
2007–2008	2025	40,334
2008–2009	2423	62,7621
2009–2010	2385	38,863
Total	8858	182,293

a part of the programme. Animals were inducted from Haryana, Punjab, Gujarat, Maharashtra, Karnataka and Tamil Nadu under the PKP. The Animal Husbandry Department invested ₹ 171.6 million in Medak district to induct milch animals in 2008–2009 and 2009–2010.[9] Table 9.3 gives the details of milch animals inducted in Medak district and Andhra Pradesh.

9.2.1 *Study Methodology*

The study starts off from a basic understanding and analysis of the local 'nondescript' buffalo population and their performance characteristics. It then analyses performance of the 164 Murrah buffaloes inducted under the *PKP* in the two Mandals. A comparative analysis of the performance of inducted animals with the local 'nondescript' buffaloes is attempted. The study uses simple statistical tests and analysis of variance for comparison.

Data were generated for all the 164 Murrah animals inducted in the two study Mandals during the period 2008–2011. The production data were collected for the year 2010–2011 based on recall method; this has covered all the Murrah buffaloes distributed under the programme and retained by the beneficiaries at the time of the survey.

Data for the nondescript buffalo population were collected from the database that a local veterinary doctor has been generating to understand and characterize the

---

[9] http://medakdistrict.net/?page_id=81.

Table 9.4 Description of the groups of local breed identified

S.no.	Group name	Description
1	Boda	Massive drooping curved horns
2	Chandravanka	Small crescent shaped horns
3	Chata	Forward apart-backward towards neck
4	Guduga	Round horns in two shapes, that is, big and small
5	Kora	Big crescent shaped
6	Mongati	Straight horns upto neck, then turned upward towards head
7	Sakkati	Straight horns like Nagpuri
8	Sonti	Two horns in different shapes; one horn in drooping and another in different shape

local 'nondescript' population. These data are analysed to see if the local population can be segregated into different (distinct) groups based on phenotypic characters and production traits. Eight specific breed groups were differentiated as per the local phenotypic descriptors as narrated by the farmers and based on their traditional knowledge.[10] The term 'breed-group' is used in this chapter to avoid any claim of these being separate breeds or breed types, an issue that needs to be resolved by animal genetic sciences. Data collected for a large unstratified sample of 517 local buffaloes were analysed to understand if the phenotypic characteristics (body size, base between two horns, length, girth, height, colour, length of left and right horns, switch of the tail, etc.) differ statistically between the groups. Except body size, all other parameters had very low p-values (0.010–0.000) indicating high levels of significance of the difference in the mean values between the groups, providing a primary basis for differentiating the local buffalo population. The analysis is not presented here. Table 9.4 gives an overview of the eight breed-groups identified by local names along with their horn type which is the main breed-group descriptor or identifier.

An analysis on the differences in the production traits among the eight groups is reported in Table 9.4. The data on milk yield in both the cases of inducted Murrah buffaloes and local ones were collected during the period 2010–2011 based on recall method. Collecting data related to milk yields, based on memory recall, has intrinsic problems. Being aware of it, the authors consider the results as a first approximation that can still provide a good basis for comparison, the method of data collection being the same for all the animals.

[10] The authors gratefully acknowledge the knowledge contribution of various buffalo farmers in the study villages of Alladurg, Bidkanne, Chaudarpally, Dosapally, Gadipeddapur, Gandlabaitanda, Gangwar, Gollagudam, Humnapur, IB tanda, Jagriyal, Kondapur, Kothlapur, Maktavenkatapuram, Pocharam, Pothirddypally, R.Itkyal, Saipeta, T. Lingampally, Tatipally, Togarpally and Usrikapally.

A scoring matrix is used giving due weights to productivity traits to arrive at performance grading of the identified eight groups of local breeds. The scores are arrived at using the standard practice followed in the Animal Husbandry Department.[11] The production traits of the best three breeds are used in the comparison.

9.2.2 Study Results

This section presents results of the analysis. The production traits of the eight local breed-groups identified are presented initially. Three best breed-groups are separated based on weighted scores of the production traits. The data on beneficiaries' responses to the distributed Murrah buffaloes are reported in the next section giving a vivid picture of what happened post induction. The results of a comparative analysis of Murrah and the local buffaloes are presented later. The results are crosschecked with other studies available in the literature.

Table 9.5 presents results of analysis of the production traits across the eight groups. The means of all these production traits are found to be significantly different across the eight groups (p-values lower than 0.000). The milk yield data mostly represent the period, second quarter after calving. It is collected for a range with lower and upper bound values.

The performance reported above is under low-input/management system where cultivated green fodder is rare, occasional feeding of concentrates, low access to health care services, no special usage of water for body washing and above all, maintained with a combination of crop residues and open grazing. As can be seen in Table 9.5, the upper bound of milk yield is quite substantial and some of the breeds have high average milk yields.

It is interesting to note that the so-called nondescript mixed population has strikingly different breed-groups with clearly identifiable phenotypic characters and having distinct productivity traits that are statistically different between different groups. As we later see, these productivity traits are comparable to the other recognized breeds like Nagpuri and even (to some extent) field level expression of the Murrah buffaloes inducted from other states.

An exercise was done to identify the best of these local eight buffalo breed-groups. A score card was developed in consultation with veterinarians having long-standing field experience. The weights are based on perceived importance of production traits by the veterinarians. Average milk yield is given a weight of 35 while lactation length, age at first calving and inter-calving period is given equal

---

[11] The authors wish to acknowledge inputs from Venkateswarlu, a retired senior official of the Animal Husbandry Department, Andhra Pradesh.

Table 9.5 Means of the production traits of local buffalo groups

Group	*N*	Milk production (litres per day)				Pre-weaning calf mortality*		Age at first calving*		Inter-calving period*		Lactation period*	
		Lower bound*		Upper bound*									
Units		l/day		l/day		Number		Months		Months		Months	
Boda	88	3.6	(0.49)	5.023	(0.15)	1.03	(0.18)	49.88	(2.12)	19.45	(2.89)	10.9	(1.22)
Chandravanka	32	3.00	(0.00)	5.00	(0.00)	1.44	(0.50)	48.25	(0.67)	18.56	(2.86)	10.38	(1.24)
Chata	42	3.11	(0.34)	4.988	(0.08)	1.19	(0.40)	51.95	(3.32)	17.77	(2.67)	10.33	(1.65)
Guduga	18	3.00	(0.30)	4.917	(0.26)	1.17	(0.38)	51.78	(3.42)	19.5	(0.97)	11.22	(1.52)
Kora	92	3.05	(0.22)	4.978	(0.16)	1.27	(0.45)	47.54	(2.68)	20.61	(3.16)	11.67	(3.22)
Mongati	14	2.93	(0.39)	4.75	(0.26)	1.07	(0.27)	47.43	(1.91)	19.14	(1.17)	11.29	(1.27)
Sakkati	135	3.08	(0.37)	4.985	(0.12)	1.02	(0.15)	43.79	(4.73)	21.31	(2.01)	12.35	(2.35)
Sonti	95	3.24	(0.50)	5.032	(0.18)	1.02	(0.14)	43.85	(4.60)	21.19	(1.75)	11.64	(2.20)
Total	516	3.18	(0.43)	4.991	(0.16)	1.11	(0.32)	46.83	(4.68)	20.33	(2.65)	11.5	(2.30)

Figures in parenthesis are the standard deviations

N Number of observations

p-value for between groups means: * = 0.000; highly significant in all cases

Table 9.6 Matrix of scores and relative ranking based on production traits

S.no.	Group of local breed	Milk yield	Lactation length	Age at first calving	Inter-calving period	Calf mortality	Total	Ranks
	Max weight◇	35	20	20	20	5		
1	Boda	35.00	9.00	6.00	12.00	4.50	66.50	1
2	Mungati	11.00	16.50	18.00	14.00	4.50	64.00	2
3	Sonti	23.00	15.00	12.00	6.00	5.00	61.00	3
4	Sakkati	11.00	20.00	18.00	4.00	5.00	58.00	4
5	Kora	11.00	15.00	10.00	6.00	2.50	44.50	5
6	Chata	11.00	3.00	2.00	18.00	3.50	37.50	6
7	Guduga	7.00	10.80	2.00	12.00	3.50	35.30	7
8	Chandra Vancha	11.00	3.00	8.00	10.00	0.50	32.50	8

weight of 20. Calf mortality is given a low weight of 5 as it depends substantially on management. The weights are allotted in decreasing scale (higher weight for maximum sample mean value) for average milk yield and lactation length. For the traits—age at first calving, inter-calving period and calf mortality—weights are given in inverse order, that is, minimum sample mean value gets higher weight.

The weighted scores based on productivity traits presented in Table 9.6 show the breed-groups Boda, Mongati and Sonti in that order as the best performing. Though average milk yield is given high weight in the above exercise, in practice the preferences may vary as the choice of inter-calving period, lactation length and even calf mortality, may well be a natural way of balancing or managing livestock densities to be in tune with the seasonal or cyclical biomass scarcities.

While the livestock are left to fend for themselves by open grazing only roughages are fed when the animals are tied up. The limited concentrates are selectively fed to the milking and working animals. The animals respond to this feed fluctuation across seasons and scarcity cycles by infertility traits (high age of first calving and calving intervals; infertility) and disease susceptibility. However, the indigenous breeds are remarkable in obtaining the required metabolic energy from the poor quality and quantity of roughages fed to them. Though this is a common knowledge, it is hard to find literature on these mechanisms of adjustment and adoption to scarcity periods.

Table 9.7 Status of inducted Murrah buffaloes

	Number	Percentage
Retained	42	26
Died	3	1
Disposed off	119	73
Total purchased	164	100
Calf mortality		
Total calves born	164	
Calf survived till weaning	101	57
Pre-weaning calf mortality	71	43

9.3 Performance of the Inducted Murrah Buffaloes in Field Situation

In the study area 164 Murrah buffaloes and eight crossbred cows were inducted under the PKP during 2007–2010. One animal per beneficiary was given first and after 6 months the beneficiary was expected to go for second animal procurement with subsidy. Only Murrah buffaloes inducted in the PKP were considered in the study.

The department has spent ₹ 2.46 million on subsidy for the purchase of 164 Murrah buffaloes and banks provided the total loan of ₹ 3.1 million. A total of ₹ 0.19 million additional expenditure was allotted for feed. Total cost per animal (loan and subsidy) approximates to ₹ 16,169. Of the total 164 beneficiaries, 88 have taken feed on subsidy.

Out of the total 164 Murrah buffaloes in Alladurg and Regode Mandals purchased under the programme, 119 (73 %) were disposed off by the beneficiaries within 4–6 months of procurement and three animals died; 42 Murrah buffaloes were retained by the beneficiaries at the end of year 2010. The animals disposed off went to various places; 13 animals were sold to brokers from Hyderabad, 20 to nearby villagers for rearing and 68 of them were sold in local shandy. Data for eight animals are not available. As per the local veterinary doctor, 92 animals (56 %) were disposed early without undergoing first pregnancy diagnosis, that is, within 6 months after grounding (Table 9.7). It is not clear whether the animals disposed off are still in production.

Only 28 farmers (17 %) have availed the option of subsidized procurement of a second animal after 6 months and procured the second animal. This might indicate the farmers' choice towards high-yielding Murrah buffaloes purchased from distant places like Haryana (Table 9.8).

It is understood that only pregnant Murrah buffaloes were procured. Of the total 164 calves born to these inducted animals during 2007–2010, 43 % mortality is recorded as 71 calves died. Murrah buffaloes require nearly two to three times of body washing every day. Water being a scarce resource in these Mandals, 57 % of the beneficiaries could only wash the animal once per day and only 28 % reported washing twice a day (Table 9.9).

Table 9.8 Body washing frequency

No. of washings/day	No. of animals(CB cows + Murrah)
Once	93 (57 %)
Twice	46 (28 %)
Thrice	6 (4 %)
Not daily	17 (10 %)
Alternate days	2 (1 %)
Total	164

The local buffaloes are normally taken to the village tank or other water bodies while the Murrah buffaloes are washed at home bringing water, requiring additional work. Government has spent total amount of ₹ 1,924,139 on subsidy on the 119 Murrah buffaloes that were disposed off. Farmers on the whole could get back about 58.5 % from selling off the animals. The bank loan is repaid to an extent of 4.81 % while 94 % still remains as debt. The accumulated debt load of ₹ 2.97 millions still remains on the farmers (Table 9.9). The programme, thus, has left the farmers who could not keep their animals into a debt trap, the same crisis the programme intended to solve! The amount of debt left on farmers is much higher than what the government has spent as subsidy. Such instances are also reported in Vidarbha region of Maharashtra.[12]

9.4 Comparison of Inducted Murrah Buffaloes with Local Breed

9.4.1 Comparative Analysis

This section first presents a comparative analysis of 42 inducted Murrah buffaloes with nearly equal number of local buffaloes drawn from the sample. Subsequently, the data are compared with the official figures of some of the officially recognized Nagpuri and Pandharpuri breeds and reported figures from other districts obtained from the literature.

Results of a comparative analysis of milk yield of the 42 inducted Murrah buffaloes retained finally by farmers with the local 'nondescript' breeds are presented in Table 9.10. Data refers to the period 2010–2011 and it compares data in the four quarters starting 1 month after calving. The local buffaloes were mixed in breed-groups and were under low-to-average management condition. The Murrah animals were under average management conditions with some farmers feeding concentrates supplied along with the animal. About 31 of the 41 farmers rearing

[12] Sainath's (2011) article "Cowed down by the Prime Minister" gives a graphical account of this process in Vidarbha region.

Table 9.9 Investments and debt burden on account of disposed animals (amount in rupees)

Breed	Total number of animals disposed	Total subsidy	Bank loan	Amount earned out of sale of animal	Loan repaid to bank	Outstanding debt on account of disposed animals
Murrah	119	1,924,139	3,120,000	1,825,500	1,50,000	2,970,000

Table 9.10 Comparative performance of milk yield (litres per day) between inducted Murrah and local buffaloes

Months	Breed	*N*	Mean	Std. error	Minimum	Maximum	*F*	*p*-value Sig.
Months 1–3	Murrah	41	4.18	0.13	2.50	6.00	24.07	0.00
	Local	42	4.85	0.04	4.50	5.00		
	Total	83	4.52	0.08	2.50	6.00		
Months 4–6	Murrah	41	3.15	0.08	2.00	4.00	146.13	0.00
	Local	42	4.43	0.07	3.50	5.00		
	Total	83	3.80	0.09	2.00	5.00		
Months 6–9	Murrah	41	2.59	0.09	2.00	4.00	6.16	0.02
	Local	42	2.83	0.04	2.50	3.00		
	Total	83	2.71	0.05	2.00	4.00		
Months 9–12	Murrah	41	0.40	0.08	0.00	1.00	232.02	0.00
	Local	42	1.80	0.05	1.02	2.50		
	Total	83	1.11	0.09	0.00	2.50		

The low *p*-values indicate high significance of the difference of mean values between and within Murrah and local buffaloes

Murrah were paddy farmers feeding paddy straw and 12 farmers of the total took up nonleguminous green fodder such as Sweet Sudan Grass (SSG).

Figure 9.1 presents the difference graphically. It is rather striking to see that in all the quarters local buffalo milk yield is more than inducted Murrah buffalo. Of the 41 Murrah buffaloes for which data are collected, in 24 cases milk yield was reported to be zero in the fourth quarter which has depressed the mean values increasing the relative difference. For the first and third quarters, the maximum milk yield in Murrah is higher than that of the locals. In general, the standard deviation in mean milk yield is higher in the case of Murrah indicating high variability in production.

Murrah is a recognized indigenous breed with native breeding tracts around Haryana with proven potential. It responds well to good management and intensive inputs. However, the analysis reinforces the significance of environmental effects over the genotype; milk yields in the constrained field situation are in reality determined not only by the genetic make up of the animal but also by environment,

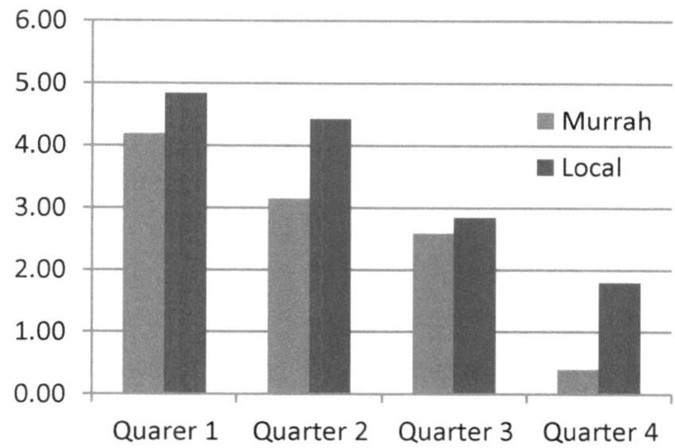

Fig. 9.1 Comparative milk yield (litres per day) of inducted Murrah and local buffaloes

adaptability, local nutrition base and management inputs. The important question here is whether livestock development by government takes these important production parameters into its purview or only focuses on promoting intensive systems even in harsh environments? Such an approach also excludes many rearers whose production system is heavily dependent on grazing based fodder resources and concentrate livestock into well-endowed households, as the study reveals.

Above being the generic situation prevailing in the study area, do the study results hold good in the case of average-to-better management situations? To understand this question, an attempt is made to study the relative performance of Murrah in a local average-to-better management situation. One farmer in the study area who has aggregated 12 Murrah buffaloes (purchased by others) into his farm and maintains them under better management conditions provided an opportunity to have insights on this issue. Performance of this flock of Murrah is compared with the best 12 in the sample of 42 local animals. The Murrah farmer was providing green fodder, drinking water, concentrates, two times body wash and stall feeding. In local buffaloes, the management conditions were: providing limited green fodder, locally made feed (i.e. mix of rice bran, oil cakes, broken maize or sorghum and salt), drinking water, one/two times body wash, open grazing(semi-intensive) of animals. Table 9.11 provides a comparative picture.

The mean of the milk yield in Murrah ranged from 3.2 to 6.4 l/day across all the quarters and in the local animals 2.6 to 5.0 l/day. The standard deviation in case

Table 9.11 Comparison of performance of Murrah under good management system with local buffaloes (under average management system)—Milk yield in litres per day

Breed		1–3 months	3–6 months	6–9 months	9–12 months	Total period
Local	Mean	5.0	4.7	3.8	2.6	4.0
	Std. Dev.	0.0	0.3	0.3	0.6	0.2
Murrah	Mean	6.4	4.7	3.8	3.2	4.5
	Std. Dev.	2.28	1.97	1.39	0.80	1.4

All the mean values have high statistical significance with p-values below 0.000

Table 9.12 Comparative performance of local breeds with recognized breeds in semiarid areas of Maharashtra. (Source: http://www.cirb.gov.in/buffalo/nagpuri.htm (given breed wise))

Production traits	Boda	Mungati	Sonti	Nagpuri[a]	Pandharpuri[a]
Milk yield (litres per day)	4.3	3.84	4.13	4.44	4
Lactation period (days)	327	339	350	270	350
Age at first calving (months)	49.88	47.43	43.85	55.8	44.83
Total milk yield (litres)	1406	1301	1445	1198	1400
Inter-calving period (months)	19.46	19.14	21.19	14.33	15.5

[a]recognized indigenous buffalo breeds of Maharashtra

of Murrah is higher than that of the locals indicating higher variability. The results show better performance of Murrah in the first and fourth quarters, whereas the difference in milk yield remains insignificant in the second and third quarters. There is also a marginal advantage for Murrah over the local breed in milk yield for the entire period in good management situation. It might be that the fodder and feed provided to Murrah even in this local-intensive system may fall substantially short of the requirement for expression of genetic potential of Murrah.

It can also be seen from comparison of the results from Tables 9.10 and 9.11 that the local breeds respond to even slightly better management conditions, an opportunity that can be exploited for overall increase in milk production. This leads us to a public policy choice of improving the management conditions of the large number of highly adapted local buffaloes to exploit their genetic potential without introducing external breeds like Murrah. The local breeds can be further upgraded through local selections.

The local breeds compare well with the recognized indigenous breeds—Nagpuri (4.44 l/day) and Pandharpuri (4.0 l/day)—of Maharashtra in terms of milk yield as reported by Central Institute for Research on Buffalo, Hissar (Table 9.12).

9.5 Discussion and Policy Implications

Subject to further confirmation and validation of the results presented here, the analysis makes a primary case for recognizing the local breeds. The study brings out that (a) there are clearly distinguishable types within the 'nondescript' local buffalo population with clear differences in phenotypic characters and productivity traits; (b) these compare well with the externally introduced breeds in terms of productivity in constrained situation; (c) these are comparable to identified local breeds in similar situations outside the state. The local population also shows significant scope for productivity improvement through further selections and improvements in management situations. Given these factors, it makes a good policy sense to institute processes to officially recognize the local breed types. Unfortunately, without such

Table 9.13 Productivity traits of Milch animals in three arid and semiarid districts. (Source: Mishra et al. 2009)

Parameter	Anantapur	Mahabubnagar	Tumkur	Average	Desired
Nondescript buffaloes					
Milk yield (litres per day)	3.65	3.75	3.39	3.6	5.0
Age at first calving (months)	46.8	47.7	46.3	46.9	32–36
Lactation length (days)	303	307	314	308	300
Inter-calving period (months)	22.9	23.4	22.7	23.0	15–17
Graded Murrah					
Milk yield (litres per day)	4.7	4.32	4.15	4.39	10
Age at first calving (months)	46.2	45.3	45.7	45.7	32–36
Lactation length (days)	309	314	310	311	300
Inter-calving period (months)	21.4	22.7	21.3	21.8	15–17

official recognition, these breeds cannot come under the purview of the programmes of Animal Husbandry Department (except for changing their breed).

The results of the study may not entirely be location specific. The results of a field study taken up by Central Research Institute for Dryland Agriculture (CRIDA) in three districts in arid and semiarid regions are reproduced in Table 9.13. The local 'nondescript' buffaloes were compared with Graded Murrah. The substantial underperformance of even Graded Murrah compared to the 'desired' levels in field situation corroborates the significant environment effects on the genotype. The local breed-groups in the study area compare favourably with the data given for the 'nondescript' population in the three districts.

The question then is what must be the strategic direction for livestock development in resource-constrained dry land areas? Public investments directed towards promoting intensification of dairy through external breeds, as brought out in the study, result in increased debt burden on the farmers as they tend to dispose off these animals. This also leads to concentration of the distributed animals in the farms of few farmers well endowed in water and land. To what extent even such small intensive units with high-yield breeds sustain in times of scarcity and drought is a serious question, when districts like Medak face drought in 30 out of 51 years, with high rates of depletion of groundwater. The anticipated rise in temperature between 2.3 and 4.8 °C; resulting from climate change is likely to aggravate heat stress in dairy animals further affecting their productive and reproductive performance.[13] Locally adapted buffalo breeds assume larger significance in this fast unfolding scenario.

[13] See Sirohi and Michaelowa (2007) for details. In buffaloes, access to water is the only way of managing heat stress and availability and access to water is under a serious threat in dry lands.

It is ironical that the state's response to farmers' distress in crisis-ridden rainfed areas further exacerbates the crisis, distress and debt burden! The strategic framework implicit in the policies and government actions in dry land areas needs to be seriously relooked at.

The study also brings out the potential for increasing productivity of the local breeds in dry land areas through larger public investments on improving the management systems. Several such experiences are available across India.[14] Such focused public investments will deconcentrate dairy opportunities as many farmers without access to secured water/irrigation and many more small, marginal and landless households with access to commons can participate in the dairy industry bringing a sense of inclusive growth. Investing and increasing productivity of nearly 70 % of the buffalo population through management interventions may give a better economic and social dividend in resource-constrained situations. It also provides a better fit into the local environmental conditions and provides good resilience to dairy industry. Such a shift towards local breeds is much needed in poverty programmes to prevent these very programmes from further accentuating vulnerabilities of the poor and their debt traps.

At the core of the issues is the question of breed. The scoring system used in the selection of breeds/animals (even the one employed in this study) is highly skewed towards high-milk yields under good management rather than stable milk yields under harsh environment. The official insulation of the government investments against inclusion of the so-called nondescript breeds, combined with inaction in recognizing the local breed types in effect, negates the potential for any public action on local animals. The study has brought out from the indigenous knowledge the possibilities of further recognizing breed types from the 'nondescript' population and they may have statistically significant variation in their phenotypic and productivity traits. It is essential that the policies must change, that is, recognize the local breed types or bring nondescript animals into the purview of public investments by changing policies, at least in resource-constrained environments.

Investments on community-managed participatory breeding programmes, involving selections within the local populations[15] combined with well-structured investments on improving biomass, water supply and management in common and private lands, can give the best returns on public investments and much better incomes to the poor, on whose name everything happens.

---

[14] SAPPLP program of FAO has made a detailed documentation of such examples. See Good Practices Brief on the subject at http://sapplpp.org/goodpractices/CPR-Livestock/SAGP02-innovations-in-common-land-development. A larger experience of regenerating 80,000 acres of common lands in arid district of Anantapur also exemplifies the potential; see http://appsatp.org/natural_regen.htm. See Parthasarathy Rao and Hall (2003) for an overview of increasing the potential of crop residues in crop–livestock systems.

[15] Sharma et al. (2005) provides a detailed account of in situ improving the indigenous breeds that shows large potential.

References

Akter S, Farrington J, Deshingkar P, Freeman A (2008) Livestock, vulnerability, and poverty dynamics in India. Discussion paper no. 10. International Livestock Research Institute, Nairobi

Das AK, Sharma D, Kumar N (2008) Buffalo genetic resources in India and their conservation. Buffalo Bull 27(4):265–268

Dept of Animal Husbandry (2010) Integrated sample survey, report on estimates of major livestock products for the years 2008–2009 & 2009–2010 (provisional) and veterinary infrastructure facilities, statistical wing. Department of Animal Husbandry, Andhra Pradesh

Köhler-Rollefson I, Mathias E, Rathore HS (2008) Local breeds, livelihoods, and livestock keepers' rights in South Asia. Trop Anim Health Prod. published on line 22 November 2008

Köhler-Rollefson I, Rathore HS, Mathias E (2009) Local breeds, livelihoods and livestock keepers' rights in South Asia. Trop Anim Health Prod 41:1061–1070

Misra AK, Rama Rao CA, Subrahmanyam KV, Ramakrishna YS (September 2009) Improving dairy production in India's rainfed agroecosystem: constraints and strategies. outlook agric, 38(3):284–292(9)

Mishra AK, Rao CAR, Shankar KR (2010) Analysis of potentials and problems of dairy production in rainfed agro-ecosystem of India. Indian J Anim Sci 80(11):1126–1133

Parthasarathy Rao P, Hall AJ (2003) Importance of crop residues in crop–livestock systems in India and farmers' perceptions of fodder quality in coarse cereals. Field Crop Res 84(2003):189–198

Sainath P (2011) Cowed down by the Prime Minister. The Hindu, May 13

Sharma KC, Goel R, Singh MR, Kumar S, Singh HN (2005) Breeding strategies for conservation of Bhadawari buffaloes in Uttar Pradesh. VIIIth national conference on animal genetics and breeding, 8–10 March, Central Institute for Research on Goats, Mathrua UP

Sirohi S, Michaelowa A (2007). Sufferer and cause: Indian livestock and climate change. Clim Change 85:285–298

Chapter 10
Soil Fertility Management in Semiarid Regions: The Sociocultural, Economic and Livelihood Dimensions of Farmers' Practices—*A Case of Andhra Pradesh*

B. Suresh Reddy

10.1 Introduction

India is predominantly an agrarian economy and more than 70 % of its population directly or indirectly depends on farming. Indian agriculture is often discussed in relation to the Green Revolution and its mixed record of successes and failures. Yet, what most Indian farmers (over 60 %) practise is rainfed agriculture, a farming system entirely different to that of irrigated areas. Green Revolution technology gains in agricultural productivity and food security were widely associated with irrigated lands, where the benefits of improved seeds and increased use of inorganic fertilisers could be realised. However, there are now widespread problems associated with the use of chemical fertilisers, mismanagement of surface water and overexploitation of groundwater, the most important source for irrigation (GoI, Economic survey 2008). The potential for expanding irrigated agriculture is decreasing as it becomes more expensive or risky to further exploit water resources. It has been fully realised that even if the full irrigation potential of the country were to become operative, 50 % of the net sown area would continue to be rainfed. Hence, its contribution is vital to the food security and livelihoods of the poorest farming families and communities who do not have access to irrigated land. Moreover, dryland agriculture holds the key for future food security in the light of Green Revolution reaching its saturation in enhancing land productivity. It is shown that returns on investments in rainfed agriculture are greater than those on investments in irrigated agriculture (Fan and Hazell 2000).

B. Suresh Reddy is an associate professor at the Division for Sustainable Development Studies (DSDS), Centre for Economic and Social Studies (CESS), Hyderabad. This chapter is based on my doctoral research work at CESS. An earlier version of this chapter has been published in *Economic and Political Weekly (EPW)*, dated 15 January 2011.

B. S. Reddy (✉)
Division for Sustainable Development Studies, Centre for Economic and Social Studies, N.O. Campus, Begumpet, Hyderabad 500016, India
e-mail: Srihithasuresh@yahoo.com

N. Ghosh et al. (eds.), *Nature, Economy and Society*,
DOI 10.1007/978-81-322-2404-4_10

In semiarid regions,[1] dryland farmers have developed cropping practices ranging from summer ploughing to crop rotations that are suited to the harsh agro-climatic constraints (Pionetti and Reddy 2002). These practices are derived from the farmers' deep understanding about the interactions between climate, soil, crops and insects, and enable the regeneration of soils and the optimal use of scarce moisture. As a major provider of organic manure, livestock are crucial to the stability of dryland agriculture and farmers also depend greatly on them for income (Reddy 2001; Sagari 2004, Reddy 2010a, Reddy 2010c). It is estimated that by 2025 India will use about 30 million t of plant nutrients per annum, 8.1 million t of which will come from organic manures. Soil fertility management (SFM) not only affects those who own or cultivate agricultural land, it is also important for the landless poor, especially women who often gather, process, transport and apply organic material to supplement their livelihoods. Cattle owners trade manure not only to be applied in fields, but also as a valuable source of fuel. Shepherds barter manure of their sheep and goats for the right to graze on particular plots of land (Adolph and Butterworth 2002). Through these multiple dimensions, SFM is a key to the sustainable livelihood of the rural people in semiarid tropics (SAT) in India. Based on the fieldwork done in Andhra Pradesh (AP), this study explores the local farmers' in-depth knowledge on SFM, and the cultural and socioeconomic web woven around these practices. It also looks at how policy interventions threaten this knowledge base and the sustainable practices it supports.

The objectives of this chapter are to identify and record the SFM strategies adopted across different size classes of farmers in dryland regions, to examine the ecological, economic, social and livelihood significance of SFM practices and contribute to the overall policy discourse on SFM in semiarid regions.

This chapter consists of seven sections including this Introduction. Section 10.2 describes the methodology followed for the study. The third section presents the findings of the study. The comparison of farmers' SFM practices with recommended practices has been done in the fourth section. Section 10.5 discusses the technical efficiency (TE) of major SFM methods. Section 10.6 presents the determinants of SFM, and the final sections make concluding observations with recommendations.

10.2 Study Area, Data Collection and Methodology

The state of AP was chosen for the study. Over 80 % of those involved in agriculture are small and marginal farmers and landless labourers who own a mere 35 % (3.5 million ha) of the total 10 million ha of cultivated land. AP has the distinction of much diversified livestock resources. In AP, agriculture has been undergoing many changes over the past two to three decades. The increasing intervention of the state in agriculture, and the Green and Yellow Revolutions, have prompted agricultural

[1] Semiarid refers to a climatic classification of typically dry areas with rainfall ranging from 500 to 950 mm and evaporative demand in excess of rainfall for a greater part of the year.

changes throughout the semiarid regions, especially in land ownership, cropping patterns, irrigation, credit and extension, agricultural productivity, prices and marketing. The use of fertilisers too was high in AP. In the period 2004–2005, as much as 11.57, 5.39 and 2.92 lakh t of nitrogen (N), phosphorus (P) and potassium (K) were used, respectively. In 2004–2005, the total NPK per hectare consumption was 158.80 kg (CMIE 2006) as against the India's 88.11 kg/ha (Fertiliser Association of India 2006). All the above-mentioned aspects have a huge bearing on the farmers' SFM practices, especially in the semiarid regions. It was in this context that AP was selected for the study on SFM focusing on the socioeconomic, ecological and livelihood dimensions of farmers' practices in semiarid regions.

The selected districts were Mahbubnagar, Anantapur and Prakasam from three different socioeconomic and ecological regions of AP, namely, Telangana, Rayalaseema and Coastal Andhra, respectively. Districts having the least percentage of net irrigated area in their respective regions with an annual average rainfall ranging between 500 and 900 mm falling under semiarid conditions were selected for the study. Two mandals from each district were selected. In each selected district, one mandal with the least percentage of net irrigated area falling under semiarid regions (those receiving rainfall between 500 and 900 mm) and another mandal with the highest net irrigated area falling under semiarid regions were selected. Though both mandals come under semiarid conditions, one mandal with high net irrigated area and another with the least net irrigated area was selected in order to observe the dynamics involved in the SFM under irrigated conditions and dryland conditions. A total of six mandals were selected from three different semiarid districts of Telangana, Rayalaseema and Coastal Andhra regions of AP. Similar criteria were followed even for the selection of villages from mandals. A total of six villages were selected from three different semiarid districts. In this way, a total of 60 farmers, 20 from each size class were selected from each village. A total of 360 farmers covering the three districts were interviewed for the research. The study used both qualitative and quantitative methods for understanding the farmers' SFM practices and the conditions under which they adopt such practices. Personal interviews were conducted with a structured interview schedule. The study used an ex post facto research design coupled with case studies, participatory rural appraisal (PRA) methods and focused group discussions (FGDs). Secondary data on rainfall, net irrigated area and demographic features of the district were collected from the Bureau of Economics and Statistics, CMIE reports, Fertiliser News and Economic Survey reports. Fertiliser recommendations were taken from the package of practices suggested by the Acharya NG Ranga Agricultural University of Andhra Pradesh.

The data analysis was basically done in two ways. One was comparing the various size classes of large, medium and small farmers; the other analysis was done comparing the irrigated and unirrigated (less irrigated) villages. The results of the study are discussed at two levels: one at the household (hh) level and the other at the plot level. The data gathered were analysed using different statistical tools. Averages, frequency and percentages were used to analyse the various information-related SFM. The model developed by Battese and Coelli (1988) for estimating stochastic frontier production was used for paddy, groundnut and sunflower crops.

Table 10.1 Study area in AP. (Source: Chief Planning Office of the selected districts)

District	Mandal	Village	No. of sample households
Telangana region			
Mahbubnagar	Peddakothapalli (3.16 % of net irrigated area and 625 mm rainfall)	Maredudinne (0.50 % of net irrigated area)	60
	Chinnachintakunta (91.57 % of net irrigated area and 565 mm of rainfall)	Dhupalle (89.20 % of net irrigated area)	60
Rayalaseema region			
Anantapur	Puttaparthi (0.36 % net irrigated area and 609 mm rainfall)	Brahmanpalle (10.16 % of net irrigated area)	60
	Tadipatri (39.30 net irrigated area and 667 mm rainfall)	Chinnapolamada (29 % of net irrigated area)	60
Coastal Andhra region			
Prakasam	Korisepadu (9.33 % net irrigated area and 865 mm rainfall)	Ravinuthala (0.48 % of net irrigated area)	60
	Darsi (75.12 % net irrigated area and 733 mm)	Darsi (82.06 % of net irrigated area)	60

Using the frontier production function approach, the average TE was calculated to find the TE of paddy, groundnut and sunflower with respect to size classes, fertility level of soils as perceived by farmers, number of SFM methods and types of soils. However, in this chapter we focus only on the TE of paddy and groundnut with respect to different SFM methods (Table 10.1).

10.3 Findings of the Study

10.3.1 Socioeconomics of Sample Households

In all the sample villages across all size classes, the percentage of backward classes (BCs) was the highest followed by other castes (OCs) in the sample hh. Among the total sample farmers, 38.61 % are illiterate and 61.39 % are literate to various educational levels. The illiterates depend on their neighbours and peers for useful knowledge on SFM. Due to self-help groups and occupational institutions, small farmers had higher social participation compared to other size classes. The highest average land-holding size with respect to large farmers (14.33 acres), medium farmers (6.90) and small farmers (3.35) was seen in Chinnapolamada, Maredudinne/Chinnapolamada and Dhupalle, respectively. Farmers from all farm size categories were keen on using organic practices, particularly Farm Yard Manure (FYM). The population of cows and bullocks is going down. Taking care of livestock is coming down slightly due to difficulties in accessing fodder, drinking water and easy labour. Livestock is taken care of by both men and women. Even today, on the whole, across all size classes more than 11.39 % of the children from the sample hh's take care of livestock.

The sample farmers depend on multiple livelihoods. Farming was found to be the main livelihood activity across all size classes, followed by dairy. For the majority of the small farmers, agricultural labour was an important source of livelihood. Dependence on small ruminants was high in unirrigated villages. Large farmers spent more on inputs which are external in nature, followed by medium and small farmers. Small and medium farmers depend on internal resources like human and bullock labour for timely agricultural operations. The average annual net income of the sample hh's was more for large farmers followed by medium and small farmers. This is attributed to bigger farm size with large farmers. The majority of the sample farmers (52.50 %) have indebtedness ranging between ₹ 1–20,000 followed by 35.83 % of them having ₹ 20,001–40,000. The majority of the sample farmers accessed loan from various sources at an annual interest rate of 12–24 %. Large farmers with better personal transport travelled greater distances for procuring agricultural inputs. Farmers' management of soil nutrients depended on a range of socioeconomic factors. Access to livestock, labour, credit and markets are of particular importance in explaining which farmers are able to maintain and improve the fertility of their soils. The hh remains central to the management of farm and the mobilization of resources, such as labour and capital. However, other social institutions are also of great value in enabling the farmers to negotiate access to obtain resources such as draught power, transport and credit.

10.3.2 Farmers' SFM Practices

Farmers actively manage soil fertility and other soil properties through a wide range of practices. These include the so-called "modern" methods like combining chemical and organic fertilisers and other practices based on long experience of farmers and a rich knowledge of the locally specific conditions; despite constraints, such practices are alive and vibrant. There are diverse SFM practices which are commonly used by farmers in AP (Butterworth et al. 2003, Reddy 2010a, b, c). Similarly, the practices adopted by sample farmers of the present study are explained in Table 10.2.

To get an idea of how various practices are being adopted and in what proportion, an attempt has been made to examine the SFM practices across size classes of the farmers. In total, there were 664 field plots distributed among 360 sample hh's covering all size classes in the six villages. Out of these 79 plots were lying fallow (current as well as permanent). Information from the remaining 585 plots is presented in the Table 10.3.

Table 10.2 Soil fertility management practices adopted by sample hh's. (Source: Field study)

Practice	Procedure	Cost and lasting effect
Farm yard manure	Gathering dung and other organic materials and dumping them in a heap. Materials are added regularly as and when available. Green leaves of pongamia, cassia tora, neem and dry leaves of various trees are added to enrich the compost. Decomposed compost is transported to field for spreading and incorporation into soil before ploughing	Ranges between ₹ 2000 and 4000/acre excluding transportation cost. The effect lasts for 3–4 years
Sheep penning	The pastoralists pen their flock in the field overnight. The faeces and urine of the flock are deposited on the farmer's field. This is rich manure and is incorporated into the soil by shallow ploughing	₹ 1000–1200/acre. Effect lasts for only one or two seasons
Legume cultivation	Growing of leguminous crop loosens soil, adds organic matter to soil through leaf fall and fixes nitrogen into the soil	Seed cost of crop. Generally ₹ 50–200/acre. In case of ground nut it is ₹ 1800–2500/acre. Effect lasts for one season
Oil cakes (neem and castor)	Neem and castor cakes are used for manuring crops. They give good colour and shine to fruits in the case of horticultural crops. They are also used to avoid certain soil-borne pests	₹ 250/50 kg bag. Cost varies depending on the crop and dosage applied. Effect lasts for one to twoseasons
Crop rotation	Crops which use different nutrients are grown alternately to keep the nutrient balance	Does not involve any cost except the decision to change the crop. Effect lasts for two seasons
Intercropping/mixed cropping	Growing a mixture of different crops in the same field. Small and marginal farmers often grow 5–6 crops in an acre	Cost varies depending upon the crops and their required quantities per acre
Tank silt application	Fine tank silt from the village tank is excavated and transported to the fields and applied to top soil	₹ 800–1000/acre. Effect lasts for 2 years
Chemical fertilisers (including micronutrients)	Chemical fertilisers are concentrated forms of nutrients that can be easily applied and are readily available to plants	Numerous brands are available in the market. Cost of most commonly used urea is approx. ₹ 250/50 kg bag and DAP is ₹ 550/50 kg bag. Effect lasts for 2–3 months only

It is important to note that multiple practices were followed in some plots. In a few plots, two to three practices were followed at a given time for a given crop. Hence, the percentages were calculated for each practice with respect to the area of the total number of sample plots in each size class and presented in Table 10.3. As

Table 10.3 Size class-wise adoption of soil fertility management practices in sample plots during the kharif 2005–2006 (in percent) ($N = 585$) (Source: Based on primary survey)

Soil fertility management practice	Large farmers ($N = 245$)	Medium farmers ($N = 183$)	Small farmers ($N = 157$)	Mean of all
Not followed any practice	1.63 (57)	4.37 (42)	10.19 (36.5)	5.40 (135.5)
FYM + chemical fertilisers	77.14 (1127.65)	63.93 (435)	57.96 (230)	66.34 (1792.65)
Only chemical fertilisers	15.51 (214)	24.84 (147.75)	17.83 (51)	19.39 (412.75)
Sheep penning	2.86 (44)	0.0	5.10 (22.5)	2.65 (66.5)
Sheep penning + chemical fertilisers	0.0	0.0	0.64 (2)	0.21 (2)
Tank silt	0.0	0.0	0.64 (3)	0.21 (3)
Neem cake	6.12 (95)	5.46 (43.5)	0.64 (0.5)	4.07 (139)
Castor cake	6.94 (99.5)	2.73 (22.5)	5.10 (15)	4.92 (137)
Legume cultivation	45.71 (606.75)	33.88 (228.5)	23.57 (92)	34.39 (927.25)
Intercropping/mixed cropping	40.41 (619.25)	38.80 (316)	40.13 (154.5)	39.78 (1089.75)
Others	00.0	00.0	1.27 (3)	0.42 (3)

Figures in the parentheses indicate the actual acres in which the practice was used. More than one practice is adopted in a given season in a given plot and the above table is based on multiple responses

more than one SFM practice was adopted in some sample plots, the total percentage exceeds 100.

Looking at the mean column in Table 10.3, it is evident that the practice of combination of FYM and chemical fertilisers (1792.65 acres) was predominant across all size class farmers followed by intercropping/mixed cropping (1089.75 acres) and legume cultivation with an area of 927.25 acres. A unique feature of farm yard manure is that it supplies humus (which can never be substituted by inorganic fertilisers). This humus is a major source of food for soil microorganisms and is essential for their wellbeing and development. The physical properties of the soil like soil porosity, water stable aggregates, water-holding capacity, infiltration rate and hydraulic conductivity are significantly increased with the conjunctive use of organic manures and inorganic fertilisers (Reddy 1999b). It was found that only chemical fertilisers were used in 412.75 acres, and no practice was followed at all in 135.5 acres. It is surprising to note that the practice of applying only FYM was not seen across any size class among all the study villages. This is due to lack of sufficient quantities of manure due to reduction in livestock population which has in turn been affected due to shortage of fodder and water crisis in summer. It was interesting to find that the practice of sheep penning still exists and is in great demand among farmers. Shepherds are the source of best quality FYM (Butterworth et al. 2003).

Organic fertilisers like neem cakes are being used for paddy to increase the efficiency of nitrogen uptake and to increase the resistance to pest attack. Similarly, another organic fertiliser, castor cake, is being used mostly for horticultural crops in Anantapur district along with neem cake. As revealed by farmers in FGDs and PRAs, organic cakes are more suitable for orchard crops than the chemical fertilisers as they serve several purposes such as adding to soil fertility, reducing pest incidence, increasing water-holding capacity of soil, giving good colour and shape to fruit and having longer shelf life which helps in transporting to distant places. It was surprising to note that vermicompost, which was becoming popular of late in some pockets of AP, was not used even in a single plot. This indicated that it is still in the early stages and is not widely spread. Efforts must be made to introduce the vermicompost technology and establish a good number of units, making it accessible to more small and marginal farmers. The empirical data of the present study clearly point out the fact that farmers attach immense significance to the practice of legume cultivation.

Contrary to the popular belief that farmers mostly use chemical fertilisers, the present study brings out the fact that across all size classes, farmers use several other methods of SFM including organic fertilisers. According to them, it is being done keeping in view the long-term sustainability of their soils. From Table 10.3 it is evident that FYM + chemical fertilisers is very predominant among the sample farmers across all size classes. This proves the fact that farmers clearly understand the prominent role of farm yard manure and other organic manures in the long-term sustainability of soil and crop yields. There is considerable diversity in farmers' conditions and SFM strategies and there is no blanket solution for improving soil fertility (Hilhorst and Muchena 2000 and Reddy 2010b).

10.3.3 Soil Type

The soils of the study area varied from deep black cotton soils to light sandy soils. The kind and depth of soils also influenced the soil fertility. Generally, it is seen that soils with greater depth will be more fertile than shallow soils. Similarly, black cotton soils and red sandy soils are more fertile compared to others. Table 10.4 indicates that the majority of the sampled plots were of depth ranging between 1.1 and 2 ft. The data show that more than 39 % of the soils belonging to large farmers have a depth of more than 2 ft and are likely to be fertile, whereas medium farmers had 12.92 % and small farmers had 13.84 % of their soils with depth greater than 2 ft. Similarly, small farmers had the highest percentage (33.05 %) of plots with a depth of less than 1 ft.

10.3.4 Agro-biodiversity

Farmers in semiarid regions developed diversified cropping systems to ensure that the most essential natural resources such as sunlight, wind, rainfall and soil are

Table 10.4 Size class-wise distribution of sample households according to their soil depth (in percent). (Source: Based on primary survey)

Soil depth	Large farmers	Medium farmers	Small farmers	Total
Up to 1 ft	17.10	31.88	33.05	23.19
1.1–2 ft	43.63	55.20	53.11	48.03
2.1–3 ft	27.94	7.78	8.47	19.91
3.1–4 ft	6.00	1.58	3.11	4.49
4.1 ft and above	5.33	3.56	2.26	4.38
Total	100.00	100.00	100.00	100.00

Table 10.5 Agro-biodiversity in study villages. (Source: Based on primary survey)

Kharif crops in 2005–2006	Rabi crops in 2005–2006
Unirrigated villages	
Groundnut, maize, red gram, green gram, black gram, cotton, chillies, chickpea, subabul, jowar, variga, bajra, sunflower, chillies, paddy, dry sown paddy, cow pea and field bean	Chickpea, groundnut, paddy
Irrigated villages	
Paddy, black gram, green gram, castor, fodder, jowar, pillipesara, red gram, maize, groundnut, sunflower, field bean, cowpea, orchards of ber, guava, lemon, mango and sapota	Paddy, groundnut, sunflower, rabi, jowar

optimally utilised throughout the year. Crops that were developed over centuries were specifically bred to suit local soils, nutritional needs of people, livestock needs and climatic conditions. A large number of farmers, especially women have been nurturing the agro-biodiversity and soil fertility without any sort of support from the government. The lands of the sample farmers of the study villages have hosted a wide range of crops (Table 10.5).

There was more crop diversity in the unirrigated villages than the irrigated villages during the kharif season. This is to insure themselves against the vagaries of nature and incidence of pests and diseases. It is noticed that millet crops like jowar, bajra and variga are still grown in unirrigated villages.

10.3.5 Crop Rotation

The growing of different crops on a piece of land in a preplanned succession is called crop rotation. The decision of rotating the crops has a huge bearing on SFM and hence it is being discussed in detail in this section. According to farmers, if we grow the same crop continuously on the same patch of land we do not get good yields. They have been advised by their elders that crops have to be rotated in order to preserve fertility of soils. Compared to monoculture cropping practices, multicrop

Table 10.6 Adoption of crop rotation by sample farmers during the last 2 years in their irrigated and dryland plots. (Source: Based on primary survey)

Village	Particulars	Irrigated plots			Dryland plots		
		Large farmers	Medium farmers	Small farmers	Large farmers	Medium farmers	Small farmers
Darsi	Total area in acres	249.5	116	60	6.5	2	2.5
	% of area under crop rotation	8.61	22.41	14.16	0	0	100
Ravinuthala	Total area in acres	–	–	–	267.5	128.5	48
	% of area under crop rotation	–	–	–	63.17	36.18	39.58
Chinnapo-lamada	Total area in acres	163.75	60.5	32.5	130.25	73.25	31
	% of area under crop rotation	27.78	38.01	6.16	31.86	39.59	59.67
Brahman-palle	Total area in acres	60.5	14.5	3.5	237.5	115	63
	% of area under crop rotation	57.02	62.06	28.57	19.36	13.04	0
Dhupalle	Total area in acres	212.9	121.5	54.5	28.5	11.5	9
	% of area under crop rotation	41.56	22.63	25.68	8.77	0	11.11
Maredudinne	Total area in acres	8	1	2.5	279	131	52.5
	% of area under crop rotation	0	0	0	32.97	30.53	32.38
Total	Total area in acres	694.65	313.5	153	949.25	461.25	206
	% of area under crop rotation	27.35	27.27	16.66	36.97	28.29	28.15

rotations with two or three crops in a year can result in increased soil organic carbon content (Purakayastha et al. 2008). This is because of the addition of large amount of biomass above the ground as well as underground. Such crop planning is practised in dryland regions. The complexity and diversity of such microenvironments created by the farmers are often undervalued (Chambers 1995). To get an idea of how crop rotation is being followed in the sample plots, data of crops grown in four crop seasons (i.e. kharif and rabi) in 2004–2005 and 05–06 were studied.

High crop rotation was seen in Chinnapolamada village across various size classes, but the area was less. Among the irrigated plots of the study villages, less crop rotation was seen in all size classes in Darsi of Prakasam district (Table 10.6).

Within the size classes, crop rotation was high (22.41 %) with medium farmers, followed by small farmers (14.16 %) and large farmers (8.61 %). Among the villages having larger area under irrigation, Dhupalle of Mahbubnagar had high crop rotation percentage when compared to Darsi (Praksam) and Chinnapolamada (Anatapur). In Dhupalle crop rotation was highest with large farmers (41.56 %) followed by small farmers (25.68 %) and medium farmers (22.63 %). There was no crop rotation in Maredudinne village in irrigated plots. However, the area under irrigation is negligible (10 acres) in this village. On the whole in the irrigated plots of sample farmers of six villages the crop rotation was significant among large farmers (27.35 %) followed by medium farmers (27.27 %) and small farmers (16.66 %). This is due to the importance of crop rotation in SFM. Crop rotation also controls weeds and diseases and helps a good variety of natural predators to survive on the farm.

Among the dryland villages, high crop rotation could be seen in Ravinuthala followed by Chinnapolamada and Maredudinne. In Brahmanpalle of Anantapur less crop rotation was seen as farmers were mostly cultivating groundnut continuously as they found it relatively more economical and suitable to their soils and climate. Among the size classes no crop rotation was seen in the case of small farmers. This might be due to small holdings coupled with dependence on other sources of income. Many farmers from this category find employment at the Satya Sai Trust, adjoining to their village.

The total area under crop rotation across all size classes was seen to be higher under dryland plots than in the irrigated plots (Table 10.6). Similarly, the overall percentage of crop rotation was high under rainfed conditions than under irrigated conditions. The probable reason could be lesser feasibility of crop rotation under irrigated conditions (canal irrigation) where water release is not under the control of farmers.

10.3.6 Farmers' Perception of Soil Fertility

The present research tried to assess the level of soil fertility of sample plots according to farmers' own perception. For this purpose, the sample plots were compared with the best fertile plot in the respective village (based on FGDs). The soils of the farmers were evaluated on a scale of continuum consisting of very bad, bad, average and good. More than 50 % of the sample plots were perceived to be less fertile. Soils with very bad fertility recorded were 5.02 % while 47.6 % had bad soil fertility status (Table 10.7). The low fertility status could be due to gradual decline in organic manure application and intensive cultivation. The bad fertility status of soils as perceived by the farmers could be the probable reason for the low yields in spite of higher NPK application than recommended quantities. Similarly, 37.13 % of the plots were perceived to be of average fertility status. Only 10.25 % had good soil fertility.

Among the size classes, the small farmers had more unfertile area. This could be due to more dependence on agricultural labour for their sustenance and also

Table 10.7 Size class-wise distribution of sample plots according to farmers' perception of level of soil fertility (percentage area) ($N = 585$). (Source: Based on primary survey)

Level of soil fertility	Large farmers	Medium farmers	Small farmers	Total
Very bad	4.17	4.86	9.07	5.02
Bad	50.61	41.86	46.17	47.6
Average	33.09	45.01	38.81	37.13
Good	12.13	8.27	5.95	10.25
	100.00	100.00	100.00	100.00

migration to distant places, leaving the soils unattended during certain years. Lack of resources for investing on inputs could also be one of the main reasons for the low fertility of the plots belonging to the small farmers. Another reason is that most of the lands distributed to SCs are far away from the village and are highly infertile and degraded. Medium farmers had more area under average soil fertility. Medium farmers have lesser amounts of very bad and bad soils and more of the soils with average soil fertility.

This makes it clear that unless the overall soil fertility status of soils is improved by way of applying good quantities of organic manures, adopting other tested traditional SFM practices, and controlling of soil erosion, the overall improvement in the yields may not be realised. The perception of farmers regarding their soil fertility also demands that the unirrigated villages be paid serious attention for improving the soil fertility in these villages which supports the livelihoods of millions of people.

10.3.7 Organic Versus Inorganic Inputs

An in-depth analysis was done to understand how the farmers are distributed with respect to the use of organic and inorganic inputs in the sampled plots (Table 10.8). For this purpose, the total amount spent by the sample farmers on organic manures and inorganic fertilisers, together in the period 2004–2005 and 2005–2006 was considered. Different ratios of both organic manures and inorganic fertilisers were considered for easier and better understanding of how money is being invested on organic and inorganic fertilisers and by what percentage of farmers. The analysis did not include cultural practices like crop rotation, intercropping/mixed cropping, as there is no separate investment for these practices, except the seed cost which the farmer has to invariably bear for raising the crop.

The distribution of sample farmers revealed that only 16.94 % used exclusively chemical fertilisers in their plots. There was no one exclusively using only organic practice. But, it was found that more than 32.50 % of the total sample farmers spend more than 50 % of their money on organic manures/practices, out of their total investment on soil fertility. This is a positive sign for the emerging organic market. This large group of farmers can be encouraged to become totally organic and

Table 10.8 Size class-wise distribution of sample hh's according to the ratio of use of amount in rupees for organic and inorganic fertilisers together in 2004–2005 and 2005–2006 (in percent). (Source: Based on primary survey)

Size class	OF:CF 0:100	OF:CF 10:90	OF:CF 20:80	OF:CF 30:70	OF:CF 40:60	OF:CF 50:50	OF:CF 60:40	OF:CF 70:30	OF:CF 80:20	OF:CF 90:10	Total
LF = Large farmer	5.83 (7)	15.83 (19)	23.34 (28)	17.50 (21)	18.33 (22)	5.0 (6)	7.5 (9)	6.67 (8)	0 (0.00)	0 (0.00)	100.0 (120)
MF = Medium farmer	20.0 (24)	5.83 (7)	12.5 (15)	20.83 (25)	9.17 (11)	10.0 (12)	9.17 (11)	5.0 (6)	3.33 (4)	4.17 (5)	100.0 (120)
SF = Small farmer	25.0 (30)	4.17 (5)	7.5 (9)	8.33 (10)	8.33 (10)	8.33 (10)	9.17 (11)	14.17 (17)	10.0 (12)	5.0 (6)	100.0 (120)
All	16.94 (61)	8.61 (31)	14.45 (52)	15.56 (56)	11.94 (43)	7.78 (28)	8.61 (31)	8.61 (31)	4.44 (16)	3.06 (11)	100.0 (360)

Figures in the parenthesis are the actual number of farmers
OF organic fertilisers, *CF* chemical fertilisers

take advantage of the growing organic market and better utilise the encouragement by the government towards organic agriculture. Among the size classes, 46.67 % of small farmers invest more than 50 % of their money on organic manures/practices.

10.4 Recommended Practices Versus Farmers' Practices

In the earlier sections, various SFM practices adopted by the total sample farmers were discussed, irrespective of the crops they were growing. In this section, an attempt is made to understand the farmers' practices with reference to agricultural university's recommendations regarding the SFM for the selected crops. Out of the total crops being cultivated by the sample farmers, only those crops which had good distribution (minimum sample of 30 farmers) were selected for analysis. Paddy, maize, sunflower, castor and groundnut (irrigated and rainfed) grown during kharif 2005–2006 were selected for comparing farmers' practices against the recommended practices. However, we are restricting our discussion to only paddy and groundnut in this chapter. It is a known fact that not only major nutrients like nitrogen (N), phosphorus (P) and potassium (K) but even micronutrients play a critical role in crop growth and yield. However, in this analysis and discussion, we are restricting ourselves to NPK and FYM.

10.4.1 SFM in Paddy

Acharya N. G. Ranga Agricultural University (ANGRAU) has divided the state of AP into nine agro-climatic zones based on physiography, soil types, crops and cropping pattern. It is interesting to note that the package of practices was given only for paddy crop based on the agro-climatic zones. Just like other package of practices, SFM recommendations were also made separately for each of the agro-climatic zone. Paddy has been one of the major food crops grown under irrigated conditions in AP. There is no specific recommendation made for the dosage of FYM in the "agricultural almanac" for paddy. But interestingly, FYM recommendations were made for other crops (viz., ground nut, sunflower and castor). The reason could be the assumption by scientists that farmers invariably use the available manure from their compost pits and hence it is not essential to mention it. But one cannot ignore the enormous importance of FYM in maintaining soil health and soil life. Hence, it is essential to make a specific mention of FYM dosages too for all the crops in the state.

Farmers across all size classes in Darsi of Prakasam district applied two and a half to thrice the recommended dosages of nitrogen and phosphorus supplying fertilisers per acre (Table 10.9). This is due to cultivation of three crops in a year due to which the soil has to be replenished either with FYM or inorganic fertilisers. The availability of FYM is comparatively less and hence excessive application of nitrogen and phosphorus fertilisers is seen when compared to recommended dosage. Coastal farmers are proactive in gaining information related to agriculture. Private companies are active in this region, and the farmers are motivated to use more of chemical fertilisers through company advertisements, which claim that they can give higher yields, whereas the applied potassium dosage is less than the recommended. This could be due to the good quantities of native potassium in the soil of Darsi Mandal (Table 10.11) and Prakasam district (Table 10.10) as a whole. Among the size classes, large farmers are seen to apply higher quantities of NPK nutrients followed by medium and small farmers. On the contrary, FYM is being applied in larger quantities per acre by small farmers (1919 kg) followed by medium farmers (1744 kg) and large farmers (980 kg). This is due to less land and more dependence of small and medium farmers on dairy, which eventually gives them some manure.

Excessive use of inorganic fertilisers over longer periods of time is a cause of concern, as it leads to gradual degradation of soils having access to assured irrigation. This is more so in cases where inorganics are not combined with good quantities of organic manures. Depletion of organic matter in the soil leads to deterioration in soil structure, reduced soil water and nutrient-holding capacity and reduced microbiological activity (Wommer et al. 1994; Reddy 1999a).

In Ravinuthala village there is no paddy crop at all. In Chinnapolamada of Anantapur district, the nitrogen dosages applied by the farmers are one third lesser and phosphorus and potassium are 50 % lesser than the recommended dosage, respectively. This is due to the lack of sufficient water, which reduces the dosages as it might have a negative effect on the crop yield under such moisture stress conditions. The fertility index in Tables 10.10 and 10.11 indicates that the soils in the study mandals and study districts have medium to high native nutrients of phosphorus and potassium. This could be one of the reasons for lesser use of phosphorus and potassium than recommended. Another reason is that the recommendations made by the scientists do not hold good for the conditions under which the sample farmers work. In this village, medium farmers were seen to apply the highest dose (21.5 kg)

Table 10.9 Soil fertility management inputs used per acre by sampled households in paddy crop during the kharif 2005–2006. (Source: based on primary survey)

Particulars	Large farmers	Medium farmers	Small farmers	Recommended nutrient dosage by Agricultural University of A.P(ANGRAU)	Recom-mended yield in kg/acre	Yield obtained by sample farmers in kg/acre
Darsi (Prakasam)						
Farm yard manure in kg	980.00	1744.53	1919.13	No specifications	2800	2031.16
Nitrogen in kg	84.91	81.68	75.86	24–32		
Phosphorus in kg	52.62	52.625	48.51	16		
Potassium in kg	9.34	9.43	8.93	12–16		
Chinnapolamada (Anantapur)						
Farm yard manure in kg	332.50	875.00	2705.00	No specifications	2500	1100
Nitrogen in kg	18.50	21.5	17.8	64		
Phosphorus in kg	17	15.00	12.60	32		
Potassium in kg	17	12.50	9.70	32		
Brahmanpalli (Anantapur)						
Farm yard manure in kg	879.90	1164.37	1531.25	No specifications	2500	1022.17
Nitrogen in kg	21.09	21.50	18.5	64		
Phosphorus in kg	18.36	18.50	16.75	32		
Potassium in kg	16.54	16.5	12.50	32		
Dhupalle (Mahbubnagar)						
Farm yard manure in kg	2061.94	1948.81	1184.33	No specifications	2500	1991.02
Nitrogen in kg	76.77	71.75	61.80	40–48		
Phosphorus in kg	42.27	37.50	27.26	24		
Potassium in kg	21.83	21.31	20.08	16		
Maredudinne (Mahbubnagar)						
Farm yard manure in kg	1075.00	0	541.66	No specifications	2500	1219.44
Nitrogen in kg	20	21	21.66	40–48		
Phosphorus in kg	16.4	16	17.33	24		
Potassium in kg	16.8	15	17.33	16		

No paddy was grown under irrigated conditions in Ravinuthala village of Prakasam

Table 10.10 Fertility index for macro nutrients (NPK) in the study districts during the year 2005–2006. (Source: Soil Testing Laboratories, Department of Agriculture, Govt. of Andhra Pradesh)

District	Fertility index		
	Nitrogen	Phosphorus	Potassium
Prakasam	1.10	1.88	2.75
Anantapur	1.18	2.78	2.81
Mahbubnagar	1.61	1.78	1.85

Fertility index rating: 0–1.66 = low, 1.67–2.33 = medium, > 2.33 = high

Table 10.11 Categorisation of soils in study mandals of selected districts based on the fertility index (2005–2006). (Source: Soil Testing Laboratories, Department of Agriculture, Govt. of Andhra Pradesh)

District	Name of mandal	Fertility index		
		Nitrogen	Phosphorus	Potassium
Prakasam	Darsi	1.08 (low)	1.89 (medium)	2.65 (high)
	Korisepadu	1.21 (low)	1.95 (medium)	2.50 (high)
Anantapur	Tadipatri	1.37 (low)	2.93 (high)	2.90 (high)
	Puttaparthi	1.07 (low)	2.46 (high)	2.84 (high)
Mahbubnagar	Chinnachinta kunta	1.81 (medium)	1.86 (medium)	186 (medium)
	Peddakothapalli	[a]	[a]	[a]

Fertility index rating: 0–1.66 = low, 1.67–2.33 = medium, > 2.33 = high
[a]Information is not available for this mandal

of nitrogen followed by large farmers (18.50 kg) and small farmers (17.8 kg). This can be attributed to the efforts of medium farmers to maximise the profits from the available land. The per acre FYM dosage being applied by small farmers in this village is the highest (2705 kg/acre) among all the study villages, whereas medium and large farmers are applying 875 and 332 kg per acre, respectively. This is quite low and the probable reason could be the lesser livestock maintained by these groups due to lack of fodder and water. However, small farmers still manage to keep the livestock as it helps them earn livelihood.

In the village of Brahmanpalle of Anantapur, the situation was found to be almost similar to that of Chinnapolamada but with slightly higher doses. FYM is applied in higher quantities by small farmers (1531.25 kg) followed by medium (1164.37 kg) and large farmers (879.90 kg). The reason is the higher livestock number with small farmers and also better maintenance by them.

In Dhupalle village of Mahbubnagar district, the NPK quantities being applied are one and a half to two times higher than the official recommendation. A significant observation made in this village was that the farmers across all size classes are applying higher doses of even potassium (it was lesser than recommended in other villages). The reason could be the medium fertility index of potassium in the villages of this mandal (see Table 10.11). Another reason is the extensive cultivation of paddy in this village in the past a couple of years due to new lift irrigation scheme and the cultivation of two crops per year. Application of complex fertilisers, which also contribute potassium, could be another reason for this. Moreover, the varieties being cultivated are long-duration BPT-5204, which require fertilisers for a longer period than short-duration varieties. Contrary to other villages, in Dhupalle, large farmers are seen to apply slightly higher quantities of FYM (2061.94 kg) per acre followed by medium farmers (1948.81 kg) and small farmers (1184.33 kg). The reason for this is the availability of FYM due to a good number of livestock and the capacity to purchase.

It can be observed from Table 10.9 that the actual yields obtained by the farmers is lesser than the recommended yield for that agro-climatic zone. Yields are lesser, both in case of those villages where nitrogen, phosphorus and potassium nutrients are being applied in excess, and also in those villages where they are applied in higher quantities than the recommended quantities. A possible reason could be that the recommendations might be based on the data obtained on research designs of 10–15 years back, which is not in line with the farmers' present condition. However, it indicates that excess quantities of chemical fertilisers are being applied in the coastal village Darsi, which may eventually lead to soil degradation. In case of the other villages, farmers are applying lesser nutrient dosages after their experience proved that it is the optimum they can apply in the given situation and higher amounts may not increase yields any more. On the other hand, it could only lead to soil degradation.

10.4.2 SFM for Rainfed Groundnut

The quantities of nitrogen and phosphorus being applied are higher than the recommended dosage (Table 10.12). Only 25 % of the recommended dosage of potassium is applied across all size classes. This is due to native potassium in the soil (Table 10.10), especially in Anantapur district where groundnut is predominantly grown under rainfed situations.

Gypsum is not at all being applied by small and medium farmers. Large farmers are applying only 12.14 kg per acre as against the recommended 200 kg/acre. This is due to difficulties in accessing the gypsum supplied by government on subsidy. The process of procuring gypsum on subsidy basis from the agricultural department is very complicated and is not within the reach of small and marginal farmers. All categories of farmers are applying very less FYM quantities for rainfed groundnut as compared to the recommended dosage. The reason attributed is the concentration of nutrients on irrigated plots to get better and assured income. Yields are quite lower than excepted. The reason is less FYM application, pest incidence and lack of gypsum application.

Table 10.12 Soil fertility management inputs used by sampled farmers per acre of groundnut crop under rainfed conditions during theperiod 2005–2006. (Source: Based on primary survey)

Particulars	Large farmers	Medium farmers	Small farmers	Recommended nutrient dosage by Agricultural University of AP (ANGRAU)	Recom-mended yield in kg/acre	Yield obtained by sampled farmers in kg/acre
Farm yard manure in kg	498.22	786.48	1053.65	4000–5000	480–600	387.24
Nitrogen in kg	18.05	15.68	10.57	12		
Phosphorus in kg	22.62	21.34	20.03	16		
Potassium in kg	5.72	5.41	4.46	20		
Gypsum in kg	12.14	0.00	0.00	200		

irrigated village, unirrigated village

10.5 SFM and TE

Production function is the average function. Each and every farmer's practice is much more important than the average production of all the farmers. Hence, it is essential to calculate the TE index of each farmer, which determines the farmers' yield, given the input bundle, and helps us find out whether the increase in TE index is because of SFM practices or some other practices adopted by these farmers. In this way, TE analysis enables us a more detailed analysis.

The frontier production function has the maximum feasible or potential output that can be produced by a farm with given level of inputs and technology.

Consider the frontier production function

$$Y_{it} = f(x_{it},\ \beta) + E_{it} \tag{10.1}$$

$$\text{and } E_{it} = V_{it} - U_i, \tag{10.2}$$

where Y_{it} denotes the appropriate function of the production for the ith sample firm ($i = 1,2,.\ N$) in the tth time period ($t = 1,2,.\ T$).

$\mathbf{x}_{it}$ is a (1 X k) vector of appropriate functions of the inputs associated with the ith sample firm in the tth period.

$\boldsymbol{\beta}$ is a (k X 1) vector of the coefficients for the associated independent variables in the production function.

V_{it} is a random variable assumed to be independent and identically distributed (*iid*) as N(0, σ_v^2) and independent of U_i random variable.

U_i is firm-specific TE-related variable and non-negative defined by the truncation (at zero) of $N(\mu, \sigma_u^2)$.

It is assumed that the random variables V_{it} and U_i are distributed independently of the input variables in the model. The density function for U_i is defined by

$$f u_i(u) = \frac{\exp[-½(u-\mu)^2/\sigma_u{}^2]}{\surd 2\Pi\sigma_u[1-\phi(-\mu/\sigma_u)]}, \qquad u > 0.$$

By the convolution formula, the joint density function of e_i, i.e. (v_{it}—u_i) can be written as

$$f_y(y) = \frac{[1-\phi(Z)]\exp\left[\frac{-\sum(e_{it}+\mu)^2}{2(1-\mu)\sigma^2}\right]\exp\left[\frac{T^2\gamma(\ddot{e}_i+\mu)^2}{2(1-\gamma)\sigma^2(1+(T-1)\gamma)}\right]}{\{(2\pi)^T\sigma^2[1+(T-1)\gamma]\}^{1/2}\left[\sigma^2(1-\gamma)\right]^{T-1/2}\left[1-\phi(-\mu\sqrt{\gamma\sigma})\right]}$$

$$Z = \frac{\sigma_u{}^2 T\ddot{e}_i - \sigma_v{}^2\mu}{\sigma_u\sigma_v\sqrt{(\sigma_v{}^2+T\sigma_u{}^2)}}$$

$$\ddot{e}_i = \frac{1}{T}\sum_{t=1}^{T} e_{it}$$

$$M = \exp\left[-1/2\sum(e_{it}+\mu)^2/\sigma_v{}^2\right]$$

$$N = \exp\left[\frac{\sigma_u{}^2T^2(\ddot{e}_i+\mu)^2}{2\sigma_v{}^2(\sigma_v{}^2+T\sigma_u{}^2)}\right]$$

and ϕ (.) denotes the distribution function of the standard normal random variable.

The density function of Y_i is written as follows:

$$f_y(y) = \frac{[1-\phi(Z)]\exp\left[\frac{-\sum(e_{it}+\mu)^2}{2(1-\mu)\sigma^2}\right]\exp\left[\frac{T^2\gamma(\ddot{e}_i+\mu)^2}{2(1-\gamma)\sigma^2(1+(T-1)\gamma)}\right]}{\{(2\pi)^T\sigma^2[1+(T-1)\gamma]\}^{1/2}\left[\sigma^2(1-\gamma)\right]^{T-1/2}\left[1-\phi(-\mu\sqrt{\gamma\sigma})\right]},$$

$$\text{where}\quad \sigma^2 = \sigma_u{}^2 + \sigma_v{}^2$$

$$\gamma = \sigma_u{}^2/\sigma^2.$$

The distribution of the non-negative firm effect *ui*, which is the generalization of the half-normal distribution in which $\mu = 0$. Log–likelihood function for the model is given by

$$\begin{aligned}\ln L_i = {} & (-NT/2)\ln(2\pi) - (N(T-1)/2)\ln[(1-\gamma)\sigma^2] \\ & - (N/2)\ln\{\sigma^2[1+(T-1)\gamma]\} - N\ln[1-\phi(-\mu/\sqrt{\gamma\sigma})] \\ & + \sum^N \ln[1-\phi(Z)] - (2(1-\gamma)\sigma^2)^{-1}\sum^N\sum^T(e_{it}+\mu)^2 \\ & + T^2\mu\sum^N(\ddot{e}_i+\mu)^2\{2(1-\gamma)[1+(T-1)\gamma]\sigma^2\}^{-1},\end{aligned}$$

$$\text{where}\quad Z = \frac{(T\ddot{e}_i+\mu)\gamma-\mu}{\sqrt{\{\gamma(1-\mu)[1+(T-1)\gamma]\}\sigma^2}}.$$

Further, a joint test on the significance of the random variable U_i in the frontier function is obtained from the generalised likelihood ratio. If the random variable is absent from the model (i.e. $\mu = \gamma^2 = 0$), then the ordinary least squares (OLS)

[2] γ is the ratio of individual variation to total variation in output, i.e. $\gamma = \sigma^2{}_u/\sigma^2{}_v + \sigma^2{}_u$.

Table 10.13 Frontier production function results for paddy

Independent variable	Yield in paddy	
	Frontier regression (half normal)	
Seed in kg	– 1.1173	a
	(0.2273)	
Farm yard manure in rupees	0.0196	a
	(0.0056)	
Total organic manures in rupees	0.0240	a
	(0.0065)	
Nitrogen (N) in kg	0.2226	a
	(0.0522)	
Phosphorus (P) in kg	0.2217	a
	(0.0460)	
Potassium (K) kg	0.0515	
	(0.0333)	
Other nutrients in rupees	0.0281	a
	(0.0076)	
Pesticide in rupees	– 0.0806	b
	(0.0329)	
Human labour in rupees	0.6479	a
	(0.0845)	
Bullock labour in rupees	0.1151	b
	(0.0540)	
R^2/sigma-squared	0.0199	a
	(0.0025)	
Gamma	0.9479	a
	(0.0228)	
log–likelihood function	202.8354	
LR test of the one-sided error	27.0134	
No. of observations	190	

Figures in the parentheses indicate standard errors
LR likelihood ratio
[a] 1 % level of significance
[b] 5 % level of significance

estimators of the remaining parameters of the production function are ML estimates. Thus, the negative of twice the logarithm of the generalised likelihood ratio has approximately Chi-Square (χ^2) distribution with parameter equal to the number of restrictions, i.e. if $\mu = 0$ then it follows the χ^2 distribution with parameter equal to one.

Table 10.13 presents the results of tests of significance conducted for paddy. In the frontier regression analysis it could be seen that the independent variables like FYM, total organic manures, nitrogen, phosphorus, other nutrients and human labour were positively significant at 1 % level, whereas bullock labour was positively significant at 5 % level of significance. This means that for every 1 rupee increase with respect to independent variables like FYM, organic manures, human labour and bullock labour, there was an increase in yield of 0.019, 0.024, 0.647 and 0.115 kg, respectively. Similarly, with every kg increase in independent variables like N, P and other nutrients, there is a yield increase of 0.222, 0.221 and 0.028 kg, respectively. The probable reason for the increase could be due to addition of more nutrients or synergetic effect which helped in higher yields, and the investments in human and bullock labour which might have helped to take up timely agricultural operation, leading to higher yields. Seed was negatively significant. With increase in every kg of seed, there will be a decrease of 1.11 kg of yield. This happens because as the seed rate increases and plant population per acre increases, competition for nutrients, moisture, sunlight and air also increases leading to lesser yields.

In the case of groundnut crop, the model showed neither positive nor negative significance for any independent variable. This means that there could be other variables such as rainfall and climate, which could have an impact on the yield, but could not be covered in this analysis.

TE After the regression analysis using frontier production function, the average TE was calculated for paddy and groundnut. TE was calculated using all inputs, which also included fertility-enhancing organic inputs.

TE is the ratio of the actual output represented in original units to the potential output measured in original units that is obtained when the firm is technically efficient ($U = 0$). Thus, the TE of the ith firm is defined by

$$TEi \quad = \frac{E(Y_{it}{}^*/U_i, x_{it}, t = 1{,}2\ldots)}{E(Y_{it}{}^*/U_i = 0, x_{it}, t = 1{,}2\ldots)},$$

where Y^*_{it} denotes the production in original units for the ith firm in the tth time period.

If the frontier production function (Eq. 10.1)–(Eq. 10.2) defined for the logarithm of the production, then the measures of individual firm specific technical efficiencies are calculated from the conditional distribution of U_i, given the joint distribution of E_{it}. The conditional distribution of U_i, given $\ddot{E}_{it}$

$$TE_i = \exp(-U_i/E_i) = \frac{1 - \phi[\sigma * -(M_i * /\sigma *)]}{1 - \phi(-M_i * /\sigma^2 *)} . \exp(-M * +{}^1\!/_2\,\sigma *),$$

where ϕ (.) is the standard normal distribution function

$$M_i{}^* = (-\sigma_u{}^2\ddot{E}_i + T^{-1}\mu\sigma_v{}^2)(\sigma_u{}^2 + T^{-1}\sigma_v{}^2)^{-1}$$

$$\ddot{E}_i \quad = T^{-1}\sum_{t=1}^{T} E_{it}$$

$$\sigma^2 * \quad = \sigma_u{}^2\sigma_v{}^2(\sigma_v{}^2 + T\sigma_u{}^2)^{-1}$$

Table 10.14 Average technical efficiency in different SFM practices. (Source: Based on primary survey)

SFM practices	Paddy	Groundnut
Chemical + FYM	0.90	0.88
Only chemical fertilisers	0.89	0.87
Sheep penning	0.92	[a]
Neem cake	0.92	0.79
Castor cake	0.85	0.87

[a]Practice was not used for this crop in 2005–2006

The details of the analysis results with respect to paddy and groundnut are discussed below.

SFM Methods and Average TE After assessing the crop-wise cost of cultivation and TE, an effort was made to see which methods were most effective for paddy and groundnut (Table 10.14). For this purpose, major SFM practices, excluding cultural practices like mixed cropping and legume crop cultivation were considered for analysis. The practices which are input-based were taken for analysis. For paddy, a combination of sheep penning and neem cake application realised the highest TE (0.92 %). This emphasises one argument that policy-makers and agricultural scientists should look at some of these traditional practices, and also practices which have been recently picking up like application of neem cake. There is a need to support such strong useful practices in various ways, so that farmers can enhance their soil fertility. It is also seen that a combination of FYM and chemical fertilisers had TE of 0.90 %; and chemical fertilisers ranked third with 0.89 %. Least TE was seen in the case of the practice of application of castor cake.

In the case of groundnut, the highest (0.88 %) average TE was seen with reference to the practice of combination of FYM plus chemical fertilisers. This is the practice being followed by a large number of sample farmers. Least TE was seen in the case of neem cake application (0.79 %). However, it must be noted that the number of farmers using this practice in groundnut crop is less.

10.6 Determinants of SFM

The SFM practices employed by farmers are determined by a wide variety of factors. An attempt was made to list out the determinants of SFM and their influence. Probit analysis was done to see what type of characters influence between the SFM adopting farmers and non-SFM adopting farmers. Regressional analysis was done to find out the variables which have influence on soil fertility. This regressional analysis was done with respect to paddy crop for which the TE was calculated earlier. The results are discussed below.

Table 10.15 Derivative probit analysis: Correlates of soil fertility management practices and paddy crop

(Dependent variable: Adoption of SFM = 1, Non-adoption = 0)			
Independent variables	Coeff.	Std. error	Sig.
Red sandy soil (= 1, Red loamy soil = 0)	0.1379	0.0873	
Black soil (= 1, Red loamy soil = 0)	– 0.0481	0.0907	
Gravel soil (= 1, Red loamy soil = 0)	– 0.1126	0.2389	
Average soil fertility level (= 1, bad soil fertility level = 0)	– 0.3119	0.2497	
Good soil fertility level (= 1, bad soil fertility level = 0)	– 0.3972	0.1971	a
Very good soil fertility (= 1, bad soil fertility level = 0)	– 0.4587	0.3592	
Rainfall in mm	0.0052	0.0015	b
Education: I–V (= 1, others = 0)	0.1366	0.0686	
Education: VI–VII (= 1, others = 0)	– 0.1374	0.1995	
Education: VIII–X (= 1, others = 0)	0.1681	0.0781	
Intermediate and above (= 1, others = 0)	0.1812	0.0753	b
Family size	– 0.0424	0.0165	a
Ratio of family labour to total labour used	0.0136	0.0061	a
Large ruminants	0.0241	0.0244	
Market distance	0.0082	0.0196	
Credit	– 0.0000005	0.0000012	
Medium farms (= 1, Big farms = 0)	– 0.2789	0.1724	
Small farms (= 1, Big farms = 0)	– 0.7660	0.1975	b
No. of observations	189.0000		
Pseudo *R*-square	0.6750		
Log-likelihood	– 41.6300		

[a]5 % level of significance
[b]1 % level of significance

10.6.1 Probit Analysis

Probit analysis is done to see what type of characters influence the SFM adopting farmers and non-SFM adopting farmers. This was done in the case of paddy. In the probit analysis, adoption of SFM practices by farmers (= 1) and non-adoption of SFM (= 0) was taken as dependent variable. It can be seen from Table 10.15 that the variables education level of intermediate and above and rainfall are positively significant at 1 % level, whereas the variable ratio to family labour is positively significant at 5 % level. This means that increase in education level, rainfall and ratio to family labour to the total labour used influences the adoptions of SFM practices positively, whereas the variable farm size is negatively significant at 1 % level. The variables' good soil fertility level and family size are negatively significant at 5 % level. With increasing rainfall, there is increase in adoption of SFM practices in paddy. This could be due to the use of only diverse practices like neem cake application, FYM, castor cake, sheep penning and chemical fertilisers due to assured moisture availability. These practices in turn help the farmer to achieve good crop yields. Similarly, with smaller farm size, the chances of adopting diverse practices is reduced, whereas farmers with bigger farm size have chances of adopting more number of SFM practices. Similarly, in soils with good fertility level, there is lesser need of adopting SFM practices as the fertility is excellent and can give good crop yields with the existing fertility level.

It is seen that the independent variables red sandy soil, black soil, gravel soil, shallow soils, medium depth soils, average soil fertility level, very good soil fertility, education: I–V, education: VI–VII, education: VIII–X, scheduled tribe, backward communities, OCs, large ruminants, market, credit and medium farms did not influence the adoption of SFM practices.

10.6.2 Regression Analysis

In this section, an attempt has been made to identify the hh, soil and climatic characteristics which influence the dependent variable percentage of organic manure value to total fertiliser values in paddy crop. For calculating the production function results of paddy crop, yield was taken as dependent variable. Hence, in this regressional analysis, the percentage of money spent on organic manures to the total amount spent on chemical fertilisers is taken as the dependent variable as soil health is directly influenced by organic manure application, which consequently gives better crop yields when crops are grown. Applied organic manure acts as a food source for soil life which in turn plays a key role in long-term sustainable soil health. Hence, the percentage of the amount spent on organic manures by farmers during the period 2004–2005 and 2005–2006 together with the total amount spent on chemical fertilisers during the same period was taken as the dependent variable.

The variables selected for the regression analysis are likely to influence the percentage of organic manure value to total fertiliser values and thereby soil fertility. The following lines describe the justification for selected variables in regression analysis and how these variables influence SFM. Generally, the soils with greater depth are more fertile and among soils black cotton soils are more fertile compared to the others. Based on farmers' perception, farmers' soils were evaluated on a scale of continuum consisting of very bad, bad, average and good. Soils with lesser fertility levels contain lesser organic matter content due to adoption of practices, which contributed lesser organic matter. Hence, the fertility level influences the ratio of organic manure used to total soil fertility inputs used per acre. Rainfall is another factor which influences soil fertility. Good amount of rainfall with proper distribution over the crop season influences higher use of chemical fertilisers. The long dry spells between two good rain showers discourages the use of fertilisers as there will be scorching effect on the crop. SFM practices which contribute good quantities of organic matter content are preferred by farmers as they hold more moisture and help to overcome the moisture stress periods. Literacy always helps farmers to get better information related to SFM from the print media. The variable family size influences the farmers' SFM through provision of family labour. Practices such as tank silt application, FYM transportation, green leaf manuring and mixed cropping require more manual labour inputs. Hence, those families with large family size contribute more labour helping easier adoption of labour-oriented SFM practices. The ratio of family labour to the total labour used in cultivation also influences SFM. Sample hh's with higher ratio of family labour to total labour used are likely to adopt more of organic practices. Caste largely determines the people's perceptions, values and knowledge, which in turn influence the adoption practices, which are either organic or inorganic in nature. Farmers from backward communities like "Gollas" own lots of sheep and hence adopt sheep penning practice in their agricultural fields. The variable "large ruminants" influences the ratio of organic manure value to total fertiliser value and thereby soil fertility by way of providing drought power for agricultural operations and supply of organic manure. Smaller market distance improves access to inorganic fertilisers and therefore, the percentage of organic manure value to total fertiliser values. Credit requirement of farmers is met from several sources. The credit provided by fertiliser dealers influences the use of inorganic fertilisers. Farmers with bigger farm sizes have lesser possibilities of applying organic manure in sufficient quantities as compared with those of small farm sizes.

In order to identify the factors that determine the percentage of organic manure value to total fertiliser values of the sample farms, a multiple linear regression model was used. The variables included in the function are given below.

$$Y = \text{Constant} + \sum\nolimits^{6}_{J=1} B_j X_j + \sum\nolimits^{17}_{j=1} C_j D_j + E_i,$$

where

Y is percentage of organic manure value to total fertiliser value

where

X_1	is rainfall in mm in the village
X_2	is family size
X_3	is family labour
X_4	is large ruminants
X_5	is market distance
X_6	is credit
$D_{1=1}$	if farmer cultivated in red sandy soil, otherwise 0
$D_{2=1}$	if farmer cultivated in black soil, otherwise 0
$D_{3=1}$	if farmer cultivated in gravel soil, otherwise 0
$D_{4=1}$	if farmer cultivated in shallow soil, otherwise 0
$D_{5=1}$	if farmer cultivated in medium soils, otherwise 0

$D_6 = 1$ if farmer cultivated in soils of average fertility level, otherwise 0
$D_7 = 1$ if farmer cultivated in soils of good fertility level, otherwise 0
$D_8 = 1$ if farmer cultivated in soils of very good fertility level, otherwise 0
$D_9 = 1$ if farmer has education level up to I–V, otherwise 0
$D_{10} = 1$ if farmer has education level up to VI–VII, otherwise 0
$D_{11} = 1$ if farmer has education level up to VIII–X, otherwise 0
$D_{12} = 1$ if farmer has education level up to intermediate and above, otherwise 0
$D_{13} = 1$ if the farmer belonged to scheduled tribe, otherwise 0
$D_{14} = 1$ if the farmer belonged to backward caste, otherwise 0
$D_{15} = 1$ if the farmer belonged to other caste, otherwise 0
$D_{16} = 1$ if the farmer belonged to medium farm size, otherwise 0
$D_{17} = 1$ if the farmer belonged to small farm size, otherwise 0

Dummy variables were used for soil types, soil depth, fertility level of soils, education, caste and farm size.

The results of the analysis in Table 10.16 reveal that in paddy crop, the variables rainfall, large ruminants and small farms are positively significant to the percentage of organic manure value to the total fertiliser value at 1 % level of significance, whereas the variable market distance was positively significant at the 5 % level. This means that with increase in rainfall and population of large ruminants the application of organic manures increases.

Similarly, the smaller the farm size, higher is the amount of organic manure used which helps in improving soil fertility. The nearer the market distance, higher is the use of organic manures. This indicates that organic manures/cakes have to be made easily accessible to farmers. The independent variables red sandy soil, black soil, gravel soil, shallow soils, medium depth soils, average soil fertility level, good soil fertility level, very good soil fertility, education: I–V, education: VI–VII, education: VIII–X, intermediate and above, family size, scheduled tribe, backward communities, OCs, ratio of family labour, credit and medium farms were not significant for paddy.

10.7 Conclusion

Soils are critical to agriculture, food security and livelihoods. With a large proportion of India's population being dependent upon small-scale, rainfed agriculture, sustainable management of soil is a high-priority issue. This chapter has brought out clearly that diverse SFM practices are being followed by farmers so as to effectively manage the health of their soils and to get good crop yields. It could be clearly seen that the contribution of organic matter by traditional SFM practices has been instrumental in maintaining soil fertility for generations. Probit analysis on paddy revealed that increase in education level, rainfall and ratio of family labour to total labour used, positively influenced the adoption of SFM practices. Without livestock,

Table 10.16 Correlation between percentage of organic manure value and total fertiliser value and household, soil, climatic characteristics in paddy

Independent Variables	Coeff.	Std. error	Sig.
Red sandy soil (= 1, red loamy soil = 0)	7.038	7.148	
Black soil (= 1, red loamy soil = 0)	5.674	3.702	
Gravel soil (= 1, red loamy soil = 0)	– 5.638	6.549	
Shallow soils (= 1, very shallow = 0)	– 6.729	3.934	
Medium depth soils (= 1, very shallow = 0)	– 7.298	5.394	
Average soil fertility level (= 1, bad soil fertility level = 0)	2.450	7.102	
Good soil fertility level (= 1, bad soil fertility level = 0)	– 4.776	7.094	
Very good soil fertility (= 1, bad soil fertility level = 0)	– 1.110	8.606	
Rainfall in mm	0.100	0.031	a
Education: I–V (= 1, others = 0)	1.451	4.632	
Education: VI–VII (= 1, others = 0)	7.375	5.158	
Education: VIII–X (= 1, others = 0)	4.374	4.788	
Intermediate and above (= 1, others = 0)	0.098	5.879	
Family size	0.468	0.629	
Scheduled tribe (= 1, scheduled caste = 0)	– 22.593	11.773	
Backward communities (= 1, scheduled caste = 0)	– 7.187	10.887	
Other caste(= 1, scheduled caste = 0)	– 7.689	11.427	
Ratio of family labour to total labour	0.077	0.126	
Large ruminants	3.013	0.635	a
Market distance	1.068	0.431	b
Credit	0.000	0.000	
Medium farms (= 1, big farms = 0)	7.340	4.202	
Small farms (= 1, big farms = 0)	15.589	5.545	a
_cons	– 51.409	27.612	
No. of observations	186.000		
r-square	0.288		

[a] 5 % level of significance
[b] 1 % level of significance

farming in semiarid regions would not be possible. Results of the regression analysis revealed that the variable "large ruminants" was positively significant in paddy, indicating the importance of livestock for sustainable SFM.

We could see from this study how farmers' SFM options are being undermined by government policies that give more priority to chemical fertiliser-based strategies. These include promotion and credit for packages of practices that include chemical fertiliser-responsive seeds, lack of recognition to mixed cropping system, and subsidising chemical fertilisers rather than organic inputs. Policies that encourage fertiliser and irrigation subsidies may discourage soil conservation and encourage depletion of groundwater (Lutz et al. 1994; Reddy 2001). The empirical results of this study call for an argument to be made for an approach to supporting SFM by farmers which is more attuned to the range of circumstances which are found on the ground and best suits the long-term productivity of soils. Hence, the study makes the following recommendations:

- Subsidies and credit policies should allow farmers to purchase whichever form of fertilising input they want.
- Such policies are needed which encourage and support low external input and labour-intensive practices such as sheep penning, composting, vermicomposting, tank silt application, or incorporation of green manure crops.
- Development of appropriate credit programmes that enable farmers to obtain crop loans for diverse mixed farming systems, including organic inputs such as farm yard manure.
- Local animal breeds are important for livelihoods and sustainable agriculture should be conserved in situ by strengthening integrated farming and indigenous systems of land use in which livestock play a key role in nutrient cycles and the maintenance of soil fertility.
- Crops that take care of fodder needs of the livestock have to be provided with good market price.
- All agricultural development projects, especially those focusing on soils, should focus on increasing supplies of organic matter to the soils to nourish soil organisms.
- INM recommendations should take into account the integrated manner in which farmers combine SFM with other aspects of their farming system.
- A more participatory approach to SFM is essential, to take account of the diversity of management techniques. It generates energy and creativity both within the farming community and among researchers and extension workers.

Acknowledgements This chapter is based on my doctoral research work. I am grateful to my Ph.D. supervisor V. Ratna Reddy, Livelihoods and Natural Resources Management Institute (LNRMI), Hyderabad, for his insightful comments and great support. Thanks to C. S. Sambamurthy and Jyothis Sathyapalan of CESS for their valuable comments on the thesis draft. I am grateful to S. Galab, Director CESS, and M. Gopinath Reddy, CESS, for their strong support. I am indebted to the farmers of study villages who contributed a lot to my research.

References

Adolph B, Butterwoth, J (2002) Soil fertility management in semi-arid India: its role in agricultural systems and the livelihoods of poor people. Natural Resources Institute (NRI), United Kingdom
Battese GE, Coelli TJ (1988) Prediction of farm-level technical efficiencies with a generalised frontier production function and panel data. J Econom 38: 387–399
Butterworth J, Adolph B, Suresh RB (2003) How farmers manage soil fertility. A guide to support innovation and livelihood. Andhra Pradesh Rural Livelihoods Programme/Chattam: Natural Resources Institute, Hyderabad
Centre for Monitoring Indian Economy (2006) CMIE Energy, Bombay
Chambers R (1995) Paradigm shifts and the practice of participatory research and development. In: Nelson N, Wright S (eds) Power and participatory development. Intermediate Technology Publications, London, pp 30–42
Fan S, Hazell P (2000) Should developing countries invest more in less favoured areas? An empirical analysis of rural India. Economic and Political Weekly, April 22, pp 1455–1464
Fertiliser Association of India (2006) Annual report, Delhi
Government of India (2008) Economic survey 2007–2008, Ministry of Finance, Economic Division, New Delhi
Hilhorst T, Muchena FM (2000) Nutrients on the move—soil fertility dynamics in African farming systems. International Institute for Environment and Development, London
Lutz E, Pagiola S, Reiche C (1994) The cost and benefits of soil conservation: the farmers' view point. World Bank Res Obs 9(2):273–295
Poinetti C, Suresh RB (2002) "Farmers' perceptions of crop diversity in the Deccan Plateau", SEEDLING, Quarterly Newsletter of Genetic Resources Action International (GRAIN), Spain
Purakayastha TJ, Swarup A, Singh D (2008) Strategies to manage soil organic matter for carbon sequestration—Indian perspective. Indian J Fertili 4(3):11–16, 19–22
Reddy RV (1999a) Valuation of renewable natural resources: user perspective. Econ Polit Wkly XXXIV(23):1435–1437 5th June
Reddy UR (1999b) "Integrated nutrient management in tomato-onion cropping system", unpublished PhD thesis, Acharya N.G. Ranga Agricultural University, Andhra Pradesh
Reddy RV (2001) User valuation of renewable natural resources: the user perspective. Nova Science Publishers, Inc., New York
Reddy SB (2010a) "Assessment of economic and ecological returns from millet-based bio-diverse organic farms vis-à-vis conventional farms", CESS Monograph series No. 8, Centre for Economic and Social Studies, Hyderabad
Reddy SB (2010b) "Soil fertility management in semi-arid regions: the socio-cultural, economic and livelihood dimensions of farmers' practices—a case of Andhra Pradesh", unpublished Ph.D. thesis, Centre for Economic and Social Studies, Dr. B. R. Amebedkar University, Hyderabad
Reddy SB (2010c) Organic farming: status, issues and prospects—a review, Journal of Agricultural Economics Research Review, Vol. 23, July–December, pp 197–208. New Delhi
Sagari (2004) Livelihoods and gender: equity in community resource management, overcoming gender barriers: local knowledge systems and animal health in Andhra Pradesh and Maharashtra. Earth Scan Publication Ltd., London
Woomer PL, Martin A, Albrecht A, Resck D, Scharpenseel H (1994) The importance and management of soil organic matter in the tropics In: Wommer, PL, Swift MJ (eds) The biological management of tropical soil fertility. Wiley-Sayce, Chichester

Chapter 11
Issues on Forest Governance in Contemporary Odisha

Kailas Sarap and Tapas Kumar Sarangi

11.1 Introduction

This chapter is an attempt to understand forest governance in general and in the context of Odisha, in particular, with a view to understand its working. It also points out the weaknesses present in the institutions. A number of factors, including access to resources, organisational efficiency, characteristics of a community and its involvement, affect the governance of natural resources like forests. The evidence (both historical and case studies) provided in this chapter shows that forest governance in the state of Odisha lacked tenurial rights for a long time for forest dwellers; this has affected their motivation for conservation and proper use of forest resources on a long-term basis. Forest protection communities, both joint forest management (JFM) and community forest management (CFM), have been working in the state over the years. However, these groups are organisationally weak in decision-making as they lack rights on the forests that they protect. Further, there is an intragroup problem. The weaker members (e.g. women and poor) are excluded from decision-making due to elite capture within the groups. As a result the distribution of benefits is not equitable among the members. This reduces the motivation of these excluded members. Further, the policy of sharing the final output between the JFM and the forest department is also not just for the JFM members. The Forest Rights Act (FRA) 2006 has been implemented in the state since 2008. There has been considerable progress in regularising the land under possession under individual rights but the progress in the implementation of conferring community forest rights is very low and other provisions of the Act have not been implemented. If the FRA is implemented effectively, it will motivate forest dwellers to conserve and develop forests.

K. Sarap (✉)
National Institute of Rural Development (NIRD), Rajendranagar, Hyderabad 500030, India
e-mail: sarap_k@rediffmail.com

T. K. Sarangi
Institute of Economic Growth (IEG), New Delhi, India
e-mail: sarangi.tapas@gmail.com

N. Ghosh et al. (eds.), *Nature, Economy and Society*,
DOI 10.1007/978-81-322-2404-4_11

This chapter is structured in the following manner: The first section discusses some analytical issues related to natural resource governance. The second section provides a discussion on the implementation of forest policies in the state that restricted the rights on forests and forestland of forest dwellers and alienated them from the process of decision-making related to the governance of forests in pre- and post-Independence periods. The third section analyses the nature and outcome of participatory forest management (PFM) as it works in the state of Odisha. The fourth section discusses the problems associated with the malfunctioning of forest institutions and is followed by a brief summary of the chapter.

11.2 Analytical Issues Related to Natural Resource Governance

The past century has experienced much degradation of forests due to increased biotic pressure in terms of increased population, technological revolution and unsustainable exploitation of natural resources. In India, as in most other developing countries, environmental degradation has manifested itself in rapid rates of natural capital depletion, exemplified by forest degradation and soil erosion. Nationalisation of forest and other natural resources without effective monitoring of traditional common property institutions has made the resources de facto open access resources and consequently faced the 'tragedy of commons'. Inefficient management of state-owned resources, market failure, increasing interdependence between the livelihood system and natural resources and the widespread concern for sustainable development have led to the evolution of alternative institutions for management of natural resources.

Decentralisation is an increasingly popular theme in the management of natural resources in recent times, all over the world. Countries have devolved and decentralised the use of resources and management system to the users. There has been effort to involve the local people and build participatory institutions to manage natural resources. In fact, a large body of case studies has demonstrated that local user groups can devise institutions to mange resources sustainably (Baland and Platteau 1996; and Ostrom 1990). A large number of countries are currently experimenting with some form of community resource management by transferring some of their power to the community for using and managing such resources. However, empirical outcomes of such devolution of resource use and management are mixed and the reasons for differences in performance of outcomes are not fully understood.

The crucial role played by institutions in the context of resource governance is increasingly being recognised in development studies. Institutions, viewed as a set of rules actually used (Ostrom et al. 1992) or 'rules of the game in society' (North 1990), are important transaction cost-minimising arrangements. In situations, such as those widespread in the developing tropics, where human beings and forests co-exist in an intricately intertwined web of interdependence, the sustainability of resource extraction largely depends upon the existence of, as well as adherence to, rules governing the common property resource (CPR) (Gibson et al. 2005).

As rule structures, community-based institutions minimise transaction costs because people themselves develop their rules suitable for particular situations. Realising the shortcomings of the traditional top–down state forest and biodiversity management, developing countries are increasingly embracing participatory approaches to natural resource management (NRM). The goal is to promote local people's active involvement in the management of protected areas and other natural resources (Kiss 1990). The same trend has manifested in India with the adoption of JFM, which aims to involve user groups on a large scale, marking an important shift in official forest policies. In parallel, the recent literature on CPR management emphasises the ability of user communities to effectively manage collectively owned natural resources through informal institutional arrangements (e.g. Wade 1988; Ostrom 1990). An example of this variety is found in the state of Odisha, where local communities have been protecting forests of their own (Sarap 2007).

With the introduction of JFM in 1990s, a dramatic shift took place in the approach of the government towards forest-dwelling communities. It changed the expectations as well as the relationship between the communities and the forest department. Much before the JFM became a programme of the government; however, community-initiated, collective action-based resource management had emerged sporadically throughout the country.[1] Studies in different parts of the country (Gadgil and Berkes 1991; Gadgil and Guha 1992; Gadgil and Chandran 1992; Ghate 2000, 2003, 2004; Pathak and Gour-Broome 2001) point to the existence of communities that were consciously maintaining and managing forests within their village boundaries, with or without tenure rights. Thus, local community participation in forest management and forest ownership has been increasing (White and Martin 2002).

It is essential to ensure rule compliance by community members as well as protection from poaching by outsiders in order to ensure effective management of degraded and dense forests. Monitoring is a necessary condition for the long-term sustainability of participating CPR regimes in order to guard against conditions that tempt individuals to cheat and gain benefits to the disadvantage of others (Ostrom 2000). When sanctions are strictly enforced they prevent the spread of free-riding behaviour, thereby instilling a sense of trust in the community. It is essential to provide conditions that facilitate a sense of justice and fair play in the participants, by ensuring that all individuals who break the rules will be sanctioned irrespective of their position in the community.

A significant fraction of local communities who are dependent on the forests have developed de facto arrangements for use and management of forests over long periods of time (Gadgil and Guha 1992; Gadgil and Chandran 1992). It is only recently, however, that participation of communities in forest management has received de jure acceptability. Participatory policies are now being considered relevant and consistent with India's overall development strategy of reducing poverty and protecting the environment. But this understanding has come after a prolonged experience

[1] See also Bhattachrya et al. 2010

of dwindling forest cover under centralised forest management (CSE 1982; World Bank 2006). Now we discuss the evolution of forest policies and governance and how these have alienated forest dwellers from the forests, which is their primary source of livelihood.

11.3 Implementation of Forest Policies in Odisha

11.3.1 Evolution of Forest Policies and Governance in India

11.3.1.1 Pre-Independence Period

Keeping in view the importance of natural resources and commercial significance of forest resource, certain regulations were formulated and implemented by colonial administrators to appropriate revenue from forest-based resources. The beginning of a forest policy in the pre-Independence India started in 1855 when the then Governor General, Lord Dalhousie, issued a memorandum on forest conservation restricting the customary rights of forest dwellers on the use of forest resources through a ban on their movement in the forest. Further, the 1865 Act empowered the government to declare authority on such resources for national interests. It was noticed that for all purposes the State seems to have played a dominant role over the right of the individuals and communities. Later, during 1878 the Indian Forest Act classified all forests of India into three categories: reserve forest, protected forest and village forest. The first ever forest policy came into existence in 1894. The primary objectives were to maintain adequate forest cover and to assume preservation of climate by emphasising on the physical conditions of the forest. Therefore, the policy regulated the rights and put restriction on privileges previously enjoyed by the local inhabitants. Since then, there has been a ban on shifting cultivation and protected the hill slopes resulting in conflicting situation for the forest dwellers with the forest department. The Indian Forest Act 1927 and Government of India Act 1935 consolidated the power of the Government on forest, emphasising on the revenue yield aspects and resource requirement of the British economy.

11.3.1.2 Post-Independence Period

The national forest policy formulated during 1952 mainly focused on forest as the source of timber, but neglected the village commons (see Appendix 1 for details). The State restricted the common people from having agricultural operations within forestland and also in the periphery areas of reserved forest. Free grazing of forests and free enjoyment of private forests were controlled, and tribal people were denied from practising shifting cultivation (Sarap and Sarangi 2010a). Due to the abolition of the *Zamindari system* in 1952, the Government of Odisha took over the management of forests and formulated a number of legislations to reduce freedom of the tribals over the use of forest and its resources. Apart from this, cultivation, hunting and fishing were also prohibited inside the reserved and protected forests. These measures increased the deprivation of people from forest resources, while assuming greater use of forest produce by the neighbouring communities. The emphasis was laid more on national interest, often, interpreted as commercial interest by reducing the access of forest dwellers to forest resources.

Subsequently, during 1976 the Government of India formulated the National Commission on Agriculture and social forestry was recommended for the creation of forest corporation to improve commercial feasibility. According to the recommendations, many conservation-oriented production forestry programmes were implemented.[2] More restrictions were made for entry of the tribals into the forests. In addition to this, the culture, tradition and ethos of the forest dwellers were also not given proper attention by the commission. Again, no special programmes were implemented for enhancing the economy of the tribals. Instead, programmes were essentially drawn for developing forest resources benefiting the tribals indirectly through wage earnings.[3]

Deprivation of tribals along with degradation of forests influenced policymakers to look forward to a new forest law. The Government of India enacted the Forest Conservation Act, 1980. It further restricted the rights of the state governments over forests. However, the law expanded the definition of 'non-forest purposes' that included the cultivation of cash crops like tea, coffee, spices, rubber plants, oil bearing plants, horticultural crops and medicinal plants. This law bill initiated a debate with respect to policies, legislations and also on the role of different stakeholders including activists, scientists, forest department contractors and industrialists. Consequently, it resulted in creating a forest department by separating it from the agriculture department and named it as Ministry of Environment and Forest (MoEF). The new department deals with forestry issues with a kind of pragmatic approach, so that the forest-related issues, both for the benefit of government and people, could be dealt with properly. Accordingly, various forest issues and related matters concerning people participation, forest revenue, deforestation, ecology etc. could be taken care of by this ministry as and when necessary.

11.3.1.3 Forest Management in the Context of Odisha

The Odisha Forest Act 1972 is based on the Indian Forest Act 1927. The formulation of this Act has been the first major attempt to bring uniformity in forest administration and management in the state. The objectives of this Act include revenue

---

[2] However the commission hardly focused on conservation- it said the only purpose of forests was to generate revenue and setup forest development corporations to cut down natural forests to replace them with commercial, fast growing plantations.

[3] There was less to do with tribals than with commercial exploitation of forests, including replacing mixed natural forests with fast growing mono-cultural plantations- 3 million dense forests were felled for the purpose.

maximisation and meeting industrial and commercial demands considering forest as a 'state property'. In a sense, the Act only formalised the process, which the state had been following since Independence. The rights and access of local communities to forest and forest products further got restricted with the enactment of policies such as Wildlife Protection Act (1972).

11.3.1.4 National Forest Policy, 1988: A Paradigm Shift

A wide discussion at national- and international-level forums suggested various ways and means to formulate a package of programmes to ensure sustainable forest development and ensure the livelihood of forest-dependent population. Similarly, there was a lot of criticism of many provisions of the Forest Conservation Act 1980. These provided inputs to the Government of India's National Forest Policy 1988 that modified a number of provisions of earlier Acts for the benefit of the poor. For the first time recognition of non-market and ecological benefits was emphasised in the Seventh Plan Document (1985–1990). It was made clear that raw materials for forest-based industries would be provided only after meeting the needs of the local people. The Central Board of Forestry recommended a ban on commercial exploitation of degraded forests and regeneration of national forest in order to reduce the growing pressure on forest resources.

Thus, the new forest policy seems to have been planned for protection, conservation and management of the forest and its resources. It also honoured the customary rights of the people; replaced the contractors with tribal co-operatives, co-operative government undertakings and corporations. It suggested suitable alternatives for shifting cultivators such as engagement of these people in forest-based industries. With the adoption of the National Forest Policy 1988, the colonial forest policy establishing straight control over forests by the forest department was relatively weakened in comparison to earlier years.

11.3.2 Changes in the Policy and Legal Framework: Governance Issues

The policy and legal regime in the forestry sector will keep focus on poverty alleviation through forestry, increasing productivity, enabling environment for private sector to grow more trees, ecological security of the nation, empowerment of communities along with their capacity building and biodiversity conservation in 2000.

The ecological security became the prime objective and was given importance for providing livelihood to the forest-dependent communities in the country. JFM mode was chosen in the country including in the state of Odisha by the State to encourage the support of the forest dwellers along with the forest department for growth and conservation of forest. There was rapid growth of JFM throughout the country.[4] The Environment Protection Act was enacted in 1986 for improving the environment in the country. There has been some progress during the past two decades for enhancing contribution of forests towards poverty alleviation through empowering people with the ownership of NTFP, participation in decision-making relating to use of forest. The traditional lifestyles of tribes and their recorded rights have been respected and embedded in the forest management practices as well as in subsequent policies to some extent.

11.3.3 Implementation of PESA

The provisions of the Panchayats Extension to the Scheduled Areas (PESA) 1996 Act was passed by the Indian Parliament to extend the provisions of the 73rd Constitutional Amendment 1993 to the Schedule fifth areas of the country. The PESA Act specifies that state governments will endow *panchayats* in the scheduled areas with such powers and authorities as considered necessary to enable them to function as institutions of self-governance.[5]

The PESA inter alia empowers the *Gram Sabhas* and *Gram Panchayats* in scheduled areas to safeguard and preserve the traditions and customs of the people, their cultural identity, community resources and customary methods of dispute resolution and more specifically to provide for endowing *Panchayats* with appropriate ownership of minor forest produce. By and large, the PESA provides a lot of useful guidelines and directives to the state government for formulation of appropriate Forest Laws and Acts for the larger benefit of the forest dwellers living in and around the forests. Decentralised management of resources by the people would not only provide incentive to them for the development and management of resources, but also supposed to solve some of the problems faced by them through participation in the decision-making.

Following the Central guidelines, Odisha has amended its *Panchayati Raj* Act in December 1997. However, practically, the state government has not transferred the power as per the provision of the Central Act, 1996. The PESA Act devolved power to the local self-government to preserve, protect and manage forest resources

[4] This programme has covered more than 22 million ha of forests with the involvement of approximately 21 million people by the end of 1990s. In case of Odisha it has cover about 14 % of the total forest area involving nearly 17 million families by the end of 2010–2011.

[5] Under this provision of the Constitution of India, the Governor is empowered to administer in the Fifth Schedule Area and also he can make, repeal or amend any act of Parliament or of the state legislature or any existing law, if he thinks them to be detrimental to the interests of the tribals. The Governor may make regulations for the good government in the Scheduled Area, he can prohibit or restrict the transfer of land of the tribal people and regulate the carrying on of business of money lending in this area.

in regards to traditional rights of forest-dwelling communities. The Act also empowered the *Gram Sabha* to preserve and protect the traditional rights and ownership of minor forest produce at the village level. But it is to be noted that the Odisha state Act does not give any power to *Gram Sabha* on matters related to land acquisition, minor minerals, planning and management of minor water bodies. Instead it has entrusted all these power to *Zilla Parisad*, which is not required to consult *Gram Sabha* while exercising all these power. Over the years it has been found that the *panchayat,* because of adhering to a particular mode of functioning, could not grow as units of self-governance, as per the provisions of the PESA Act.

In the management of minor forest products, there have been various shortcomings in the state Act as well as in the functioning of *Gram Panchayat*. In Odisha the *Gram Panchyats* are now empowered to register the traders in their territorial jurisdiction for trading of 68 items. However, they have not been legally empowered to take any penal measures against traders who do not pay fair prices (prices fixed at the regional level) to primary collectors. In case they want to take penal action they have to approach the divisional forest officer for further action.

Further, contradiction has emerged between the Odisha Forest Policy and the PESA Act. The Odisha Forest Act 1972 Section 77 invests the powers with the divisional forest officers to enter upon any land to survey, demarcate and prepare a map of the same, power to hold an inquiry into forest offences and in the course of such inquiry to receive and record evidence. The criminal procedure authority has been lying under the jurisdiction of forest officials (Orissa Forest Manual 2005). Taking the advantage of the 1972 Act, the local level forest officials assert their power to control forests and forest products. As a result the traditional rights of the tribal people over forests have been eroded.

The *Panchyati Raj* institutions in the present forms have neither any capacity to control the traders, who misuse their power, nor have funds to buy MFPs from the primary collectors. Further, they also do not have any capacity to store the procured products and to sell the same to buyers at reasonable prices.[6]

11.4 Nature and Outcome of Participatory Forest Management in Odisha

11.4.1 Participatory Forest Management (PFM) in Odisha

This section provides a discussion on the development of PFM in the state of Odisha. It is noteworthy that the self-initiated forest protection committees took the initiative well before the forest department, and it has spread widely across the state since the 1960s. The forest department took the initiative in the formation of *Vana Samrakhyan Samitee (VSS)* during 1993 and later.

[6] However, in the administrative perspective, no effective coordination was found between PRIs and Forestry institution over controlling forest resources.

The PFM approach, in its present form, that is, JFM came as an outcome of the National Forest Policy in the year 1993 in Odisha. Theoretically, the JFM resolution looked upon local communities as equal partners with forest department for protection and management of forests.

From 1993 to 1997 the process of formation of JFM was slow but later it picked up. As a result, 11,915 *Vana Samrakhyan Samities* (VSSs) and 398 Eco-Development Committees (EDC) had been formed by the end of 2010–2011. These VSSs have been managing around 14 % of total forest area in the state and nearly 17 million families were involved in this programme of which around 0.7 million families belonged to tribal communities (Government of Odisha 2011–2012). Similarly, it has been estimated that about 10,000 self-initiated forest protection committees (FPCs) have been functioning in the state. These groups are also protecting the forest, of course, without the support of the forest department. But it has been found that many of the JFM and CFM Committees are dormant in the sense that the committees have been formed, but they are not active in activities relating to protection or management of forests.

About 29,000 villages have forest as recorded land. The number of FPCs, both, JFM and self-initiated forest protection groups (SIFPGs) constitute more than 20,000 villages. In any case some of the committees of both JFM and CFM are dormant. In any case more than three fourths of the villages with forest have some sort of FPCs to look after the forest.

11.4.1.1 Forest Development Agency (FDA)

With the introduction of the FDA scheme during the year 2002–2003, a renewed drive for formation of VSS was started by the forest department in the state. The scheme was launched in 2002–2003 by MoEF to implement the National Afforestation Programme (NAP). The NAP was introduced during the Tenth Five-Year Plan and has been formulated by merging four centrally sponsored afforestation schemes of MoEF.[7] The National Afforestation and Eco-Development Board (NAEB) of the MoEF is in-charge of operation of this 100 % centrally sponsored scheme and its tenure has been kept till the end of the 10th plan period.

11.4.1.2 Progress Under FDA in Odisha

FDA was formulated for the regeneration of the depleted forest. However, the objective of FDA formation has limited success. Up to 2006–2007, 1565 VSSs constituting 16 % of the total VSS (9776) have been covered under FDA. These VSSs

[7] i.e., Integrated Afforestation and Eco-Development Projects Scheme (IAEPS), Area Oriented Fuel Wood and Fodder Projects scheme (AOFPS), Development of Non-Timber Forest Produce including Medicinal Plants Scheme and Association of Scheduled Tribes and Rural poor in regeneration of Degraded Forests (ASTRP).

are protecting 55,000 ha of forest area, which constituted hardly 6.15 % of area protected by VSSs in the state during this period. Clearly, very few VSSs have been included under FDA and the selection is often based on the subjective judgment of forest department officials. The structure of general body and executive body is designed to favour the government officials, who control the decision-making of the FDA. Active participation by VSS representatives is rarely found. Major decisions are taken by the forest department officials.

Though some developmental works have been undertaken in the entry point phase of NAP, transparency has not been maintained in the use of funds. Further activities are selected mainly by the forester and the president. People participation in the micro planning is found to be marginal. People are used as labourers in nurseries and other forestry activities. The performance of FDA has not been satisfactory and participatory (see Sarap 2007; Sarap and Springate-Baginski 2013).

Though FDA has provided a few days of employment to the poorer sections of the society, it has left out communities depending on timber, fuelwood, bamboo and charcoal trade for their livelihood. National Afforestation Programme has left problems like encroachments, salinity, desertification, vulnerability of members, etc. In some VSSs, plantation has been done without a micro plan. Cooperation between the villagers and the forest officials has not been encouraging (RCDC 2004). Clearly, the working of FDA has discriminated against many weaker communities in its operations and the decision-making process has become centralised.

11.4.2 Odisha Forestry Sector Development Project (OFSDP)

The Odisha Forestry Sector Development Project (OFSDP) aims at promoting sustainable forest management in the state with the larger goal of supporting rural livelihoods. Supported by the Japan International Cooperation Agency (JICA), OFSDP is a 7-year project (2006–2013) and is executed by an autonomous society under the Forest and Environment Department, Government of Odisha. The project is being implemented at selected project villages in 14 forest divisions of the state, through active participation of the village community organised as *Vana Sarakshana Samities* (VSSs) or Eco-Development Committee (EDC). The project targets to cover 2275 VSSs/EDCs in the forest fringe villages in the targeted forest divisions. Local-level evidence shows that working of this programme is similar to that of the JFM programme working in the state and beset with several problems relating to management of FPCs.

11.4.2.1 Emergence of Forest Right Acts 2006: A New Hope for the Tribals and Forest Dwellers

The Central Government passed the Forest Right Act in 2006. It provides the legislative basis to redress historical injustice meted out to forest dwellers in the country for long years and thus has major implications for them, both in promising a more secured form of livelihoods as well as the legal provisions necessary to defend them in the future. This law recognises the rights of occupation of forests by tribes and other forest dwellers and empowers them for management of forests used by them as CPRs. It conforms to the policy prescription of PFM, the accepted principles of biodiversity conservation. The land title will be given to those forest dwellers that are under the possession of the forest land till December 2005. It is to be noted that the vested forest rights are heritable, but not alienable or transferable (Government of India 2006).

By the end of March 2013 about 300,000 of individual land titles (including hardly 879 community forest titles) constituting 5.39 lakh acres of forest land have been given to the claimants (Government of India 2013). The average area of land given per claimant was 1.62 acre and it was 64.25 acre in case of community forest title.[8] Effective and comprehensive implementation of the FRA will have significant impact on the livelihood of forest dwellers and conservation of forest (Sarap et. al. 2013). It will reduce the tenure insecurity and provide a number of benefits under the anti-poverty programmes. It will also increase the bargaining power of the tribals vis-à-vis the others in the decision-making process leading to better governance of forests.

11.5 Problems Associated with Malfunctioning of Forest Institutions

11.5.1 Problems in the Participatory Forest Management

This section discusses the problems associated with the PFM institutions. Though both the CFM & JFM institutions serve the same purpose of involving the local communities and their concerns in the management of forest, there are several cases of conflict between these institutions. At the administrative level JFM is patronised by the forest department where as CFM initiatives do not get the official recognition from any government authority and are often treated as illegal. There are several instances of conflicts between the forest department and CFM groups in many forest divisions of the state. This is mostly due to difference in the approach and perspective of the stakeholders. It has been observed that CFM institutions are characterised

[8] At the all India level (as on 31st March 2013) 32,45,369 number of claims have been filed. Out of this 12,81,926 titles (39.5 %) have been issued to the claimants.

with higher level of participation as compared to JFM. But JFM institutions have certain advantages due to the patronisation of the government (ibid). In view of this, many self-initiated CFM groups have converted them into JFM to avail the benefits given under the JFM programme and to seek the government recognition (Sarap 2005).

Even though local forest institutions have been working in the state for the past several years, they are besieged with a number of problems in their functioning. The PFM in the state of Odisha has been a haphazard affair reflecting the weakness of the Odisha Forest Department as well as forest protection committees as an institution. There has been lack of proper participatory process, either at the outset or post-formation; local people's participation in the preparation of the 'micro-plan' is generally marginal, as the forester exerts major control over this. In VSS executive committee and general body meetings, important decisions are being taken by elites, including the forester (who is the secretary). Self-initiated groups were found to be relatively more participatory than the VSS, although few women are involved in either types of management. Further, women have little power in decision-making (see Sarap 2007).

The study of Sarap and Springate-Baginski (2013) reveals that the organisational structure of JFM has several types of lacuna. These include exclusion of poorer members including women from the decision-making process, resulting in weak bargaining power of these members vis-à-vis the elites in the FPCs. It has lead to the capture of elites in the governance of the organisation. This has also led to weakening of FPCs in elimination of several types of conflicts faced by them.

11.5.1.1 Organisational Set-up

There are many reasons for the failure of JFM, the foremost being the emphasis on a formal and uniform organisational structure. The JFM framework prescribes for the constitution of a committee termed as VSSs with defined membership. The recently enacted JFM resolution 2000 by the Government of India talks about facilitating a uniform structure for JFM committees i.e. Society in all the states and registration of all JFM committees under the Society Registration Act, 1860. This is in contrast to diverse institutions and organisational arrangements under CFM, which undergo changes in response to internal dynamics, local situations and context. Though their organisational structures differ, they are essentially democratic bodies reflecting the ground realities of the area. On the other hand, appointment of local forester in the position of secretary replacing the natural leadership virtually puts the power in the hands of forest officials in JFM. Since the forester had responsibilities of a number of committees at the same time, he failed to perform the duty of a functional leader and is unable to give adequate time to the affairs of the committee.

11.5.1.2 Benefit Sharing

The 1993 resolution of JFM by Government of Odisha provides for 50 % share in major or final harvest and a 100 % share of intermediate produce to the VSS members. There is a feeling among the community that 50 % of the benefits are taken away from it. As such there is problem of incentive for hard work to the members of PFM. The Non-timber Forest Produce (NTFP) policy in the state is regressive in comparison with policies of other neighbouring states. Up to late 1990s most of the marketable items (28 items) were leased out to private traders, alias a joint sector company. Thus, even on supposedly jointly managed forest land, the co-managers are treated as mere labourers who are to gather NTFPs and handover to state-appointed agents at the prices fixed by the state (Sarap and Sarangi 2009, 2010b).

11.5.1.3 Financial Transparency

There is no robust organisation structure among the forest protection committees both JFM and CFM for management of finance is concerned, even though the scale of finance available to the CFM is low as compared to the JFM. Evidence shows that there is lack of financial transparency in most of the FPC. Complaints of misutilisation of funds by president/secretary in some of the JFM/CFM studied villages were reported by the members of the FPCs. Further, there is no regular external auditing of financial account, in especially the CFM and in the case of JFM secretary (forester) is doing the audit (see Sarap and Springate-Baginski 2013).

11.5.1.4 Issue of Tenure Security

Under the existing JFM framework, villagers have hardly any secure rights over forest. As such the incentive to the members to ensure the growth of forest products on a long-term basis is minimal. The forest department has been the senior partner in the control of forest and powers over forests as well as the systems of management. Provision of community forest rights over the forest would likely change the power structure of the forest dwellers in future. But this is yet to happen in the state.

11.5.1.5 Inter and Intra Conflicts

Much of the conflicts around PFM were concerned with the local institutional level and the nature and functioning of institutions (Sundar 2012). Conflicts of various natures, including intra-village and inter-village are found in the forest area. For instance, Sarap (2007) study revealed several types of conflicts present in the study area. These include sharing benefits, usufruct rights, illegal felling, forest boundaries and with forest mafias. Mining, mostly located in forest areas, has led to conflict between forest-fringe communities and mining leaseholders. The VSS or the CFM institutions are unable to resolve such conflicts and when the conflicts are resolved, it is temporary in nature and occurs again after a point of time (see Sarap 2005). The recent industrialisation policies in the state, which are mostly based on mining, and generally located in forest areas, have aggravated the conflicts between the forest-dependent communities and mine contractors significantly due to displacement of local people from the forest area. As the traditional livelihood options of local people are affected because of this policy the conflict is becoming recurrent (see Mishra 1998; Sundar 2012)[9].

Further, there is no coordination between *Panchayati Raj Institution*s (PRIs) and other local institutions in the states. This has weakened the conflict management in forest fringe villages (World Bank 2006)

11.5.1.6 Equity Issues

Equity in the distribution of benefits from the PFM between different sections of participant households within a community is another important issue that is likely to affect household participation in the PFM activities as well as the sustainability of PFM institutions. It has been found that poorer members in the group, whether in JFM or CFM, are unable to realise fully the benefits accruing from the forest. The participation of women is also weak. Such situations lead to exclusion of many poor from benefit sharing and forest management efforts.[10]

These forest management systems were meant to include and empower the community, but the nature of empowerment remained very limited (Saxena 2003). Joint forest management has been working in the state since the past two decades, but the progress in terms of institutional development as well as impact on the livelihood of local people is marginal.[11] Furthermore, JFM has been used as a strategy to co-opt CFM and to enable the forest department to establish and expand its control over the forest areas, which are under 'de-facto-control' of local communities.

Local communities find the VSS institution uncomfortable since it tends to the erosion of the decision-making authority at the community level. It also disregards the traditional knowledge system of the community and instead has introduced a situation where the forest department plays an important role in decision-making related to forests.

[9] As per the JFM plan the VSS, through its executive committee, is to execute an MoU with the concerned Divisional Forest Officer (DFO) for protecting, regenerating and management of forest area, VSSs constituted prior to 1993 have not been registered as VSS in full. Further, VSS institutions, in the absence of legal authority, have failed to resolve many interpersonal conflicts prevalent in the groups. In such situations VSS committee find it difficult to take strong action against erring individuals or the state for non- compliance. It also becomes difficult to ensure equitable distribution of usufructs given the prevailing socio economic inequalities in the rural communities.

[10] See also Baland and Platteau 1999, pp. 774.

[11] A study conducted by Bhattarchrya et al. 2010 based on a preliminary assessment of the status of JFM in the forest tract of central and central-eastern part of India including Odisha found that due to lack of or inappropriate conflict resolution mechanism, fragile institutional arrangement, inadequate people's participation, inefficient accountability mechanism, and poor collaboration between the state forest department and the communities, the program has suffered.

Clearly, the functioning of PFM in the state is not satisfactory. As a result these institutions are unable to perform their role properly in the distribution of benefits to the members and in resolving the conflicts present in the forest fringe villages.

11.6 Summary

On the whole, it is clear that a number of factors including clear tenurial rights and provisions of incentives to the people are dependent on natural resources and protecting them is important for the proper functioning and governance of forest institutions. There have been systematic efforts on the part of the state, both in pre- and post-Independence period, to curtail the participation of forest dwellers in the process of decision-making in the forest institutions, thereby alienating them from not only conservation efforts, but also on livelihood supports.

The PFM through state efforts has led to some improvement in the forest dwellers' participation in the process of governance. But the forest institutions are characterised by several types of problems that constrain effective participation of weaker members. This affects equitable distribution of benefits of forest products. Proper functioning of PFM and effective implementation of FRA in its wider perspective and especially through community forest rights will go a long way in providing incentives to the forest dwellers for proper governance of forest institutions and distribution of benefits.

11.7 Appendix 1 (Table 11.1)

Table 11.1 Phases of forest governance during the post-Independence period

Phases	Major policies	Highlight points
Phase- I 1947–1970	National Forest Policy (1952)	Commercial exploitation of forest for industrial development
Phase- II 1971–1988	National Commission on Agriculture (1976); Forest Conservation Act (1980)	Conservation through powerful legislation such as Wildlife Conservation Act & Forest Conservation Act
		No place for forest dwellers and tribals in protection and management of local forest resources
Phase- III 1988 onwards	National Forest Policy (1988)	Based on three major components:
		Emphasis on participation of forest dwellers
		Increasing access to forest products
		Enhancing livelihoods

References

Baland JM, Platteau JP (1996) Halting degradation of national resources: is there a role for rural communities? Clarendon Press, Oxford

Baland JM, Platteau JP (1999) The ambiguous impact of inequality on local resource management. World Dev 27(5):773–788

Bhattachrya P, Pradhan L, Yadav G (2010) Joint forest management in India: experience of two decades. Resour Conserv Recycl 54(2010):469–480

CSE (1982) The first citizen report. Centre for Science and Environment. New Delhi

Gadgil M, Berkes F (1991) Traditional resource management systems. Resour Manag Optim 8(3/4):127–141

Gadgil M, Chandran MDS (1992) Sacred groves. India Int Cent Q 19(1/2):183–187

Gadgil M, Guha R (1992) This fissured land: an ecological history of India. Oxford University Press, New Delhi

Ghate R (2000) The role of autonomy in self-organising process: a case study of local forest management in India, Working Paper. Workshop in Political Theory and Policy Analysis, Indiana University, Bloomington.

Ghate R (2003) Ensuring collective action in 'Participatory' forest management, Working Paper No. 3. South Asian Network for Development and Environmental Economics, Nepal.

Ghate R (2004) Uncommons in the commons: community initiated forest resource management. Concept Publishing Company, New Delhi

Gibson C, Ostrom E, Williams JT (2005) Local enforcement and better forests. World Dev 33(2):273–284

Government of India (2006) Scheduled tribes and other traditional forest dwellers: recognition of forest rights act, 2006

Government of India (2013) Status report on implementation of the scheduled tribes and other traditional forest dwellers. (Recognition of Forest Rights) Act, 2006 [for the period ending 31st March 2013], Ministry of Tribal Affairs, New Delhi

Government of Odisha (2011) Economic Survey of Orissa, 2011–2012. Bhubaneswar

Kiss A (1990) Living with wildlife: wildlife resource management with local participation in Africa. Washington, DC: World Bank Technical Paper 130

Mishra R (1998) Capacity building for participatory management of degraded forest in Orissa, India. Sandia Consult for Nature, AB/Asia Forest Network, November

North DC (1990) Institutions, institutional changes and economic performance. Cambridge University Press, Cambridge

Orissa Forest Manual (2005) Orissa Law Review. Cuttack.

Ostrom E (1990) Governing the commons: the evolution of institutions for collective action. Cambridge University Press, New York

Ostrom E (2000) Collective action and evolution of social norms. J Econ Perspect 14(3):137–158

Ostrom E, Walker J, Gardner R (1992) Covenants with and without a sword: self-governance is possible. Am Polit Sci Rev 89:404–417

Pathak N, Gour-Broome V (2001) Tribal Self-Rule and Natural Resource Management. Kalpavriksha, Pune

RCDC (2004) Community forestry 3(4), May Issue

Sarap K (2005) The forest-livelihood context in Orissa. Phase-II, Project Report submitted to Centre for Economic and Social Studies (CESS), Hyderabad

Sarap K (2007) Forest and livelihood in Orissa. In: Springate-Baginski OS, Blaikie P (eds) Forests, people and power- the political ecology of reform in South Asia. Earthscan, London

Sarap K, Springate-Baginski O (2013) Participatory forest management in Odisha- implementation, outcomes and livelihood impacts. Bookwell, New Delhi

Sarap K, Sarangi TK (2009) Malfunctioning of forest institutions in Orissa. Econ Polit Wkly 44(37):18–22

Sarap K, Sarangi TK (2010a) Factors contributing to erosion of access to forest and private land in contemporary tribal Orissa. In: Sarap K (ed) Institutions, livelihood and natural resource management. Bookwell, New Delhi

Sarap K, Sarangi TK (2010b) An analysis of working of forest institutions in Orissa. J Soc Econ Dev 12(2):193–210

Sarap K, Sarangi TK, Naik J (2013) Implementation of Forest Rights Act 2006 in Odisha: process, constraints and outcome. Econ Polit Wkly (September 7) XLVIII(36):61–67

Saxena NC (2003) Livelihood diversification and non-timber forest products in Orissa: wider lessons on the scope for policy change? Working Paper No. 223, London, UK. Overseas Development Institute.

Sunder N (2012) Violent social conflicts in Indian's forests: society, state and the market. In Rights and resource initiatives. Deeper roots of historical injustice: trends and challenges in the forests of India. Washington, DC.

Wade R (1988) Village republics: economic conditions for collective action in South India. Cambridge University Press, Cambridge

White A, Martin A (2002) Who owns the world's forests? Forest trends, Washington, DC.

World Bank (2006) India unlocking opportunities for forest dependent people. Agriculture and Rural Development Sector Unit South Asia Region. Oxford University Press. New Delhi.

Chapter 12
Multifunctional Benefits of Community-Based Mangrove Restoration in Gujarat: An Analysis

P. K. Viswanathan

12.1 Introduction

The importance of development and restoration as well as the multifunctionality of mangroves has received greater attention in recent years[1] especially in the context of the increasing risks of climate change-induced natural hazards. Mangroves are salt tolerant plant species and provide a wide range of ecological and economic functions and services and support a variety of living organisms and other coastal and marine ecosystems. Mangroves possess several characteristics that make them structurally and functionally unique (Alongi 2002) and mangrove forests have traditionally been utilized by local people for a variety of purposes (Choudhury 1997).

However, it is reported that 35 % of the world's mangroves have been lost in the past two decades alone due to various pressures, including natural disasters, anthropogenic interventions, including overharvesting, aquaculture as well as coastal development activities[2] (Macintosh and Ashton 2003). Undervaluation of natural products and ecological services supported by mangrove ecosystems has also been a major driving force behind mangrove destruction. In the Far East, mangroves

---

[1] The environmental and ecosystem-related benefits of mangroves and their conservation needs have been widely demonstrated in recent years (e.g. Othman 1994; Nagelkerken et al. 2000; Kathiresan and Rajendran 2005), as well as their direct and indirect contributions to the livelihoods of millions of coastal communities the world over (Sathirathai and Barbier 2001; Soontornwong 2006).

[2] Mangroves are seriously threatened ecosystems (Valiela et al. 2001), with threats coming from coastal development schemes/programmes, conversion to aquaculture, overharvesting of trees, pollution and global climate change-induced risks (Adeel and Pomeroy 2002; Alongi 2002; Macintosh and Ashton 2003).

P. K. Viswanathan (✉)
Gujarat Institute of Development Research, Ahmedabad, India
e-mail: pkviswam@gmail.com

N. Ghosh et al. (eds.), *Nature, Economy and Society*,
DOI 10.1007/978-81-322-2404-4_12

have been extensively damaged for firewood and charcoal, used in the construction of dwellings, furniture, boats and fishing gear (Ronnback 1999; Ronnback and Primavera 2000).

At the same time, there has been a realization of the importance of protection and conservation of the coastal ecosystems the world over[3] (Field 1996; Kairo et al. 2001) with greater efforts across countries and regions for the development and restoration of mangroves involving international and national development agencies as well as local communities.

A vast but growing empirical research brings out the historical significance of mangrove restoration efforts and a large number of these studies have used interdisciplinary techno-economic, social and biological perspectives in understanding the contemporary relevance of mangrove restoration activities. For instance, the environmental and ecosystem-related benefits of mangroves and their conservation needs have been widely demonstrated by some researchers (e.g. Othman 1994; Nagelkerken et al. 2000; Kathiresan and Rajendran 2005), whereas others (Sathirathai and Barbier 2001; Soontornwong 2006) have highlighted the direct and indirect contribution of mangroves to a vast number of coastal communities the world over. It has been recognized for some time that community-based coastal management shows promise. For instance, in the Philippines, the community-based coastal management projects have existed for more than 20 years (Pomeroy and Carlos 1997; Katon et al. 2000; Walters 2003; Walters et al. 2008).

The issue: "How can community management of mangroves be assessed for conservation outcomes?" has been a major concern for researchers at least for some time now. Quantitative surveys may report biological outcomes of mangrove conservation or the lack of it. But when the resource is managed by local communities, it becomes all the more important to: (a) gather information on the socioeconomic interface with mangrove conservation efforts; (b) explore the contextual factors associated with the varied outcomes of interaction between mangroves and local communities; and (c) understand the interface between conservation/restoration efforts made by the state and other agencies and the local communities. In this regard, though researchers have tried to understand the interface with respect to forests and joint-forest management systems (Kijtewachakul et al. 2004; Gautam and Shivakoti 2005), there are hardly such studies on mangroves and their interface with coastal communities and institutional and governance systems. Hence, it is important to explore how mangrove conservation strategies lead to successful socioeconomic, environmental and biological outcomes, while accounting for the needs of coastal communities who depend on this resource for their survival.

[3] To a large extent, this realization was based on several reasons, viz., (a) the coastal zone is a dynamically unstable system where natural disasters of one or the other kind like sea intrusions, cyclones, tsunami, etc., strike intermittently; (b) as global warming causes ice melt-induced rise in water in oceans and surge in tidal waves, mangrove prop roots help arresting them.

This chapter assumes relevance in this context and tries to understand how the local communities consider the importance and benefit from the mangrove restoration activities being carried out by the state in recent years in several villages in Gujarat. The chapter has four sections, including the Introduction. Section 2 provides a brief summary of the global scenario of development and restoration of mangrove ecosystems, followed by a review of empirical studies on the multiple benefits of mangroves in the Indian context. Section 3 presents empirical results on the multifunctionality of mangrove ecosystems based on the community-based mangrove restoration (CBMR) underway in Gujarat. Section 4 concludes the chapter by suggesting the need for long-term policy and institutional interventions for protecting and further expanding mangroves as important sources of biodiversity conservation.

12.2 Multifunctionality of Mangroves: Global and Indian Contexts

There was a time when mangroves were considered as mere swamps. But mangroves have become quite important in the current context as they are productive and protective systems with multiple socioeconomic, environmental and ecosystem functions and benefits. Mangrove forests filter out pollution, stabilize sediments, hold nutrients, protect the shoreline from erosion and provide food, nesting and nursery areas for many animals, including a large number of species of fish, reptiles and amphibians, mammals and birds.

An earlier inventory of the mangrove plantations indicate that there were slightly more than 18 million ha mangrove plantations inhabiting 112 countries and territories in tropical and subtropical regions (Spalding 1997). More recent estimates of the global mangroves area ranges from 110,000 to 240,000 km^2 (FAO 2007). Around 34 major and 20 minor mangrove species belonging to about 20 genera in over 11 families have been recorded globally (Tomlinson 1986). Mangroves of South and Southeast Asia form the world's most extensive and diverse mangrove systems covering 41 % of the global mangroves and the region has the greatest number of species of crustaceans (229 species) and fish (283 species) associated with mangroves (Saenger et al. 1983; Saenger 2002). Indian mangroves barely make up 4 % of the global mangroves and are distributed along all the maritime states, covering an area of about 4461 km^2 along the 7500-km-long Indian coastline. The floral diversity of mangroves of India comprise 38 core mangrove species (Kathiresan 2003a).

Since 2005, a major initiative for the development and restoration of mangrove plantations has received worldwide attention, following the devastating Asian tsunami that struck the Indian Ocean on 26 December 2004.[4] The catastrophe led to greater realization that the presence of mangrove cover might have mitigated a significant scale of the economic damages and loss of lives and properties caused by such natural disasters(EJF 2005; Kathiresan and Rajendran 2005; UNEP 2005; Vermaat and Thampanya 2006; Dahdouh-Guebas et al. 2005). Thus, there have been enormous efforts for the expansion of mangroves in the Indian Ocean region, in particular, supported by national governments, international agencies and NGOs. These efforts were attempts to replant and rehabilitate mangrove ecosystems as bio-shields or "natural barriers" to future tsunamis and other tropical storms (Barbier 2008).

12.2.1 Socioeconomic and Ecological Significance of Mangroves

As observed, mangroves entail protective and productive values, as they provide numerous tangible and intangible benefits to the coastal communities (Vannucci 2004). They also help reducing the impact of tropical cyclones and tidal surges, which frequently occur along the coasts of South Asian countries (Kathiresan 2003b, c; Das 2007, 2009; Kathiresan and Rajendran 2005). The role of mangroves in reducing the sea wave energy is also documented. For instance, during the cyclone that struck the southwest part of Bangladesh in 2007, people have observed the protective role of the Sundarbans. The mangroves also provide necessary nutrients and habitats for a range of animal species. The rich fishery resources along the coasts of these countries can be attributed to the presence of mangroves to some extent (Amarasinghe et al. 2002; Islam and Haque 2004; Islam and Wahab 2005).

Reportedly, millions in South Asia earn their livelihoods by extracting mangrove resources and working in industries that use mangrove as raw material. It has been estimated that in Bangladesh and India, around 9 million people depend on the Sundarbans for their livelihoods (PDO-ICZMP 2004; GOI 2005). In the impact zone of the Sundarbans in Bangladesh, around 18 % of the households are directly dependent on the forest (Iftekhar and Islam 2004; Awal 2010). While around 200,000 fishermen operate daily in the Sundarbans water, another 225,000 of the population, i.e. about 14 % of the people residing within 10 km buffer around the Sundarbans, depend on collection of *Penaeus monodon* fry (Hoq 2007).

By and large, the assessment of socioeconomic as well as the ecological significance of mangroves in country/region-specific contexts has been undertaken by scholars using the total economic valuation (TEV) framework as developed by Barbier[5] (1993, 1994). For instance, it was estimated that in Sri Lanka, the total value

[4] Since the 2004 Indian Ocean tsunami, there has been a general increase in the awareness of the importance of mangrove ecosystems; efforts to conserve, protect and restore them can be seen in Bangladesh, India, Indonesia, Myanmar, Seychelles, Sri Lanka, Pakistan, Thailand and Vietnam (Macintosh et al. 2012).

[5] The concept of total economic value (TEV) captures the socioeconomic and environmental functions of tropical wetlands with an emphasis on their regulatory ecological functions in support or protection of economic activities. TEV is composed of use (UV) and nonuse values (NUV).

of goods and services of mangrove forests was US$ 12,229/ha/annum. Similarly, for India, Badola and Hussain (2003) have estimated values for different functions of Bhitarkanika mangrove, such as nutrient retention US$ 865/ha/annum, offshore fishery US$ 37.97/ha, inshore fishery US$ 1.9/ha, fry collection US$ 0.2/ha, and storm abatement US$ 116.28/household (Iftekhar 2008).

Suthawan (1998) estimated the TEV of mangroves in Thailand. Using the techniques of valuation, it assessed the foregone benefits of mangroves compared to the net returns from converting the areas into shrimp farms, which is the major competitor for mangroves. It was found that although shrimp farming creates enormous private benefits for those engaged in it, the net social benefits, considering the externalities in terms of mangrove destruction and water pollution, are not so economically viable. This is especially true when the mangroves in focus are located along the coasts and serve as a nursery for small fish and marine life. In the Philippines, initial costs are estimated to be US$ 204/ha (Walton et al. 2006). On the other hand, mangrove restoration cost estimates for the USA ranged between US$ 225 and US$ 216,000/ha (Lewis 2005). These costs thus vary very widely depending on various cost components (Brander et al. 2006), site conditions and the labour required for hydrological restoration and removal of debris and weeds, and planting material types where replanting is undertaken.

Nevertheless, Barbier's conceptual framework has limitations as it does not provide a holistic perspective of the multifunctionality of mangrove ecosystems. It provides only a partial assessment of the socioeconomic, environmental and ecosystem services of mangroves. Also, there are significant variations across countries in terms of valuation of the direct and indirect benefits realized from mangroves. These variations in the values may be due to several reasons, including the geographical/locational characteristics of the mangroves and the communities' perceptions of the benefits as well as access to various use and nonuse benefits of mangroves. Yet, despite these shortcomings, the valuation studies available based on the TEV framework help raising the awareness about the multiple roles and values of ecosystem services for human wellbeing. An important study by the Conservation International (2008) on coral reefs, mangroves and sea grass, compiles the results of wide-ranging economic valuation studies on coral reef and related ecosystems around the world, with a focus on the five important ecosystem functions and benefits, viz., (a) tourism benefits; (b) fisheries; (c) coastal protection; (d) biodiversity benefits; and (e) carbon sequestration potential.

Use value (UV) is classified into three, viz., direct use value (DUV), indirect use value (IUV) and option values (OV). Similarly, nonuse value (NUV) is classified as existence value (EV) or bequest (legacy) value. The indirect use value includes support for fisheries, storm protection, etc. Although OV is defined as the value of increased information in the future, EV is considered in terms of the benefit value of resources for future generations. Using Barbier's TEV framework, many scholars have tried assessing the multiple use and nonuse values of mangroves in different country contexts. A summary of the various studies on the economic valuation of mangroves is presented in Viswanathan et al. (2011).

12.2.2 Mangrove Restoration—Role of Local Communities

Assessment of local peoples' attitudes towards conservation and participation in such activities has been one of the least explored aspects of mangrove restoration science (Kovacs 2000; Glaser 2003). There are several reasons that suggest involvement of local communities in the restoration efforts, and it was observed by Ostrom (1990) that the failure to consider local needs and values has brought a collapse in many conservation projects. However, over time many countries have initiated participatory or community-based natural resource management programmes, including mangroves (Zorini et al. 2004). Regional coastal zone development plans have also been formulated (Olsen and Christie 2000) and the local people, who were often held responsible for natural resource decline (Contreras-Hermosilla 2000), are now considered important stakeholders in such programmes (Mbilea et al. 2005).

Although community participation plays an important role in rehabilitation and management of mangroves, economic benefits for local people from mangroves seem to be necessary in order to sustain the conservation programmes. Besides, ownership of or access rights to mangroves are also most important factors in utilization, conservation and management. Since mangrove conservation requires long-term maintenance, the expectations of the local people in terms of both short-term and long-term economic benefits from mangrove rehabilitation and conservation should also be addressed. This could lead to "self-mobilized (or voluntary) participation" and sustainable management of natural resources in coastal areas (Amri 2005).

Further evidence suggests that in Southern Thailand, degradation of mangroves has prompted many communities to organize locally and do mangrove conservation collectively (Rittibhonbhun et al. 1993). Reportedly, coastal communities are also interested in gaining rights to mangroves under a community forestry umbrella (Suthawan 1998). Sudtongkong and Webb (2008) note that community management was the principal factor in protecting, managing, and conserving the mangrove ecosystem in a manner superior to the conventional state management outside of protected areas. This is an important conclusion, because most terrestrial and coastal ecosystems are outside of the protected area system, and strategies for conservation of such "unprotected" ecosystems are very much needed. Community-based mangrove management and protection, therefore, is an important mechanism to achieve the goal of mangrove ecosystem conservation.

12.2.3 Mangroves and Coastal Protection

The indirect benefits of mangroves, such as coastal protection and nonuse values (option, bequest and existence values) are more difficult to measure. Since the establishment of the Buswang mangrove, storm surge damage and coastal erosion has been negligible, whereas in some other countries around the Indian Ocean, cases about storm-associated costs have been documented (Gilman et al. 2008). In India for instance, monetary losses due to repair and reconstruction costs of personal property (including livestock and agricultural products) ranged between US$ 32/household in mangrove-protected villages to US$ 154/household in villages that were not protected by mangroves (Badola and Hussain 2005). Other indirect benefits include accretion of agricultural land. In the Sundarbans in Bangladesh, the planting of 150,000 ha of mixed mangrove species has enhanced the deposition of sediments to such an extent that the elevation of 60,000 ha is no longer suitable for mangrove, and can be used for agriculture worth US$ 800/ha/annum (Saenger and Siddiqi 1993).

There is a growing urgency for conserving and restoring protective mangrove greenbelts to lessen the dangers from catastrophes affecting the coasts, such as sea-level rise, hurricanes and storm surges. Reportedly, mangroves can buffer against the fury of such destructive storms, protecting those settlements located behind a healthy mangrove fringe. In regions where these coastal fringe forests have been cleared, tremendous problems of erosion and siltation have arisen, and terrible losses to human life and property have occurred due to destructive hurricanes, storm surges and tsunamis. In the Mekong Delta, Soc Trang province, Vietnam, extensive planting of Rhizophora species was used as a coastal protection measure (Viswanathan et al. 2011).

12.2.4 Mangroves and Mitigation of Climate Change Effects

Of late, mangroves have been considered as an important protective cover against climate change: they protect coastal areas from tidal waves and cyclones and are among the most carbon-rich forests in the tropics (Cornforth et al. 2013). In the face of rising sea levels and changing climates, coastal buffering against the negative impacts of wave action becomes critical and will play an important role in climate change adaptations. Gilman et al. (2008) review the state of knowledge of mangrove vulnerability and responses to predicted climate change and consider the adaptation options. Based on available evidence of all the climate change outcomes, it has been highlighted that the relative sea-level rise may be the greatest threat to mangroves. The Pacific Islands mangroves have been demonstrated to be at high risk of substantial reductions due to climate change-induced hazards.

Protecting and replanting mangroves (and finding incentives for governments to do this) will help protect communities from the effects of climate change. Nevertheless, more research on climate change-induced risks on mangroves is needed in terms of developing assessment methods and standard indicators of change in response to effects from climate change, whereas regional monitoring networks are needed to observe these responses to enable well-informed adaptations, including planting mangroves for carbon sequestration.

12.2.4.1 Multifunctionality of Mangroves: Evidence from India

There have been several studies examining the multiple benefits of mangroves in the Indian context. A study by the Centre for Marine Fishery Resource Kakinada in Andhra Pradesh showed that prawn catch per boat load from mangrove areas of Godavari and Krishna has been 25 % more than in non-mangrove areas. The benchmark survey conducted by the MS Swaminathan Research Foundation (MSSRF) showed that in Andhra Pradesh, the fishery resources from the Godavari mangrove wetlands supported 32,300 families from 26 hamlets in 1998. The total value of their catch was estimated at ₹ 2.53 crore/annum or an average annual income per family of about ₹ 3500. It also revealed that about 375 t of fodder grass was obtained from the mangrove area every year (Chatterji undated).

Selvam et al. (2003) tried to map the impact of restoration on the degraded areas of Pichavaram mangrove wetland in Tamil Nadu by comparing 1986 digital data (before restoration) with digital data of 2002 (after restoration). It was observed that the mangrove forest cover had increased by about 90 %. A science-based, community-centred and process-oriented approach was followed for the restoration of the Pichavaram mangrove wetland in collaboration with the Forest Department, Government of Tamil Nadu and the participation of local mangrove using communities; it was mainly responsible for the success of the restoration efforts. This study has novelty as it uses remote sensing data as a monitoring tool to assess the effectiveness of restoration and conservation programmes of mangroves, where direct and regular physical monitoring is difficult due to the marshy nature of the soil and the presence of numerous tidal creeks and canals (Viswanathan et al. 2011).

A study by Santhakumar et al. (2005) on the Sundarbans indicate that the direct benefits from mangroves included abundance of brackish water fish, shrimps, crabs, honey, beeswax and tannin, which provided for requirements of both local and urban consumption. The exports of dried fish, shrimps, crabs and honey brought in substantial gains in foreign exchange.

The study by Badola and Hussain (2003) on the mangroves surrounding the Bhitarkanika mangrove ecosystem in the Kendrapada district of Orissa examined the multiple functions and services performed by mangroves. It revealed that people appreciated the contribution of mangroves to their lives and livelihoods directly in the form of increased production of fish and prospects for better tourism. A high percentage of people (88.6 %) recognized the contribution of mangroves in cyclone and flood mitigation. People have recognized the importance of even functions such as biodiversity conservation and ground water recharge. A majority (84.3 %) feel that they have a responsibility towards the conservation of flora and fauna. Nevertheless, despite such positive attitudes there was a high degree of resource extraction by the locals. This is because local people do not have any other livelihood options other than paddy cultivation and fishing. Consequently, more and more mangrove areas are being converted into paddy fields. Intensive prawn culture has also resulted in large-scale removal of mangroves from the Mahanadi delta, situated south to the Bhitarkanika (Badola and Hussain 2003).

The study by Mitra et al. (2006) has analysed the impact of controlled mangrove regulatory regime followed by the state on the local communities surrounding the Bhitarkanika wildlife sanctuary (BWS) in Orissa. The village communities surrounding the BWS area depend directly on mangroves for fuel wood and fodder and indirectly on the fish and prawn seedlings for their livelihood. But the imposition of conservation strategies by the government, regulating the free rides on resources that the people had been enjoying for generations, was not favoured by the locals. The study observes that the restrictions for conservation led to an economic non-sustainability of the local community.

The study by Das (2007) assesses the storm protection role of mangroves based on data on human casualties, damages to houses and livestock losses suffered in the Kendrapada district in Orissa (presently Odisha) during the 1999 super cyclone. The cyclone devastated 12 of the 30 districts of the state causing 9893 human casualties and 441,531 livestock deaths, damaging 1.95 million houses and 1.84 million ha of crop. The results indicate that the mangroves significantly reduced human and animal casualties. There was significant reduction in damage due to mangroves to residential houses and to big animals like cattle and buffaloes. It also observes that if the coverage of mangrove forest was 10 % more than what it was at the time of the cyclone, human casualties would have been reduced by 12.5 %.

Another study by Das (2009) on the storm protection role of mangroves revealed that if the mangrove cover had remained at the level that it had been in the 1950s, the area would not have witnessed any completely collapsed houses. The total protection benefits of mangroves in terms of averted damages to residential property in Kendrapada were estimated at ₹ 592,647,800 (US$ 14,110,662). The study suggests that mangrove forests provided protection benefits to houses to the extent of ₹ 975,800 (US$ 23,233) per kilometre width of forests or ₹ 51,168 (US$ 1218) per hectare of forests.

A detailed study on the mangroves in Gujarat by Hirway and Goswami (2007) indicated the loss of mangrove cover as caused by various natural and human interventions and development of industries, expansion of salt pans, etc. In fact, marine national parks and wild life sanctuaries have been established in the state with the intention of protecting vital/critical habitats, such as mangroves, coral reefs and wetlands. In a study by Nayak et al. (1989) on the Marine National Park, Jamnagar, on the Gujarat coast, significant changes in the mangrove vegetation and coral reef area were observed between 1975 and 1998. Recent industrialization, development of ports, etc. have again put these ecosystems under stress, as evident from recent satellite data. The earlier estimate of mangroves was 6740 km^2. The estimate based on IRS data is 4460 km^2, which indicates large-scale degradation of mangroves (Viswanathan et al. 2011). Highlighting the importance of Gujarat's fishery sector in the country (20 %), Saravanakumar et al. (2009) observe that 90 % of fish catch in Gujarat is contributed by the high yield of marine fisheries, supported by mangroves.

The study by Hirway and Goswami (2007) may be considered as an important case study on the impact of mangroves on the local communities in Gujarat. It tries to quantify the multiple benefits (direct use and nonuse values) in physical and monetary terms. Whereas the direct use value (at 2003 prices) of mangroves has been estimated at ₹ 1603 million, the indirect use value of the current status of mangroves was ₹ 2858 million/annum. The total use value (direct and indirect) of mangroves was thus estimated at ₹ 7731.3 million/annum for the state at 2003 prices (Hirway and Goswami 2007).

Thus, the foregoing section provides a comprehensive account of the multiple socioeconomic and ecological benefits of mangroves in the global as well as the regional contexts of Indian states, including, Gujarat. The review highlights that socioeconomic linkages of mangrove ecosystem play a pivotal role in the management of mangrove ecosystems, especially in coastal areas, where major economic activities and livelihoods depend mainly on extraction of natural capital and related resources. Empirical assessments from other countries as discussed also call for the need for proper resource management strategies involving local communities to ensure long-term sustainability of the mangrove ecosystems.

12.3 Multifunctionality of Mangroves: Empirical Evidence from Gujarat

The latest assessment of the Forest Survey of India (FSI) for the year 2013 shows that mangrove cover in India is 4628 km^2, which is 0.14 % of the country's total geographic area. The very dense mangroves comprises 1351 km^2 (29.2 % of mangrove cover); moderately dense mangroves is 1457 km^2 (31.5 %), whereas open mangrove covers an area of 1819 km^2 (39.3 %). Compared with the 2011 assessment, there has been a slight decline in mangrove cover (34 km^2) during 2013 (FSI 2013).

The country has recorded an increase in mangrove cover by 181 km^2 between 2001 and 2011. Table 12.1 presents the trends and the latest status of mangrove cover in India. It shows that West Bengal tops with the largest mangrove cover (45 %), followed by Gujarat (24 %), A&N Islands (13 %) and Andhra Pradesh (7.6 %). Mangrove cover in the rest of the states has been showing slight improvements particularly in Odisha, Maharashtra and Tamil Nadu. The growth in mangrove cover in the country has mainly originated from Gujarat, which reported an addition of about 192 km^2 from 911 to 1103 km^2 between 2001 and 2013.

The observed expansion in mangrove cover in Gujarat has been mainly contributed by the Restoration of Mangroves (REMAG) project implemented by the Gujarat Ecology Commission (GEC) during 2002–2007 with financial support from the India–Canada Environment Facility (ICEF). The REMAG project was aimed at development and restoration of mangroves in the Gulfs of Kutch and Khambhat, outside the conventional forest areas, i.e. in the wastelands provided by the government of Gujarat. Although the original project was over by 2007, the GEC continued mangrove restoration activities in view of the socioeconomic benefits as well as the environmental and ecological functions and services provided by

Table 12.1 Trends in mangrove plantation in major states in India (sq. km), 2013. (Source: FSI Report 2013)

States	1987	1991	1995	2001	2005	2011	2013	% share
1. Andhra Pradesh	495	399	383	333	354	352	352	7.61
2. A&N Islands	686	971	966	789	635	617	604	13.05
3. Gujarat	427	397	689	911	991	1058	1103	23.83
4. Maharashtra	140	113	155	118	186	186	186	4.02
5. Odisha	199	195	195	219	217	222	213	4.60
6. Tamil Nadu	23	47	21	23	36	39	39	0.84
7. West Bengal[a]	2076	2119	2119	2081	2136	2155	2097	45.31
8. Others[b]	0	3	5	8	26	34	34	0.73
Total	4046	4244	4533	4482	4581	4663	4628	100.00

[a] As per the West Bengal Forest Department, mangrove area in Sundarbans is 4200 km^2 (approximately), which is almost double the area estimated by the FSI. This is mainly due to the difference in assessment methods. West Bengal Forest Department includes the intervening water in the mangrove cover; whereas, assessment of FSI takes into account the mangrove cover only, as discerned from the satellite image

[b] Others includes states, viz., Puducherry, Kerala, Karnataka and Daman and Diu

mangroves. The programme has given a major thrust for public–private partnership (PPP)-based mangrove restoration programme. The REMAG programme has been further incorporated into the World Bank-aided Integrated Coastal Zone Management (ICZM) to improve the socioeconomic conditions of the coastal villages. To a greater extent, investments by private sector companies/industries have also been encouraged with a commitment to cherish community participation and corporate environmental responsibility (CER) in mangrove development/restoration efforts. Accordingly, a total of about 12,000 ha of mangrove plantations have been established in the state during the period between 2003–2004 and 2011–2012. Almost 35 % of these mangrove plantations have been established under the PPP model. As per the latest information available, a total of 176 community-based organisations (CBOs) have been formed covering about 170 coastal villages spread across the districts of Kutch, Rajkot, Bharuch, Navsari, Anand, Ahmedabad, Surat, Bhavnagar, Jamnagar and Valsad (Table 12.2). Figure 12.1 shows district-wise locations where mangrove plantations are carried out in Gujarat.

Under this programme, the GEC has selected various voluntary organizations as its project implementation partners (PIPs), which in turn, act as facilitators for community mobilization; formation and registration of CBOs; helping the community in micro-planning, implementation of project-related activities, like seed collection, nursery development, plantation, land development, etc.; protection of the plantations through social fencing, formation and amendment of bylaws for the utilization of the corpus funds, benefit sharing and conflict resolution.

As stated, a large part of the mangrove restoration in Gujarat has been undertaken under the PPP and private sector agencies accounted for close to one-third of the

Table 12.2 Extent of community-based mangrove restoration in Gujarat, 2011–2012. (Source: GEC and ICZM Project GEC (2006)

District	Mangrove area (ha)	% share	CBOs formed (#)
1. Ahmedabad	500	4.17	1
2. Bharuch	685	5.72	3
3. Bhavnagar and Anand	1250	10.43	2
4. Jamnagar	1515	12.64	54
5. Kutch	3320	27.70	98
6. Navsari	275	2.29	5
7. Rajkot	850	7.09	9
8. Surat	3000	25.03	2
9. Valsad	590	4.92	2
Total	11,985	100.00	176

Fig. 12.1 Locations of mangrove planted areas in Gujarat. (Source: Gujarat Ecology Commission)

mangrove-planted areas, which form the second largest category after the ICEF in the state sector (Table 12.3). Among the various private sector companies, the largest mangrove area belongs to Adani (28 %), followed by Pipapav Shipyard (23.4 %), Shell Hazira (14 %) and NIKO (12 %).

Table 12.3 Distribution of Mangrove plantations by various agencies in Gujarat 2010. (Source: Estimated from data provided by GEC, Gandhinagar, March 2010)

Type of implementing agency	Area (ha)	% share
State sector agencies		
Gujarat Heavy Chemicals Limited (GHCL)	50	0.81
Gujarat Maritime Board (GMB)	120	1.95
Gujarat Mineral Development Corporation (GMDC)	170	2.76
GSPC Pipapav Power Company (GPPC)	10	0.16
India Canada Environment Facility (ICEF)	4101	66.51
Ministry of Environment and Forests (MoEF)	300	4.87
Government of Gujarat (GoG)	1415	22.95
Subtotal	6166 (74.1)	100.00
Private sector agencies		
ABG shipyard	50	2.31
ADANI	600	27.78
Ambuja cement	150	6.94
Anjan cement	25	1.16
Bayer	10	0.46
Essar	200	9.26
NIKO Resources India Ltd.	250	11.57
Petronet LNG	50	2.31
Pipavav shipyard	505	23.38
Shell hazira	300	13.89
Ultatech	20	0.93
Subtotal	2160 (25.9)	100.00
Grand total	8326 (100.0)	

Figures in parentheses indicate the percentage of the total area

12.3.1 Multiple Benefits from Mangrove Restoration: Empirical Analysis

In this study, we attempted at understanding how different communities perceive the multiple (tangible and intangible) benefits of the mangrove restoration activities and how the restoration outcomes have impacted on their socioeconomic status and livelihoods. Since the restoration activities involve active participation of various stakeholders, such as the government, local communities and the private sectors, it is also important to understand whether mangrove restoration efforts receive proper support in terms of policy and institutional interventions so as to motivate local communities and strengthen and building capacities among them towards sustainable mangrove development outcomes.

Table 12.4 Distribution of sample villages and households. (Source: Primary Survey 2009–2010)

No.	District	Taluka	Village	No. of households covered[a]
1	Kutch	Lakhpat	Lakki	17 (7.5)
		Abadasa	Ashira Vandh	18 (7.9)
2	Bharuch	Jambusar	Nada	41 (18.1)
		Hansot	Kantiyajal	50 (22.0)
3	Surat	Olpad	Dandi	47 (20.7)
		Olpad	Karanj	15 (6.6)
4	Anand	Khambhat	Tada Talav	39 (17.2)
Total				227 (100)

[a] Figures in brackets are percentage of the total sample

In this regard, a community survey was undertaken among 227 households drawn from seven villages, viz., Lakki, Ashira Vandh, Nada, Kantiyajal, Dandi, Karanj and Tada Talav, covering six talukas and four districts, viz., Kutch, Bharuch, Surat and Anand (Table 12.4). The beneficiary households selected from each village are also members of CBOs engaged in plantation and management of mangroves.

12.3.2 Mangroves and Socioeconomic and Environmental Benefits

The results of the empirical analysis as presented in Table 12.5 indicate significant socioeconomic gains and environmental benefits as realized by the households across study villages within a shorter span of 6–7 years of the CBMR activities (also see Fig. 12.2). Evidently, the household dependence on mangroves for various benefits has been quite significant across villages (34 %). While fishermen communities are present in all the villages, households also report having farm lands closer to the mangrove restoration sites (53 %), which could not be cultivated earlier due to coastal cyclones.

Another important benefit realized by the communities has been increase in employment and wage earnings. Whereas all households benefited from the employment opportunities created by development and maintenance of mangrove plantations, the percentage of women households getting such opportunities has been quite significant in all the villages, except Karanj. On an average, each household received wage employment for 134 days/annum during the 7-year period, with notable differences across villages. The resultant gain in wage earnings per household has been as high as ₹ 14,820/annum in Ashirawandh, followed by ₹ 12,577 in Lakki and ₹ 10,327 in Tada Talav.

Table 12.5 Socioeconomic and environmental benefits of community-based mangrove restoration in Gujarat 2002–2007. (Source: Viswanathan et al. 2011)

Descriptives	Lakki	Ashiravandh	Karanj	Dandi	Nada	Kantiyajal	Tada Talav	All villages
1. Household (HH) dependence on mangroves (%)	27.6	25.2	35.2	54.8	28.4	35.1	33.6	34.1
2. Presence of fishermen communities (%)	29.4	16.7	66.7	40.4	41.5	18.0	10.3	29.5
3. Farm lands protected from salinity ingress (%)	25.0	42.9	–	0	66.7	80.0	36.4	52.5
4. Women household members getting employment (%)	52.2	63.0	–	60.3	45.1	55.9	51.5	53.9
5. Annual average employment generated per household (No.)	188	243	–	67	103	85	185	134
6. Annual average earnings from mangrove work (Rs.)	12,577	14,820	–	4728	6822	3954	10,327	8735
7. Households extracting mangroves for various uses (%)	88.2	94.4	–	21.3	56.1	22	71.8	45.8
8. Women extracting mangroves (%)	26.7	76.5	–	90.0	52.2	27.3	85.7	62.5
9. Mangrove extraction for fodder use (%)	100.0	100.0	–	10.0	56.5	90.9	92.9	64.3
10. Mangrove extraction for cooking fuel use (%)	–	–	–	90.0	43.5	9.1	7.1	35.7
11. Percentage gain in fishery income after mangroves	32.8	50.1	31.3	17.2	73.0	19.1	13.5	31.5
12. Households owning livestock (%)	82.4	94.4	13.3	0.0	26.8	30.0	71.8	38.3
13. Mangroves reduced salinity ingress in crop lands (%)	58.8	44.4	–	68.1	68.3	76.9	71.4	55.4
14. Mangroves reduced crop damage due to winds (%)	82.4	55.6	–	19.1	53.7	68.0	71.8	50.1
15. Households (HHs) reporting seriousness of cyclones (%)	94.1	100.0	73.3	51.1	87.8	72.0	97.4	78.9
16. HHs reporting reduced effects of cyclones after mangroves (%)	82.4	83.3	25.0	66.0	85.4	78.0	94.9	73.6

The values are averages for the period 2002–2007

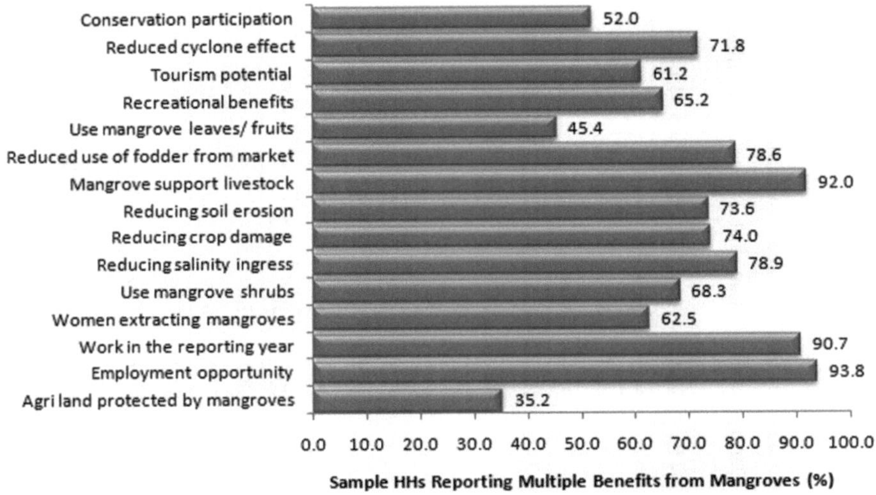

Fig. 12.2 Multiple benefits from mangroves reported by sample households in Gujarat

About 30 % of the respondents' main sources of income were fishery, followed by agriculture (25 %), agricultural labour (15 %), livestock (13.2 %) and other activities (3.1 %). About 14 % of the respondents solely depend on mangroves for income and occupation. The occupational structure of the household members seems to be very interesting, as almost 34 % of the households depend on mangroves for income and occupation as compared to other occupations, such as agricultural labour (22.6 %), agriculture (14 %), animal husbandry (11 %), fisheries (10 %), etc.

A majority of the households extract mangroves for various uses (46 %), ranging from 94 % in Ashirawandh to 88 % in Lakki, 72 in Tada Talav and 56 % in Nada. Compared to male household members, more women have been engaged in mangrove extraction for various uses. Among the various uses for which mangroves are extracted, fodder and fuel uses are the prominent ones, with highest use reported for fodder use for cattle. In fact, this provided at least two immediate benefits to the communities, viz., (a) helping them avoid purchase of fodder from the market; (b) an increase in milk production as mangrove leaves were available in plenty to be fed to the livestock.

12.3.3 Mangroves and Biodiversity Benefits

The study also undertook a detailed biological assessment to examine the vegetative growth and biodiversity impacts of mangrove plantations in the study villages. In order to do that it assessed the species diversity, i.e. the presence and abundance of species; vegetation cover and structure; and the ecological process by indirectly measuring the nutrient availability and biotic interactions.

Table 12.6 Status of biodiversity observed in the mangrove restoration villages. (Source: Viswanathan et al. 2011)

	Tada Talav	Lakki	Ashirawandh	Karanj	Dandi	Nada	Kantiyajal
Crustacean							
Crabs	++	++++	+++	+++	++	++	++++
Prawns	++	+++	++	++	+	+	++
Molluscan							
Gastropods	++	++++	+++	+++	+++	+++	++++
Bivalves	++	+++	++	++	+	+	+++
Snake	+	–	–	–	–	–	+
Birds	–	++	+	–	–	–	++
Mudskippers	+++	++++	++++	++++	+++	++++	+++
Other fish	++	+++	++	++	++	++	+++

– absent, + satisfactory, + + good, + + + excellent, + + + + quite rich

Table 12.7 Factors determining efforts of households: results of the regression analysis

Explanatory variables	Regression coefficients	*t*-stat	*p* value
Agri land protected by mangroves	0.076	3.224	0.001
Employment benefits	0.057	1.790	0.075
Women extracting mangrove stems/leaves	0.115	4.073	0.000
Increase in fish catch	0.172	6.424	0.000
Reduction in salinity Ingression	0.315	8.493	0.000
Decline in soil erosion after plantation	0.086	3.104	0.002
Mangrove benefits for livestock	0.126	3.892	0.000
Use of mangrove as feedstock	0.064	2.316	0.022
Recreational benefits	0.116	3.926	0.000
Tourism potential	0.092	3.517	0.001
Number of observations	226		
	Adjusted R square	0.796	
	Durbin Watson	1.816	
	F-stat	182.007 ($p=0.001$)	

Dependent variable—conservation participation by HHs (1 = yes; 0 = otherwise)

The biological assessment (Table 12.6) also brought about the diversity of the study villages in terms of the presence of invertebrates, mobile fauna and other species. Among the villages, Lakki, Ashirawandh and Kantiyajal have reported the highest species richness supported by the mangrove ecosystem. The mangrove areas have been found to be quite rich in terms of other species, such as mudskippers, crabs, bivalves, gastropods, fish, and habitats for other species. As mangrove areas have achieved good growth over the past few years, they have also been found providing habitat for birds and marine reptiles like snakes.

The empirical analysis also tried to capture the conservation participation outcome as revealed by the households. To examine this, we undertook a simple regression analysis of the important factors that motivate the respondents to participate in the mangrove conservation activities. Accordingly, the dependent variable considered was "participation in mangrove conservation" along with 10 important explanatory variables, expressed in binomial terms. The results of the analysis are presented in Table 12.7.

As seen from the table, the ten variables significantly explain the high level of household participation (R square being 0.796) in mangrove restoration with as many as five variables reporting higher coefficient values ($p=0.000$) and these included reduction in salinity ingress, increase in fish catch, livestock benefits, women extracting mangroves, and recreational benefits. The other two important factors reported included (a) the agricultural lands protected by mangroves from coastal waves or salinity ingress, and (b) perceived tourism potential.

12.4 Concluding Observations

This chapter provides a holistic view of the mangrove restoration efforts being initiated in Gujarat since the past one decade under the joint initiatives of the state (through the GEC) and the CBOs. Though this study is limited in coverage (7 villages of the total 22 mangrove restoration villages in Gujarat), it brings out the beneficial outcomes of mangrove restoration activities in place. It needs to be mentioned that the socioeconomic, environmental and biodiversity benefits derived from mangroves as presented in the analysis are only indicative of the vast potential that could be realized in the future when mangrove plantations attain maximum growth.

This chapter highlights the importance of evolving long-term policies and institutional intermediations required for carrying forward the development of new mangrove plantations as well as conservation/restoration of the existing plantations. In this connection, it is important to consider that the local communities and the CBOs need to be more strengthened in terms of increased awareness, skill development, capacity building, etc. so as to enable them to conserve/restore the mangrove ecosystems for the future. Though a majority of the communities (91 %) do feel that growing mangroves is important for protecting the coastal systems and livelihoods from the adverse effects of cyclones, soil erosion, etc., they still lack the motivation and incentives to conserve the resources on a sustainable basis. This is an important challenge that needs to be addressed through policies and interventions for creating motivations for conservation and restoration.

The fact that the growth of mangrove plantations is adversely affected by industrial pollution as well as garbage deposited in the coastal waters has been identified as a serious environmental issue by the communities in Ashirawandh and Karanj villages, which needs further in-depth studies. Besides, oil spill from boats and dumping of plastics were also reported. The discharge of saline water from salt pans is also reported as adversely affecting the growth of mangroves. Similarly, the closeness of a cement factory to the mangrove area is also reported as creating environmental problems for mangrove trees. The lack of inflow of freshwater (river water) due to construction of a dam is also reported to be affecting the growth of mangroves.

The provision of boats for fishing is yet another important need indicated by the fishing communities. The study reveals that only 9 of the 98 fisher communities own boats for fishing. In the absence of boats or inability to hire costly boats, almost 89 % of the fishing communities walk in deep waters for catching fish. This situation needs to be addressed through arrangement for provision of fishing boats to the communities who can own and operate fishing boats on a collective basis.

It has been observed that a majority of the households have started using mangroves as the major source for firewood, even replacing the use of conventional woods available in the local areas. Earlier, the use of local woods was also used to be supplemented with either neem or charcoal. There are also many instances in which respondents have stopped using kerosene for cooking with the abundant availability of mangroves. Such a scenario suggests the increasing pressure on mangroves for use of firewood other than fodder for livestock. This may invariably affect the existing stock and future growth of mangrove plantations as households prefer mangroves to other sources of fuel for cooking due to its easy availability and access. Perhaps, this calls for proper mechanisms and solutions through awareness creation to arrest the excessive extraction of mangroves. As imposition of policies or regulations alone will not work in such contexts, this problem needs to be addressed through promotion of planting of suitable species of wood for cooking in the villages.

Finally, sustainable development and restoration of mangroves essentially calls for more efforts for creating opportunities for collective action among the multiple stakeholders, like the state departments under various government portfolios, the local communities, private firms and industries who are increasingly investing in mangrove plantations, NGOs and local administration units like the village panchayats. This requires more frequent interactions among these stakeholders towards identifying more strategies and action plans for sustainable development and restoration of mangrove plantations in the villages. Needless to say that all these strategies and action plans should be targeted towards strengthening the capabilities of the local communities and sustaining their livelihoods without compromising on the broader goals of sustainable management of mangrove ecosystems.

References

Adeel Z, Pomeroy R (2002) Assessment and management of mangrove ecosystems in developing countries. Trees 16(2–3):235–238

Alongi DM (2002) Present state and future of the world's mangrove forests. Environ Conserv 29(3):331–349

Amarasinghe US, Amarasinghe MD, Nissanka C (2002) Investigation of the Negombo estuary (Sri Lanka) brush park fishery, with an emphasis on community-based management. Fish Manag Ecol 9(1):41–56

Amri A (2005) Community participation in rehabilitation, conservation and management of mangroves: lessons from coastal areas of South Sulawesi, Indonesia. Afr Study Monogr 29:19–30.

Awal MA (2010) Ecological pollution in Sundarbans. A H Development Publishing House, Dhaka, p 138

Badola R, Hussain SA (2003) Valuation of the Bhitarkanika mangrove ecosystem for ecological security and sustainable resource use, theme: wetlands and biodiversity, EERC working paper series: WB–1, A World Bank Aided India Environmental Management Capacity Building Technical Assistance Project. Indira Gandhi Institute for Development Research (IGIDR), Mumbai, p 101

Badola R, Hussain SA (2005) Valuing ecosystem functions: an empirical study on the storm protection functions of Bhitarkanika mangrove ecosystem, India. Environ Conserv 32:85–92

Barbier EB (1993) Valuing tropical wetland benefits: economic methodologies and applications. Geogr J, Part 1 59(Mar.):22–32

Barbier EB (1994) Valuing environmental functions: tropical wetlands. Land Econ 70(2):155–173

Barbier EB (2008) In the wake of tsunami: lessons learned from the household decision to replant mangroves in Thailand. Res Ener Econ 30(2):229–249

Brander LM, Florax RJGM, Vermaat JE (2006) The empirics of wetland valuation: a comprehensive summary and a meta-analysis of the literature. Environ Res Econ 33:223–250.

Chatterji J (undated) Conservation of mangroves along the east coast, in Utkarsh Ghate (undated): biodiversity: people's ecological planning (PEP) network and agricultural revitalization in South India.

Choudhury JK (1997) Sustainable management of coastal mangrove forest development and social needs. Proceedings of the XI World Forestry Congress, vol. 6, topic 38.6.

Conservation International (2008) Economic values of coral reefs, mangroves, and seagrasses: a global compilation. Center for Applied Biodiversity Science, Conservation International, Arlington, p 23

Contreras-Hermosilla A (2000) The underlying causes of deforestation. Occasional Paper No. 30, Centre for International Forestry Research (CIFOR), Jakarta

Cornforth A, Fatoyinbo E, Freemantle P, Pettorelli N (2013) Advanced land observing satellite phased array Type L-Band SAR (ALOS PALSAR) to inform the conservation of Mangroves: Sundarbans as a case study. Remote Sens 5:224–237

Dahdouh-Guebas F, Jayatissa LP, Di Nitto D, Bosire JO, Lo Seen D, Koedam N (2005) How effective were mangroves as a defence against the recent tsunami? Curr Biol 15(12):443–447

Das S (2007) Storm protection values of mangroves in coastal Orissa. In: Kumar P, Sudhakara Reddy B (eds) Ecology and human well-being. Sage, New Delhi, pp 197–212

Das S (2009) Can mangroves minimize property loss during big storms? An analysis of house damage due to the super cyclone in Orissa. SANDEE Working Paper No. 42–09, (July), p 36

EJF (2005) Did the loss of Mangrove forests exacerbate the impact of the Tsunami? Environmental Justice Foundation, London

FAO (2007) The world's Mangroves 1980–2005. Food and agriculture organization of the United Nations, FAO Forestry Paper 153. FAO, Rome

Field CD (ed) (1996) Restoration of mangrove ecosystems. International Society for Mangrove Ecosystems, Okinawa, p 250

FSI (2013) India State of Forest Report 2013, Forest Survey of India, Ministry of Environment & Forests, Government of India, p 282. http://www.indiaenvironmentportal.org.in/files/file/State%20of%20Forest%20Report%202013_final.pdf. Accessed 5 Sep 2014

Gautam AP, Shivakoti GP (2005) Conditions for successful local collective action in forestry: some evidence from the hills of Nepal. Soc Nat Resour 18:153–171.

GEC (2006) REMAG—Restoration of Mangroves in Gujarat—a community based approach, p 76

Gilman EL, Ellison J, Duke NC, Field C (2008) Threats to mangroves from climate change and adaptation options: a review. Aquat Bot 89:237–250

Glaser M (2003) Interrelations between mangrove ecosystem, local economy and social sustainability in Caet'e Estuary, North Brazil. Wetl Ecol Manag 11:265–272

Government of India [GOI] (2005) Ramsar COP9 national planning tool-national report format India. www.ramsar.org/cop9. Accessed 11 May 2010

Hirway I, Goswami S (2007) Valuation of coastal resources: the case of mangroves in Gujarat. Academic Foundation, New Delhi, p 170

Hoq ME (2007) An analysis of fisheries exploitation and management practices in Sundarbans mangrove ecosystem, Bangladesh. Ocean Coast Manag 50(5–6):411–427

Iftekhar MS (2008) An overview of mangrove management strategies in three South Asian countries: Bangladesh, India and Sri Lanka. Int For Rev 10(1):38–51

Iftekhar MS, Islam MR (2004) Degeneration of Bangladesh Sundarbans mangroves: a management issue. Int For Rev 6(2):123–135

Islam MS, Haque M (2004) The mangrove-based coastal and near shore fisheries of Bangladesh: ecology, exploitation and management. Rev Fish Biol Fish 14:153–180

Islam MS, Wahab MA (2005) A review on the present status and management of mangrove wetland habitat resources in Bangladesh with emphasis on mangrove fisheries and aquaculture. Hydrobiologia 542:165–190

Kairo JG, Dahdouh-Guebas F, Bosire J, Koedam N (2001) Restoration and management of mangrove systems—a lesson for and from the East African region, S. A. J Bot 67:383–389

Kathiresan K (2003a) Global policies and institution, biodiversity in mangrove ecosystems. In: Kathiresan K, Subramanian AN (eds) UNU—UNESCO international training course on biodiversity in mangrove ecosystems, Course manual. Annamalai University, India, pp 317–332
Kathiresan K (2003b) Conservation strategies for mangroves in India. Botanica 53:61–75
Kathiresan K (2003c) How do mangrove forests induce sedimentation? Rev Biol Trop 51(2): 355–359
Kathiresan K, Rajendran N (2005) Mangrove ecosystems of the Indian Ocean region. Indian J Mar Sci 34(1):104–113
Katon BM, Pomeroy RS, Garces LR, Ring MW (2000) Rehabilitating the mangrove resources of Cogtong Bay, Philippines: a co-management perspective. Coast Manag 28:29–37
Kijtewachakul N, Shivakoti GP, Webb EL (2004) Forest health, collective behaviors, and management. Environ Manag 33(5):620–636
Kovacs JM (2000) Perceptions of environmental change in a tropical coastal wetland. Land Degrad Dev 11:209–220
Lewis RR (2005) Ecological engineering for successful management and restoration of mangrove forests. Ecol Eng 24:403–418
Macintosh DJ, Ashton EC (2003) Draft code of conduct for the sustainable management of mangrove ecosystems. World Bank, ISME, Center Aarhus, p 81
Macintosh D, Mahindapala R, Markopoulos M (eds) (2012) Sharing lessons on mangrove restoration, Bangkok, Thailand: mangroves for the Future and Gland. IUCN, Switzerland
Mbilea P, Vabib M, Mebokac M, Okonc D, Arrey-Mbod J, Nkonghoe F, Ebong E (2005) Linking management and livelihood in environmental conservation: case of the Korup National Park Cameroon. J Environ Manag 76:1–13
Mitra R, Bhattacharya RN, Hazra S, Santra SC (2006) Mangrove conservation efforts and the local economy: a case study. Econ Political Wkly 41(33):3612–3616
Nagelkerken I, van der Velde G, Gorissen MW, Meijer GJ, Van't Hof T, den Hartog C (2000) Importance of mangroves, sea grass beds and the shallow coral reef as nursery for important coral reef fishes, using a visual census technique. Estuar Coast Shelf Sci 5(1):31–44
Nayak S, Pandeya A, Gupta MC, Trivedi CR, Prasad KN, Kadri SA (1989) Application of satellite data for monitoring degradation of tidal wetlands of the Gulf of Kachchh, Western India. Acta Astronaut 20:171–178
Olsen S, Christie P (2000) What are we learning from tropical coastal management experiences? Coast Manag 28:5–18
Ostrom E (1990) Governing the commons: the evolution of institutions for collective action. Cambridge University Press, New York
Othman MA (1994) Value of mangroves in coastal protection. Hydrobiologia 285(1–3):277–282.
PDO-ICZMP (2004) Areas with special status in the coastal zone. Programme development office for integrated coastal zone management plan project, Water resources planning organization, Ministry of water resources, Bangladesh. Dhaka, Bangladesh.
Pomeroy RS, Carlos MB (1997) Community-based coastal resource management in the Philippines: a review and evaluation of programs and projects, 1984–1994. Mar Policy 21(5):445–464
Rittibhonbhun N, Chansanoh P, Tongrak S, Suwannatchote R (1993) Community-based mangrove rehabilitation and management: a case study in Sikao District, Trang Province, Southern Thailand. Reg Dev Dialog 14:111–122
Ronnback P (1999) The ecological basis for economic value of seafood production supported by mangrove ecosystems. Ecol Econ 29(2):235–252.
Ronnback P, Primavera JH (2000) Illuminating the need for ecological knowledge in economic valuation of mangroves under different management regimes–a critique. Ecol Econ 35:35–141
Saenger P (2002) Mangrove ecology, silviculture and conservation. Kluwer Academic Publishers, Dordrecht, p 360
Saenger P, Siddiqi NA (1993) Land from the sea: the mangrove afforestation programme of Bangladesh. Ocean Coast Manag 201(11):23–39

Saenger P, Hegerl EJ, Davie JDS (eds) (1983) Global status of mangrove ecosystems. Environmentalist 3.3(Suppl 3):88

Santhakumar V, Enamul Haque AK, Bhattacharya R (2005) An economic analysis of mangroves in South Asia. In: Khan M (ed) Economic development in South Asia. Tata McGraw Hill, New Delhi, pp 368–437

Saravanakumar A, Rajkumar M, Sesh Serebiah J, Thivakaran GA (2009) Fishery resources in arid zone mangroves in Gulf of Kachchh, Gujarat, northwest coast of India. J Ocean Univ China (Oceanic and Coastal Sea Research) 8(3):233–240

Sathirathai S, Barbier EB (2001) Valuing mangrove conservation in Southern Thailand. Contemp Econ Policy 19(2):109–122

Selvam V, Ravichandran KK, Gnanappazham L, Navamuniyammal M (2003) Assessment of community-based restoration of Pichavaram mangrove wetland using remote sensing data. Curr Sci 85(6):794–798

Soontornwong S (2006) Improving livelihood through CBNRM: a case of self-organization in community mangrove management in Thailand. In: Mahanty S, Fox J, Nurse M, Stephen P, McLees L (eds) Hanging in the balance: equity in community-based natural resource management in Asia. East-West Center, Honolulu, pp 182–199

Spalding M (1997) The global distribution and status of mangrove ecosystems. International News Letter of Coastal Management—Intercoast Network, Special edition 1, pp 20–21

Sudtongkong C, Webb EL (2008) Outcomes of state- vs. community-based mangrove management in Southern Thailand. Ecol and Soc 13(2):27. http://www.ecologyandsociety.org/vol13/iss2/art27/. Accessed 11 May 2010

Suthawan SS (1998) Economic valuation of mangroves and the roles of local communities in the conservation of natural resources: case study of surat Thani, South of Thailand, EEPSEA Research Report Series, June 1998, 60 p. http://idl-bnc.idrc.ca/dspace/bitstream/10625/15261/39/108378.pdf. Accessed 12 May 2015

Tomlinson PB (1986) The botany of mangroves, Cambridge tropical biology series. Cambridge University Press, USA

UNEP (2005) After the Tsunami: rapid environmental assessment report. United Nations Environment Program (UNEP), Nairobi, p 140

Valiela I, Bowen JL, York JK (2001) Mangrove forests: one of the world's threatened major tropical environments. Bioscience 51(10):807–815

Vannucci M (ed) (2004) Mangrove management and conservation: present and future. United Nations University, Tokyo, p 324

Vermaat JE, Thampanya U (2006) Mangroves mitigate tsunami damage: a further response. Estuar Coast Shelf Sci 69:1–3

Viswanathan PK, Pathak KD, Mehta I (2011) Socio-economic and ecological benefits of mangrove plantation: a study of community based mangrove restoration activities in Gujarat. Gujarat Ecology Commission (GEC), Government of Gujarat, Gandhinagar, p 164. www.gec.gov.in/pdf/GIDR%2520Report.pdf

Walters BB (2003) People and mangroves in the Philippines: fifty years of coastal environmental change. Environ Conserv 30(2):293–303

Walters BB, Ronnback P, Kovacs JM, Crona B, Hussain SA, Badola R, Primavera JH, Barbier E, Dahdouh-Guebas F (2008) Ethnobiology socio-economics and management of mangrove forests: a review. Aquat Bot 89:220–236

Walton MEM, Samonte-Tan G, Primavera JH, Edwards-Jones G, Le Vay L (2006) Are mangroves worth replanting? The direct economic benefits of a community-based reforestation project. Environ Conserv 33:335–343

Zorini LO, Contini C, Jiddawi N, Ochiewo J, Shunula J, Cannicci S (2004) Participatory appraisal for potential community-based mangrove management in East Africa. Wet Ecol Manag 12: 87–102

Chapter 13
Linkages Between Environment and Globalization: A Case Study of Developing Countries

Trupti Mishra

13.1 Introduction

The relationship between the trade and the environment has become an increasingly contentious issue between economists and environmentalists. Economists have the firm opinion that trade helps the natural environment because rich countries can better afford to protect their unspoiled areas. Environmentalists counter that the motivation of increasing income, drives global environmental degradation and that free trade accelerates the process. In addition, people are increasingly becoming concerned and apprehensive about the environmental consequences of globalization. These result in clash between trade community and globalization opponents from past several years (Copeland and Taylor 2004). The debate was originated from North American Free Trade Agreement and the Uruguay round of General Agreement on Tariffs and Trade (GATT) and gained its strength with the scientific evidence of global warming, species extinction and anthropogenic activities. However, the creation of World Trade Organization (WTO) and proposals for future rounds of trade negotiations were trying to address the debate (Copeland and Taylor 2004). Nevertheless, the motivation behind the debate was never satisfying as parties lacks trust, common concern/solution and empirical evidence (Copeland and Taylor 2004).

Liang 2006 brings out the role of trade in relationship between growth and environment. The study mentions that one-hand growth increases the country's endowment or improves the technology, while trade liberalization changes relative goods prices by opening up the economy to foreign competition. At the same time, trade liberalization could cause economic growth when the competition or technology transfers from foreign firms spur innovation or capital accumulation in domestic sectors, as supported by numerous empirical works on the relationship between trade and innovation and productivity improvement. On the other hand, it

T. Mishra (✉)
Shailesh J. Mehta School of Management (SJMSOM), Indian Institute of Technology (IIT) Bombay, Mumbai, India
e-mail: truptimishra@iitb.ac.in

N. Ghosh et al. (eds.), *Nature, Economy and Society*,
DOI 10.1007/978-81-322-2404-4_13

also brings environmental degradation with increasing production activities. Thus, trade can influence environmental outcome by increasing per capita income (Liang 2006).

In this backdrop, the present chapter examines environmental consequences of globalization/liberalized trade. An attempt has been made to study the effect of globalization on environment for some of the developing countries. The chapter is organized as follows: Sect. 13.2 presents literature review and objective of the study, Sect. 13.3 discusses the database and methodology and Sect. 13.4 presents the result and discussion.

13.2 Literature Review

As per the study of Liang 2006, the economic literature on relationship between trade, growth and environment goes back to the year 1970s with some normative research. Nevertheless the pioneering work by Grossman and Krueger (1993) on trade policy and growth's impact on environmental outcomes brought the positive direction about this issue (Liang 2006). In this chapter they hypothesize an inverse U-shaped relationship between a country's per capita income and its level of environmental quality: increased incomes are associated with an increase in pollution in poor countries, but a decline in pollution in rich countries. According to the study, the effects of growth on environment can be divided into three components: *scale effect, composition effect and technique effect*. Scale effect is the change in pollution with the change in the scale of the production, holding constant the mix of goods and production techniques. Composition effect is the change in the share of the dirty goods in national income and the increase in pollution as the economy devotes more of its resources to produce the dirty goods. Technique effect is the change in the pollution as cleaner technique is used for production (Liang 2006; Grossman and Krueger 1991). Liang's 2006 study also mentioned Grossman and Krueger's (1995) study with cross-country data set covering 58 countries in the1980s and finds support for an inverse U-shaped relationship between income and pollution. Grossman and Krueger's work led to a booming literature on the Environmental Kuznet Curve (EKC), that is an inverse U-shaped relationship between growth and environmental quality (Liang 2006).

Environmental quality could decline through the scale effect as increasing trade volume (especially export) would expand the size of the economy which in turn leads to increase in pollution level (Dinda 2008). In other words, scale effect could decrease the environmental quality and it might be a cause of environmental degradation.

The studies of Wheeler 2000, Tisdell 2001 and Dinda 2008 supports the role of globalization for increasing environmental quality. According to the studies, since globalization increases competition for investment which in turn increases the income levels that helps maintaining environmental quality. However, the previous studies also bring out the point that, even if globalization has positive effect on environment, still there is always a conflict between economic reform and environmental protection.

Country-specific regulation with respect to environment also plays a role in the linkage between trade and environment. There is a tendency that the polluting industries always shift their location to the countries having weaker environmental regulation. These shifts may in turn increase global pollution or they may have a chilling effect on environmental policy, as countries will be reluctant to tighten environmental regulations because of concerns over international competitiveness (Dinda 2008).

Researchers have proposed two major hypotheses to explain the impact of trade: the pollution haven hypothesis (PHH) and the factor endowment hypothesis (FEH; Liang 2006). PHH assumes that the countries are identical except for exogenous differences in pollution policy, so it is cheaper to produce dirty goods in the country with weaker environmental policy, usually the poorer country (Liang 2006). It should be mentioned that a polluting activity in a high-income country normally faces higher regulatory costs than its counterpart in a developing country (Mani and Wheeler (1998). Under these circumstances the pollution-intensive industries will have a natural tendency to migrate to countries with weaker environmental regulations (Copeland and Taylor 1995; Eskeland and Harrison 2003). Trade induced by pollution policy differences creates a pollution haven in the poorer country (Liang 2006). Thus, developed countries are expected to benefit in terms of environmental quality from trade, while developing countries will lose. In other words, the PHH basically suggests that countries having strict environmental standard will lose all the dirty industries and poor countries (i.e. those having poorer environmental standard) will get them all (Dinda 2008).

FEH is the main alternative to the PHH. It suggests that the direction of trade is determined by the relative abundance of factor endowments each country (Liang 2006). Copeland and Taylor 2004 emphasized the role of comparative advantage in this context. According to the study, affluent countries are capital abundant countries and since the nature of polluting industry is capital intensive, the countries will have comparative advantages in these industries. Also they will export capital intensive goods. The study concludes that the differences in environmental policy and factor endowments might jointly determine the comparative advantage in trade. It is clear that effects of trade liberalization on environmental quality depend on, among other factors, jointly by differences in pollution policy and factor endowments, which leads to two competing theories in question (Dinda 2008). Liang 2006 brought up the problem with empirical tests of the two hypotheses as government of few countries are believed to make trade and environmental policy separately.

Previous literature also advocated the beneficial effect of trade in the long run (World Bank 1992). In the initial phase, although trade increases the environmental degradation, in the later phase with the increase in regulation for better environmental quality, there would be increase in pollution control innovation and technology transfer (Birdsall and Wheeler 1992; Lee and Roland-Host 1997; Jones et al. 1995; Frankel and Rose 2002). Liddle (2001) argues that trade may be good for environment and may improve the environmental quality through technological effect.

Liang 2006 mentioned the work of Frankel and Romer (1999), Antweiler et al. (2001) and Dean (2002) which deals with the issue of endogenous trade policy and the relationship between trade, growth and environment. The outcome of the studies identified factor for trade and positive effect of trade on income and negative effect on environment.

Objective of the Study In view of the issues discussed in previous section, the main objective of the present chapter is to examine the impacts of globalization on pollution level, pollution intensity and relative change of pollution for the developing country

13.3 Database and Methodology

The dataset for the study consists of 15 developing countries randomly chosen from the non-OECD (Organisation for Economic Co-operation and Development) counties list for the time period 1990–2005. The data on annual per capita real gross domestic product (GDP), capital per worker, trade intensity and annual per capita CO_2 emission taken from Penn world table and Carbon Dioxide Information Analysis Center (CDIAC). Globalization is the proxy measurement in terms of openness and trade intensity, which is measured as export plus import to GDP (Dinda 2008). Following Dinda (2008) the below estimating equation is used.

$$\begin{aligned} Eit &= \alpha + \beta \ln Yit + \beta(\ln Yit)^2 + \beta \ln\ KLit + \beta(\ln KLit)^2 + \beta\ \ln Pit \\ &\quad + \lambda OPit + \theta t + uit \end{aligned} \tag{13.1}$$

Where *Eit* is the pollution measured in country *i* at time *t;* similarly *Y, KL, P, OP* and *t* denote income per capita, capital labour ratio, population, openness and time trend, respectively. Both the fixed effect (FE) and the random effect models are estimated.

13.4 Results and Discussion

The estimated result of the analysis is presented in Table 13.1. For per capita CO_2 emission level, result shows that the coefficient of openness is positive and statistically significant which implies that the globalization is negatively related to the pollution level in the developing counties. Based on the findings of CO_2 emission intensity per dollar and relative change of CO_2 emission per capita, it is observed that globalization promotes to increase emission (pollution) intensity and relative change of emission and thereby liberalization or openness hurts the environment of developing counties. These empirical findings are in similar line of few earlier

Table 13.1 Estimated results of globalization and CO_2

Variable	Globalization and CO_2 emission level		Globalization and CO_2 emission intensity		Globalization and relative change of CO_2 emission	
	FE	RE	FE	RE	FE	RE
Constant		6.42 (9.87)		− 8.74		
Yit	− 1.46 (9.24)	− 1.62 (9.51)	− 3.2 (4.16)	− 2.9 (3.92)	3.25 (2.41)	3.19 (1.76)
$(Yit)^2$	0.24 (12. 83)	0.28 (13.21)	− 1.94 (3.85)	− 1.91 (2.46)	− 0.23 − (5.76)	− 0.21 − (5.28)
K/L	− 0.26 − (3.54)	− 0.31 − (3.89)	− 3.2 (2.12)	− 3.1 (1.92)	0.36 (2.15)	0.32 (1.87)
$(K/L)^2$	0.032 (4.57)	0.029 (4.17)	8.37 (1.24)	6.28 (1.08)	0.0008 (1.09)	− 0.0012 − (0.76)
P	0.0046 (0.73)	0.0034 (0.24)	3.47 (1.74)	2.18 (0.92)	0.302 (2.86)	− 0.12 (− 1.25)
Λ	− 8.67 (2.8)	− 4.84 (1.92)	4.37 (2.74)	3.19 (1.78)	4.79 (5.86)	3.39 (3.28)
T	− 0.0038 (1.68)	0.0012 (0.63)	− 7.61 (2.82))	− 3.67 (1.47)	− 0.0224 − 2.87()	− 0.0018 (− 1.04)

studies (Dinda 2008, Low and Yeats 1992, Agras and Chapman 1999, and Suri and Chapman 1998).

Although the finding implies the negative relation between globalization and environmental quality, however it is difficult to analyse other linkages within the scope of this study. As discussed in some previous literature, there is a need to study some other factors like country's environmental policy regulation, competitive advantages and international competitiveness to estimate the impact of globalization on environmental quality of the economy. Nevertheless, the emission share of developing countries in the world emission is growing with the increasing economic activity.

The present study is an attempt to examine the linkage of globalization and environment. The motivation of the present study comes in the backdrop of challenges of twin objectives: economic and environmental quality growth with the globalization and free trade. It is observed from the results that there is a negative relation between globalization and environment and increase in CO_2 emission in the developing countries. However, the further study is required to address the linkages on the basis of country's regulation and basic and trade profile. Since all these factors are based on government's decision, the role of government plays an important role in deciding what may be beneficial for the country. The study outcome is based on limited case of few developing countries; hence, the study outcome may not be same when we take a large set of data. The future scope of the study lies in analysis of large number of countries segmented on the basis of abovementioned factors and comparison with the developed countries.

References

Agras J, Chapman D (1999) A dynamic approach to the environmental Kuznets curve hypothesis. Ecolog Econ 28(2):267–277

Antweiler W, Copeland BR, Taylor MS (2001) Is free trade good for the environment? Amer Econ Rev 91(4):877–908

Birdsall N, Wheeler D (1992) Trade policy and industrial pollution in latin america: where are the pollution havens? In: Low Patrick ed International trade and the environment, World Bank discussion papers, no. 159. Washington, D.C., World Bank, pp 159–167

Copeland BR, Taylor MS (1995) Trade and environment: a partial synthesis. Amer J Agr Econ 77:765–771

Copeland BR, Taylor MS (2004) Trade, growth and the environment. J Econ Lit 42: 7–71

Dean JM (2002) Testing the impact of trade liberalization on the environment: theory and evidence. Can J Econ 35(4):819–842

Dinda S (2008) Impact of globalization on environment: how do we measure and analyze it? ISEE 2006

Eskeland GS, Harrison AE (2003) Moving to greener pastures? multinationals and the pollution haven hypothesis. J Devel Econ 70:1–23

Frankel J, Romer D (1999) Does trade cause growth? Am Econ Rev 89:379–399

Frankel JA, Rose AK (2002) Is trade good or bad for the environment? sorting out the causality. Mimeo, Harvard University

Grossman GM, Krueger AB (1991) Environmental impacts of a North American free trade agreement. NBER working paper 3914

Grossman GM, Krueger AB (1993) Environmental impacts of a North American free trade agreement. In: Garber PM (ed) The U.S.–Mexico free trade agreement. MIT Press, Cambridge, pp 13–56

Grossman GM, Krueger AB (1995) Economic growth and the environment. Quarterly J Econ 110(2):353–377

Jones LE, Rodolfo EM (1995) A positive model of growth and pollution controls. NBER working paper 5205

Lee H, Roland-Holst D (1997) The environment and welfare implications of trade and tax policy. J Devel Econ 52(1):65–82

Liang FH (2006) Does foreign direct investment harm the host country's environment? Evidence from China. Working paper, from faculty. Haas. berkeley, edu/fenliang/rearch/Pollution & FDI/FDI Pollution. Pdf

Liddle B (2001) Free trade and the environment-development system. Ecolog Econ 39:21–36

Low P, Yeates A (1992) Do 'dirty' Industries Migrate. In: Low P (ed). International trade and the environment, World Bank discussion paper, no. 159: 89–104

Mani M, Wheeler D (1998) In search of pollution havens? Dirty industry in the world economy. 1960–1995. J Environ Devel 7(3):215–247

Suri V, Chapman D (1998) Economic growth, trade and the energy: implications for the environmental Kuznets curve. Ecolog Econ 25:195–208

Tisdell C (2001) Globalisation and sustainability: environmental Kuznets curve and the WTO. Ecolog Econ 39:185–196

Wheeler D (2000) Racing to the Bottom? Foreign investment and air quality in developing countries. World Bank working paper

World Bank (1992) World development report. Oxford University, New York

Chapter 14
Can the Poor Resist Capital? Conflicts over 'Accumulation by Contamination' at the Ship Breaking Yard of Alang (India)
How Struggles for Environmental Justice Contribute to the Environmental Sustainability of the Economy

Federico Demaria

List of Abbreviations

ASSBY	Alang–Sosiya ship-breaking yard
DWT	Dead weight tonnage
GMB	Gujarat Maritime Board
GPCB	Gujarat Pollution Control Board
IMF	International Metalworkers' Federation
IMO	International Maritime Organization
LDT	Light displacement tons
OSHA	Occupational Safety and Health Administration

14.1 Introduction

In August 2009, a fire broke out aboard the European container ship MSC Jessica killing six workers on the Indian ship breaking beaches of Alang.[1] These kinds of tragedies are rather common. The fire erupted as they were dismantling the

An earlier version of this chapter has been published with the title 'Shipbreaking at Alang–Sosiya (India): An ecological distribution conflict' in Ecological Economics, Volume 70, Issue 2, pp. 250–260, 2010. (Special Section: 'Ecological Distribution Conflicts', edited by Joan Martinez-Alier, Giorgos Kallis, Sandra Veuthey, Mariana Walter and Leah Temper).

[1] http://www.indianexpress.com/news/six-die-in-fire-at-alang-ship-breaking-yard/498063/ (Accessed in January 2012).

F. Demaria (✉)
Institute of Environmental Sciences and Technology (ICTA), Autonomous University of Barcelona (UAB), Edifici Z, ICTA-ICP, Carrer de les Columnes, Campus de la UAB, 08193 Bellaterra (Cerdanyola del Vallès), Barcelona, Spain
e-mail: federicodemaria@gmail.com

Centre for Studies in Science Policy (CSSP), Jawaharlal Nehru University (JNU), New Delhi, India

N. Ghosh et al. (eds.), *Nature, Economy and Society*,
DOI 10.1007/978-81-322-2404-4_14

cargoship's engine room. It took place as the ship had neither been decontaminated by the original owner nor made safe by the local enterprise. The Geneva-based Mediterranean Shipping Company (MSC), the world's second-largest shipping line in terms of container vessel capacity, had profitably used the ship since its construction in 1980. MSC denied all responsibilities as, officially, the owner of the vessel was a Panamanian company and the vessel was registered in Panama while under bareboat charter to MSC; after the vessel left MSC's service it was sold to a Saint Vincent company.[2] In other words, MSC, as most shipping companies, normally uses flags of convenience, cash buyers and shell companies to bypass the international regulations. This chapter investigates, through the lens of an ecological distribution conflict related to ship breaking in Alang–Sosiya (India), how to understand the linkages among nature, economy and society.

Rich societies use large amounts of resources. Conflicts of resource extraction and waste disposal, such as the conflict over the excessive production of carbon dioxide, arise as a consequence of this. Rich societies generate large quantities of other kinds of waste, encountering opposition to local waste treatment and disposal sites, such as incinerators and landfills, (Pellow 2007) and facing rising management costs (Pearson 1987). This is the background of a rapidly changing and lucrative trade, global in nature, in which waste flows towards developing countries or poorer areas of developed countries (McKee 1996). Under a world-system perspective, the core, through unequal power relations, manages to export entropy to distant sinks in the periphery (Scott Frey 1998; Hornborg et al. 2007). These flows, legal or not (with mafias as important players), consist of urban and industrial waste, hazardous and nonhazardous waste, and waste intended for reuse, recycling and final disposal (Clapp 1994; D'Alisa et al. 2010).

In the 1970s and 1980s, scandals of toxic waste dumping in the South led to attempts to stem these flows, such as the Basel Convention on the Control of Transboundary Movements of Hazardous Wastes and their Disposal of 1989. Yet, India, among others, has been increasingly used as a dumping ground for toxic industrial waste (like asbestos and mercury) from developed countries (Singh 2001).

The issue of ship breaking is examined here as an example of toxic waste trade (Alter 1997). Ship breaking is the process of dismantling an obsolete vessel's structure for scrapping or disposal. Conducted at pier or dry dock, or directly on the beach as in Alang–Sosiya, it includes a wide range of activities, from removing all machineries and equipment to cutting down the ship infrastructure. It is the destiny of ocean-going ships like oil tankers, bulk carriers, general cargo, container ships and others like passenger ships. Depending on their interests, stakeholders will call it breaking, recycling, dismantling or scrapping (Stuer-Lauridsen et al. 2004). It is a challenging process, owing to the numerous problems of safety, health and environmental protection (OSHA 2001).

---

2 Mediterranean Shipping Company (MSC) response re allegations regarding a fire causing six deaths on MSC Jessica shipbreaking operation. 17 October 2012. Available at http://www.business-humanrights.org.

The industry provides steel at cheap prices and employment, which contribute to economic growth. On the other hand there are concerns about the health and safety of workers, and the impact on the environment. These are the premises of the debate on whether ship breaking in India falls under a Welcome Into My Backyard (WIMBY) logic or is a case of (environmental) injustice (Singh 2001) and application of Lawrence Summers' Principle (Martínez-Alier 2002). This chapter discusses the controversy under a framework of ecological economics and political ecology.

Changing social metabolism (meaning the flow of energy and material in the economy) (Fischer-Kowalski 1998; Foster 1999), driven by economic and population growth, generates growing quantities of waste. Georgescu-Roegen proposed a paradox highlighting that 'technical evolution leads to an increase in the rate at which society "wastes resources"... the economic process actually is more efficient than automatic shuffling in producing higher entropy, i.e. waste' (Georgescu-Roegen 1971). In other words, the more developed a society, the higher its rate of generation of wastes per capita (Giampietro and Mayumi 2009). It is generally accepted that under a fair allocation of responsibility, developed countries should deal with their own waste. Principles such as 'the polluter pays' and 'producer liability' appear to be legally settled. However, in some cases where countries from the North 'externalize the costs' of toxic waste disposal outside their own national borders (notably to the South) are not rare. The pollution haven hypothesis (Antweiler et al. 2001) refers to the idea that lower trade barriers will shift pollution to those countries with less stringent environmental regulations, which are normally also poorer. According to the Lawrence Summers' Principle, Southern countries have an environmental 'comparative advantage' regarding waste treatment (Pearson 1987). In an internal memo leaked to the press,[3] Lawrence Summers, then chief economist at the World Bank in 1991, wrote: 'I think the economic logic behind dumping a load of toxic waste in the lowest wage country is impeccable and we should face up to that'. Pollution should be sent to places where there are no people, or where the people are poor, since 'the measurements of the costs of health impairing pollution depend on the foregone earnings from increased morbidity and mortality. From this point of view a given amount of health impairing pollution should be done in the country with the lowest cost, which will be the country with the lowest wages'. The cost of internalizing the externalities would be the lowest.

The question is whether decisions on matters of life and death should be taken only on economic grounds. Poor people that meet 'Lawrence Summers' criteria, often complain, as several studies from political ecology document (Martínez-Alier 2002). Such ecological distribution conflicts express underlying valuation conflicts, actors deploying different languages to affirm their right to use a safe environment (Martinez-Alier 2009).

This chapter investigates ship breaking in India from the vantage point of political ecology, paying attention to the unequal distribution of benefits and burdens (already in the present generation) in a context of growing global social metabolism that

---

[3] 'Let them eat pollution'. The Economist, 8 February 1992.

leads to greater generation of waste, and with an analytical focus on the ways actors express alternative claims in the political arena and the valuation conflicts that hence emerge.

Section 14.2 describes the methods and the study region. Then Sect. 14.3 introduces the ship breaking industry, describing the process through which a ship becomes waste for the ship owner, enters the scrapping market through a cash buyer and is finally dismantled by a ship breaker. Section 14.4 presents different options for the management of the ship's toxic waste and analyzes the socio-environmental impacts resulting from current practices. The conflict in the 'Blue Lady' case at the Supreme Court of India is analyzed in Sect. 14.5 with particular attention to the valuation languages used by the different social groups. Finally, conclusions are drawn in Sect. 14.6.

14.2 Methods and Study Region

Data from interviews, official documents, direct and participant observation have been combined using the case study research methodology (Yin 2003). Fieldwork was carried out from April to July 2009. The access of researchers to the industry site is strictly regulated and workers' freedom of expression is limited. Semi-structured or in-depth interviews with 64 respondents were conducted with local villagers (10), farmers (8), fishers (9), ship breaking entrepreneurs (4), workers (11), political and administrative authorities (6), legal experts (4), academics (5) and activists (7). Interviewees were selected to represent a broad spectrum of interests and knowledge regarding ship breaking, using both random and snowball sampling methods. Moreover, focus groups have been led with farmers, fishers and workers. Interviews were conducted in English or with the help of local translators in Hindi and Gujarati. National and international documentation was researched with special focus on the 'Blue Lady' case at the Indian Supreme Court during 2006 and 2007 (Civil Writ Petition No. 657 of 1995). Official documents were examined under the guidance of the lawyer Sanjay Parikh and the petitioner Gopal Krishna. Media coverage on ship breaking has been extensively examined on the web and at the Centre for Education and Documentation in Mumbai.

The case study is located in the Gulf of Cambay, Bhavnagar district of Gujarat state in the north-west of India (Fig. 14.1). Alang and Sosiya are the two local villages that give the name to the ship breaking yard (ASSBY) (Fig. 14.2). The district, originally based on farming and fishing, is under rapid industrialization and urbanization which resulted in the degradation of the environment and decline in biodiversity (Gov. of India 2002). Gujarat state, historically a main centre of trade and commerce, has one of the fastest growing economies in India.

Fig. 14.1 Location map of Alang–Sosiya in the state of Gujarat, India

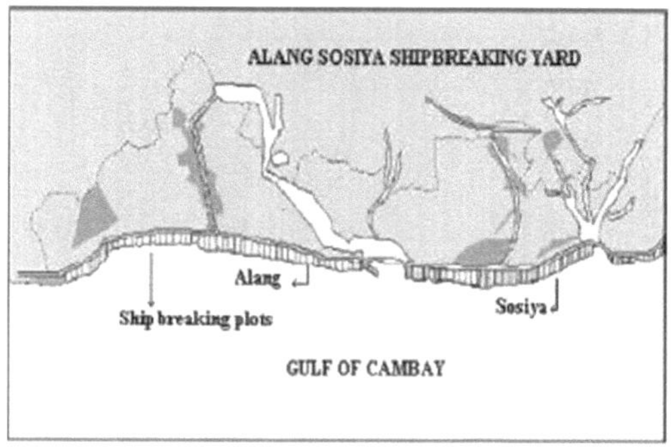

Fig. 14.2 Map showing ship breaking plots at Alang and Sosiya

14.3 The Ship Breaking Industry

14.3.1 The Shipping Industry[4]

The shipping industry (Fig. 14.3) constitutes a key infrastructure for the world's social metabolism as more than 80 % of international trade in goods (both raw materials and manufactured goods) by volume is carried by sea. Material flows resulting from international trade (direct import and export flows in terms of their weight) are part of physical accounting methods, such as material flow analysis (MFA) (EUROSTAT 2001; Vallejo 2010), used to quantify 'social metabolism' processes (Fischer-Kowalski 1998). In 2010, developed countries accounted for 34 % of goods loaded and 43 % of goods unloaded in tons, while developing countries accounted for 60 and 56 %, respectively (post-communist European transition

[4] If not diversely specified, data for this section comes from Review of Maritime Transport (UNCTAD 2011). All presented data refers to vessels of 100 gross tons (GT) and above.

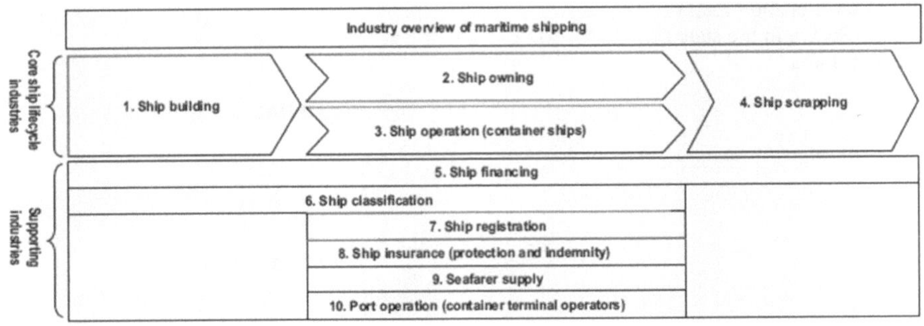

Fig. 14.3 Maritime sectors along a ship's lifecycle. (Source: UNCTAD, Secretariat)

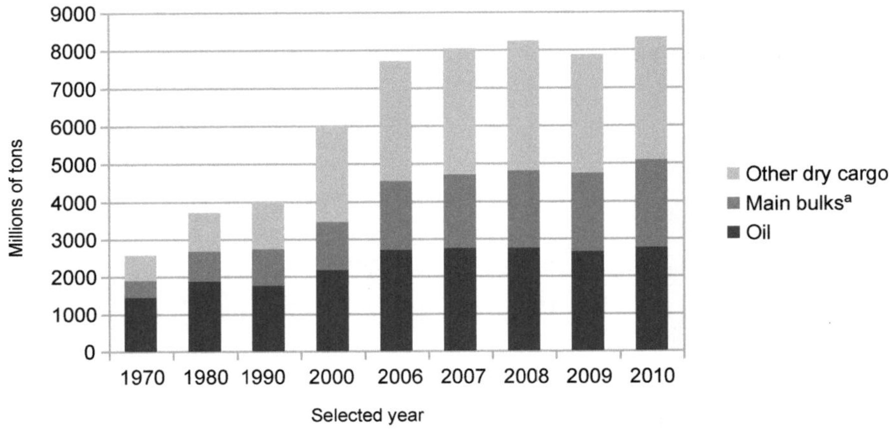

Fig. 14.4 Development of international seaborne trade, selected years (*millions of tons loaded*). **a** Iron ore, grain, coal, bauxite/aluminia and phosphate. (UNCTAD, review of maritime transport, various issues)

economies account for the rest). Some regions are characterized by a physical import surplus while others face a physical trade deficit (Giljum and Eisenmenger 2004).

Since the 1990s 'International seaborne trade' (goods loaded) increased faster than world gross domestic product (GDP), highlighting the effects of changing production processes, consumption patterns and the deepening of economic integration (globalization). In 2010 this trade reached 8.4 billion t, from 2.5 billion t in 1970. Figure 14.4 shows the historical evolution per type of cargo for selected years. Data from 2009 reflects the economic crisis.

As a direct consequence, the number and capacity of ships have significantly increased. In 1960, the world ocean-going fleet was composed by 15,000 ships (84 million of deadweight tons; dead weight tonnage (DWT) a measure of how much weight a ship is carrying or can safely carry), while in 2011 it had reached 103,392

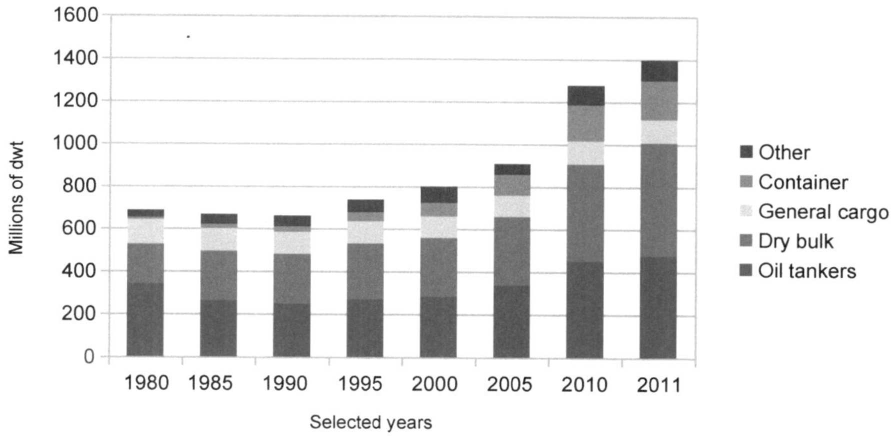

Fig. 14.5 World fleet by principal types of vessel, selected years. (Note: vessels of 100 gross tons (GT) and above; UNCTAD 2011)

(1396 million of DWT). Figure 14.5 shows the composition of the world fleet by principal types of vessel, selected years.

In 2007, developed countries controlled about 65.9 % of the world DWT, developing countries 31.2 % and economies in transition the remaining 2.9 %. In 2011, the four top ship owning economies (Greece, Japan, Germany and China) together controlled 50 % of the world fleet. Fleet ownership, however, does not always reflect ship registration. Foreign flagged ships accounted in 2011 for 68.1 % of the world total, most of them registered in the so-called states of convenience (or open registers). The top five registries (Panama, Liberia, Marshall Islands, China Hong Kong and Greece) together accounted for 52.6 % of the world's DWT. Figure 14.6 shows ship entries at Alang–Sosiya ship breaking yard, India, in 2004–2005 by ship owner's country: 82.5 % of them used a flag of convenience.

Flags of convenience, together with fiscal havens, shell companies and cash buyers, allow under-invoicing (resulting in evasion of import tax and money laundering) and facilitate ship owner's access to the ship breaking market.

This increase in the size of the world fleet does not immediately lead to a general increase in the supply of ships for scrap (Fig. 14.7).

Ship owners evaluate the expected future earning potential and the expected cost of keeping the ship in operation against the revenue obtained when the vessel is sold for scrap. This mainly depends on the price of steel. Potential earnings are more important in the decision than the scrapping price. The 2008–2009 economic crisis resulted in a boom of ship breaking because of excess shipping capacity (Fig. 14.7), with ship owners associations planning to eliminate 25 % of the world fleet. In fact, according to the data elaborated by the French NGO *Robin de Bois*, if in 2006 demolitions were equivalent to only 0.6 % of the existing fleet (293 vessels), the economic

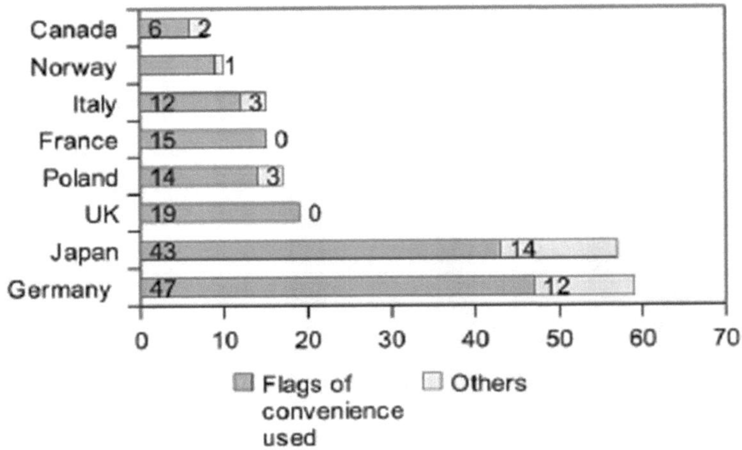

Fig. 14.6 Ship entries at Alang–Sosiya ship breaking yard, India, 2004–2005. (GMB; www.gmbports.orgs)

crisis reversed the situation (288 vessels in 2007, 456 in 2008, 1.006 in 2009,[5] 956 in 2010[6] and 1.020 in 2011[7]). The excess supply is reflected since 2009 in the spectacular fall in the Baltic Dry Index that measures the rates charged for chartering dry bulk cargoes. In 2011, strong steel prices and the recovery of maritime business increased costs for ship procurement but at the same time significantly increased the margins in the ship scrapping business.[8] In general, ship owning companies look to sell their ships for demolition at the best price.

The 2011 United Nations Conference on Trade and Development (UNCTAD) report rightly argues that 'the competitiveness of a country's scrapping industry is mostly influenced by labour costs and the regulatory environment. All major ship scrapping countries are developing countries' (pp. 151). In other words, ships go for scrapping wherever is easier to externalize social and environmental costs.

---

[5] In 2009 of 1.006 vessels (8.2 million tons), 435 were demolished in India (43 %), 214 in Bangladesh (21 %), 173 in China (17 %), 87 in Pakistan (9 %), 42 in Turkey (4 %). Robin de Bois, Information Bulletins on Ship Demolition: #17, September 2009; #18, January 2010. www.robindesbois.org.

[6] In 2010 of 956 vessels (6.5 million tons), 422 were demolished in India (44 %), 135 in Turkey (14 %), 125 in China (13 %), 90 in Pakistan (9 %), 79 in Bangladesh (8 %), 5 in Europe (1 %) and 100 in other countries (10 %). Robin de Bois, Information Bulletins on Ship Demolition: #19–22, January 2011. www.robindesbois.org.

[7] In 2011 of 1.020 vessels (8.2 million tons), 458 were demolished in India (45 %), 154 in Bangladesh (14 %), 142 in China (14 %), 108 in Pakistan (7 %) and 19 in Turkey (2 %). Robin de Bois, Information Bulletins on Ship Demolition: #23–26, February 2012. www.robindesbois.org.

[8] Article by Xu Hui, Executive Manager, China Ship Fund. Available at http://www.chinadaily.com.cn/bizchina/2010-04/08/content_9703387.htm (accessed in January 2012).

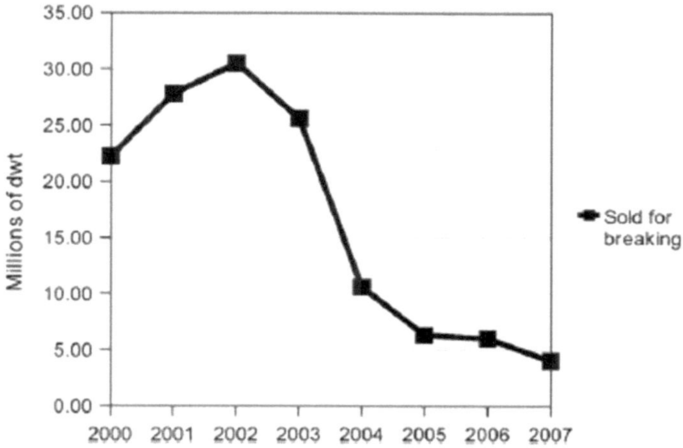

Fig. 14.7 Tonnage reported sold for breaking at the world level, 2000–2007 (millions of DWT). (UNCTAD 2007)

Fig. 14.8 Alang–Sosiya ship breaking yard (April 2009)

Companies look to sell their ships for demolition at the best price. South Asian yards are main destinations. For processing capacity ASSBY in India and Chittagong in Bangladesh are the world's biggest yards (Fig. 14.8); Chinese yards are catching up to them.

Again, according to *Robin de Bois* (Fig. 14.9), in 2011 India continues to be the undisputed leading country not only per number of units, but also for tonnage (45 %

2011	N° of vessels	%	Tons of metal	%
India	458	45	3,500,000	43
Bangladesh	145	14	1,600,000	19
China	142	14	1,000,000	17
Pakistan	108	11	1,000,000	13
Turkey	72	7	206,0003	3
USA	19	2	131,0001	1
Europe	5	1	-	-
Others	55	6	-	-

Fig. 14.9 Numbers and tonnage of ships dismantled (2011). (Robin de Bois, Information Bulletins on Ship Demolition: #23–26, February 2012)

of the total 8.2 million t) followed by Bangladesh (14 %), China (14 %), Pakistan (11 %) and Turkey (7 %).[9]

Data on ships sent for scrapping are not easily accessible.[10] The 2011 report by UNCTAD presents statistics (Fig. 14.10) based on data from the information company IHS Fairplay (Maritime Intelligence and Publications). Data differ significantly in terms of tonnage but the country ranking remains the same, where the four largest ship scrapping countries covered 98.1 % of the activity in terms of recycled DWT in 2010. This data series allows to see on which types of ships the different countries specialize: India on tankers, dry cargo and passenger ships; Bangladesh and Pakistan on tankers; and China on bulk carriers.

In Bangladesh, since 2010, the industry has only being working periodically due to an intensification of the controversy around safety and environmental concerns. The Supreme Court had suspended the authorization of beaching following an umpteenth fatal accident in 2009 and a new action by the NGO Bangladesh Environmental lawyers Association (BELA[11]) who demanded compliance with environmental and social standards (UNCTAD 2011). Attempts by shipyards to circumvent the Court's decision had been successful, but the activity was then again suspended due to new fatal accidents (at least 12 workers have died in 2011). All site activity had then been stopped pending an investigation report and dismantling authorizations for new ships were suspended (Robin de Bois 2011). Later, the

[9] Robin de Bois, Information Bulletins on Ship Demolition: #23–26, February 2012. www.robindesbois.org.

[10] Database from the French NGO Robin de bois is public and presents a lot of details for each ship sent for scrapping. Instead IHS Fairplay data might be more exhaustive, but is less transparent and detailed (therefore difficult to assess), and only accessible by paying an expensive fee. For the purpose of this analysis the two are complementary and do not contradict each other.

[11] http://www.belabangla.org/.

				Scrapped ships, percentage of total volume				
Country	Scrapped amount, dwt	Number of ships scrapped	Rank	Bulk carriers	Dry cargo / passenger	Offshore	Tankers	Others
India	9,287,775.00	451	1	9.7	32.8	5.3	46.2	5.9
Bangladesh	6,839,207.00	110	2	15.1	5.5	5.7	71.1	2.5
China	5,769,227.00	189	3	46.6	36.3	2.5	12.2	2.4
Pakistan	5,100,606.00	111	4	8.1	2.9	6.2	80.6	2.2
Turkey	1,082,446.00	226	5	24.3	48.7	0.2	14.1	12.8
United States	217,980.00	15	6	0.0	19.9	0.0	80.1	0.0
Romania	16,064.00	4	7	0.0	100.0	0.0	0.0	0.0
Denmark	15,802.00	25	8	0.0	53.4	22.7	0.0	23.9
Japan	13,684.00	1	9	0.0	100.0	0.0	0.0	0.0
Belgium	8,807.00	12	10	0.0	100.0	0.0	0.0	0.0

Fig. 14.10 Top-ten ship-scrapping nations, 2010. (Source: Compiled by the UNCTAD secretariat on the basis of data from IHS Fairplay)

industry restarted operations. In any case, this is probably not the end point of ship breaking in Bangladesh.[12]

China has overtaken Pakistan and reached the levels of Bangladesh. It keeps growing rapidly thanks to modernization of its industry, lower taxes and the complete lack of democratic control over accountability for social and environmental impacts. New large facilities have been built near Shanghai in association with major shipping and other companies (including the Peninsular & Oriental Steam Navigation Company and British Petroleum) who have guaranteed a steady supply of ships for breaking. In exchange, the Chinese firms have promised good environmental controls and safe working conditions for the workers. In fact, ship owners are under public scrutiny in their countries for being the source of alleged misery in ship breaking countries (Wayne Hess et al. 2001). Then, one could wonder why they are doing it in the less transparent country (China) which keeps strictly under control labour trade unions and environmental NGOs, a part from denying access to researchers.

Ship breaking yards in Europe and the USA receive very few ships, as the prices they can offer are close to zero, and tend to receive state-own ships, like the ones from the navy.

14.3.2 History of an Industry

Ship breaking first developed in the USA, UK and Japan during World War II (WWII) because there were many ships damaged by war, and an urgent demand for steel. In the 1960s, it moved to less industrialized European countries such as

[12] http://www.recyclinginternational.com/recycling-news/5930/ferrous-metals/bangladesh/recovery-bangladesh-shipbreaking-tonnages.

Spain, Italy and Turkey. In the 1970s, it left Europe and established itself in Asia, first in Taiwan and South Korea, and then during the 1980s, in China, Bangladesh, India, Pakistan, Philippines and Vietnam. South Asian countries have benefited from favourable natural characteristics (high tidal ranges, gentle sloping and rocky bottom beaches) which allow the vessels to be beached, turning a highly mechanized industry into a labour intensive one.

14.3.3 From the Ship Owner to the Ship Breaker Through Cash Buyers

Ship owners sell their ships through brokers operating in London, Dubai, Singapore and Hamburg. All ships are sold per ton (light displacement tonnage; LDT[13]) at a price ranging from US$ 100 to 400, depending on the ship type and on the market. In the last 10 years, 'cash buyers' have emerged as important intermediaries officially to assure fulfillment of the contract. They differ from traditional ship brokers because they acquire ship ownership, becoming themselves ship owners (although only for a limited period pending its sale or during the handing over of the ship to a recycling facility). Original (last operational) ship owners get lower prices, but this system allows them to bypass liabilities and regulations (Hillyer 2012).

14.3.4 ASSBY: Alang–Sosiya Ship Breaking Yard

The first ship, called 'Kota Tenjong', was beached in Alang on the 13th of February 1983. Alang–Sosiya ship breaking yard (ASSBY), which occupies 10 Km of coastline, became in the 1990s the world largest ship breaking yard. In 2007, India accounted for 41 % of the world's recycling capacity, 90 % of it taking place in ASSBY (Fig. 14.11; Table 14.1).

14.3.5 Ship Breaking Process

Once a ship arrives in the Gulf of Cambay, it is inspected and checked by the competent authorities which issue (occasionally with corruption) the relevant certificates. The ship is then beached by its own propulsion power at high tide and during low tide is laid down stable on its flat bottom. At this point cutters and their helpers, using simple liquid petroleum gas (LPG) gas and oxygen torches, can start taking apart the vessel structure. All operations take place directly on the beach in a

[13] LDT is the mass of the ship excluding cargo, fuel, ballast water, stores, passengers and crew.

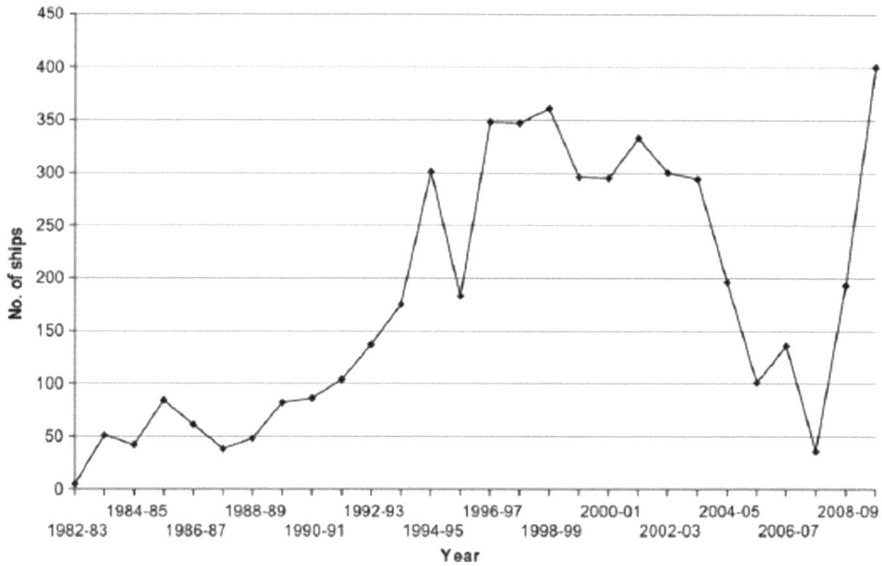

Fig. 14.11 Number of ships broken per year at ASSBY. (GMB; www.gmbports.org)

relatively small and congested area called a plot. Machinery and heavy equipment (engines, compressors, generators and boilers), together with other dismantled components (navigation equipment, life saving equipment, furniture, electrical cables, utensils, etc.) are sold to traders for reuse. These operations do not require investment in infrastructure or technology, as they are labour intensive and moving cranes and motorized winches are reused from the same ships. Depending on their size and type, scrapped ships have an unloaded weight of between 5000 and 40,000 t, with an average composition as shown in Table 14.2. It requires from 3 to 6 months for an average ship (15,000 t) to be dismantled with a variable number of workers involved at different stages (from 150 to 300). The industry requires relatively low-fixed capital (plot lease, machinery and equipment) and high-working capital. The cost of the vessel itself corresponds to more than 50 % of the total cost. Interests on investment, duties (customs, excise, value added tax—VAT, etc.) and port charges represent the second important item. Labour and energy (torch oxygen and fuel) constitute each between 3 and 6 % of the total expenditure (Upadhyay 2002; Dubey 2005). Environmental, safety and health insurance costs do not appear in the accounting.

Table 14.1 Number and LDT of ships broken at ASSBY.[a] Gujarat Maritime Board (GMB; www.gmbports.org)

Year	Number of ships	LDT
1982–1983	5	24,716
1983–1984	51	259,387
1984–1985	42	228,237
1985–1986	84	516,602
1986–1987	61	395,139
1987–1988	38	244,776
1988–1989	48	253,991
1989–1990	82	451,243
1990–1991	86	577,124
1991–1992	104	563,568
1992–1993	137	942,601
1993–1994	175	1,256,077
1994–1995	301	2,173,249
1995–1996	183	1,252,809
1996–1997	348	2,635,830
1997–1998	347	2,452,019
1998–1999	361	3,037,882
1999–2000	296	2,752,414
2000–2001	295	1,934,825
2001–2002	333	2,727,223
2002–2003	300	2,424,522
2003–2004	294	1,986,121
2004–2005	196	938,975
2005–2006	101	480,405
2006–2007	136	760,800
2007–2008	36	643,437
2008–2009	193	–
2009–2010 (estimated)	400	–
Total	5033	31,877,972

ASSBY Alang–Sosiya ship breaking yard
[a]*LDT* light displacement tonnage (is the mass of the ship excluding cargo, fuel, ballast water, stores, passengers and crew

Table 14.2 Average components (both in weight and value) obtained by a demolished ship. Interviews with ship breakers; Upadhyay 2002

	Weight (%)	Value (%)
Rerollable ferrous scrap and iron plates	75–85	65
Reconditioned machinery	10–15	25
Remelting scrap	3	2
Nonferrous metal	1	7
Furnace oil and oils	2	0.50
Wooden and furniture	2	0.50
Burning, cutting losses and waste	5–10	0
	100	100

14.4 Hazardous Waste and Socio-environmental Impacts

14.4.1 Hazardous Waste Generation and Management

Ships contain (in-built and on board) hazardous and nonhazardous substances, significant both in quantity and toxicity, which cannot (or should not) be totally reused or recycled. The waste output of the process represents between 0.5 and 10 % of the ship's total weight (see Table 14.2). Composition is diverse, mainly constituted by scrap wood, plastic, paper, rubber, glass wool, thermocol, sponge, polyvinyl chloride (PVC) pipes, oil, metals, heavy metals, paints, cement, asbestos and radioactive waste. Independent and reliable statistics on quantity and composition are not available, while estimates are difficult because there are many different types of ships, which vary considerably in their structure (Reddy et al. 2005a, b).

The controversy over ship breaking mainly concerns the disposal of hazardous waste. There are three methods of disposal:

1) Decontamination Prior to Export Decontamination is the process of removing hazardous materials contained in the ship structure (partially or totally), normally without endangering sea worthiness. This must be done by ship owners. It is a costly operation that requires expertise and technology. A totally decontaminated ship would not fall under the Basel Convention.

2) Environmental Sound Management on Site Hazardous materials are safely removed and then properly disposed once the ship has been beached. This is the option recommended by the International Convention for the Safe and Environmentally Sound Recycling of Ships adopted in May 2009 by the International Maritime Organization (IMO).

3) Dumping Hazardous materials are freely released into the environment. ASSBY, since the beginning, has used the third method (HPC 2003; Reddy et al. 2003,

2005a, b). Waste, hazardous or not, has generally been directly released into the sea from the ship or the plot, burnt on the plot or dumped during the night in surrounding villages. Some has been transported and dumped in areas (like the surroundings of the city of Surat in the Golden Corridor) where other industries undertake similar actions so that it is impossible to identify the source and enforce any liability.

14.4.2 Pollutants Discharged

Scrapping activity discharges a number of liquid, gaseous and solid pollutants which are hazardous for the environment and human beings (Islam and Hossain 1986; Zhijie 1988; Hossain and Islam 2006, Neşer et al. 2012). Most common are oil, bacteria, asbestos, heavy metals[14] and persistent organic pollutants.[15]

14.4.3 Socio-environmental Impacts

14.4.3.1 Environmental Impacts

In ASSBY, waste materials accumulate over the soil and then ramify incrementally to seawaters in a stepwise manner through tidal and subtidal zones, deep sea and their respective sediments. This has led to a deterioration of physicochemical properties of seawater and intertidal sediments. Chemical oxygen demand (COD) and biological oxygen demand (BOD), used as indicators of water quality (organic degradation and tension in the system), are present at high levels. Ship breaking activity has substantially affected the ecosystem at Alang–Sosiya (GEC 1997; Tewari et al. 2001; Reddy et al. 2003; Reddy et al. 2004a, b, 2005a, b). System stress has led to a decline in biotic structure: a decrease in biomass, abundance and species diversity has been measured. Pollutants mix with suspended solid and migrate long distances, carried by high currents (Bhatt 2004). They have been found, to a lesser extent, together with floating objects and oil, all along the 100 Km of coastline on the East and West side of Alang (Pathak 1997; Mehta 1997). The exact spatial dispersion of contaminants remains unknown as all selected control sites (10, 30 or 50 Km away from Alang) have always affected by pollution (Dholakia 1997). The intertidal zone around ASSBY has practically no vegetation. Mangroves disappeared many years ago, soon after the industry began. The sea off ASSBY has very poor biological production potential with very low phytoplankton pigment concentration, low

[14] Mercury (Hg), Lead (Pb), Arsenic (As), Chromium (Cr), Copper (Cu), Manganese (Mn), Zinc (Zn) and Nickel (Ni).

[15] Polychlorinated Biphenyl Compounds (PCBs), Dioxins, PVC), Polycyclic Aromatic Hydrocarbons (PAHs) and Organotins (Monobutyltin—MBT, Dibutyltin—DBT, Tributyltin—TBT, etc.).

zooplankton standing stock, very poor macrobenthic standing stock and low numerical abundance of fish eggs and larvae (Soni 1997; Majumdar 1997). Exotic species might have been carried in with ballast water, which represents a serious biological risk. The population and diversity of fish have decreased and species tolerant to petroleum hydrocarbons seem to have adapted better to the environmental stress (Mandal 2004). The absence of sanitation facilities for the workers has led to the presence of pathogenic and nonpathogenic bacteria (fecal and nonfecal coliforms, salmonella, clostridium and staphylococcus) in the water of the ASSBY area (surface and underground) rendering it unsafe for human consumption while marine coastal water has become harmful for fish population and unsuitable for recreation (Desai and Vyas 1997; Trivedi 1997; MECON 1997).

There is a lack of studies into the potential impacts on local terrestrial ecosystems. Gujarat Pollution Control Board (GPCB), a local government agency, claims to keep a complete monitoring, but it has not made data available. For a comprehensive environmental impact assessment one should go beyond local impacts and analyze the complete material recycling chain (ancillary industries). The furnace emissions of rerolling mills are rendered toxic by the presence of volatile organic matter of marine paints and antifouling paints (such as lead, arsenic and pesticides) which has resulted in acid rain during the monsoon season (Bhatt 2004).

The 1997 Report by the Gujarat Ecology Commission 'Ecological Restoration and Planning for Alang–Sosiya' (GEC 1997) remains the most comprehensive study to date. None of its suggestions have been followed, so that the assessment maintains its validity as confirmed by more recent studies of the 'Central Salt & Marine Chemicals Research Institute' (Tewari et al. 2001; Reddy et al. 2003, 2004a, b, 2005a, b; Mandal 2004).

14.4.3.2 Impacts on Workers

Workers in ASSBY, mainly seasonal migrants from the poorer states of India (Orissa, Bihar, Uttar Pradesh and Jharkhand), live and work in pitiable conditions (FIDH 2002; IMF 2006). They migrate as a survival strategy because with their previous jobs (at US$ 1 per day) and small farms they are unable to maintain their families. Their number varies from 5000 to 50,000. They work under contractors, on a daily basis and with no contract or rights. They work 12 h per day, 6 days per week. During the field work in Spring 2009, their daily salaries ranged from a minimum of ₹150 (US$ 3) for helpers and loaders to a maximum of 375 ₹(US$ 7) for experienced cutters. They live, without their families, in shared shanties, locally called Kholi, close to the yard with no running water, electricity or sanitation. They are continuously exposed to pollutants, from the air they breathe, the water they drink and the fish that they eat (Deshpande et al. 2012). Notably their jobs present a number of hazards. Frequent accidents find their causes in fire and explosion, falling objects, trapping or compression, snapping of cables, falls from heights, and lack of personal protective equipment, housekeeping standards, and safety signs (ILO 2004). In case of injury or death, they are rarely compensated (Rousmaniere and Raj 2007). Local fishers report that severely injured workers are sometimes dumped at sea and left to drown. The Final Report of the Technical Experts Committee (TEC), presented in 2006 to the Indian Supreme Court, offers an insight of the hazards faced by these workers. With regard to accidents, the Final Report notes that 'the average annual incidence of fatal accidents in the ship breaking industry is 2.0 per 1000 workers while the all India incidence of fatal accidents during the same period in the mining industry, which is considered to be the most accident-prone industry, is 0.34 per 1000 workers'. This is based on official data from 1995 to 2005 (roughly 40 traumatic work fatalities per year). It would be methodologically more accurate to correlate the number of fatal accidents to the number of dismantled ships, as workers do when they say 'one ship, one death'. Others say, 'one per day'. With regard to pollutants, the Final Report cites the 'Medical Examination of the Asbestos Handlers' by a team from the National Institute of Occupational Health (NIOH) which concludes, 'The X ray examination by NIOH showed linear shadows on chest X rays of 15 (16 %) of 94 workers occupationally exposed to asbestos. These are consistent with asbestosis...' There are no medical records on the short- and long-term effects of the workers' exposure to contaminants.

14.4.3.3 Impacts on Fishing Communities

The South Saurashtra coastal area has always been well known for fisheries of Bombay Ducks (Harpodon neherius), Hilsa, Prawns and other species. Fish catch in the gulf of Khambat is found to be rich on the Western side over 100 Km away. Data for fish catch for 1991 and 1995 are indicative. Table 14.3 shows the fish landing situation at Gogha, Bhavnagar Lockgate and Katpar. Gogha and Bhavnagar Lockgate are on the East side of ASSBY about 50 Km away. Katpar is on the West side of ASSBY again 50 Km away. From the data available for the commercially important fishes, a definite fall can be observed in the fish catch, apart from disappearance of certain species.

In the same area, there are about 2500 fishers living in small communities on the beach and in villages (from East to West: Gogha, Mithi Virdi, Sosiya, Alang, Talaja, Sartampar, Gopnath—Gadhula, Mahuva—Katpar). Fishing activity constitutes the main source of livelihood for about 10,000 people. Apart from Gogha where the majority is Muslim, they all belong to the Koli community. Kolis belongs to the scheduled tribes (ST), that (together with the scheduled castes, SC) are unprivileged population groups explicitly recognized by the Constitution of India.

Fishers report that, since ship breaking began, the quantity, variety and size of fish has decreased, the flavour has changed, and a number of species have disappeared. Others like mudskippers (an amphibious fish with a special air breathing system) have better adapted, but are normally less commercially valuable (apart from being contaminated).

Fishers report not to have noticed any damage to their own health due to pollution. However, a number of pollutants can bioaccumulate and enter the food chain. Heavy metals bioconcentrated in the fish have been found to be many times higher

Table 14.3 Fish landing in Kg for some species at different centres near ASSBY. Dholakia 1997

Name of fish	Ghogha		Katpar		Bhavnagar lockgate	
	1991	1995	1991	1995	1991	1995
Bombay duck	102,069	93,862	116,865	46,129	74,129	32,596
Hilsa	7020	Nil	31,762	15,860	–	–
Culpid	1860	Nil	22,905	23,390	–	–
Mullet	44,308	24,809	112,695	12,776	–	5689
Catfish	21,715	–	13,950	2250	Nil	–
Colmi(shrimp)	175,250	909,151	30,015	48,072	20,240	62,004
Medium prawn	704,179	408,121	108,534	18,690	78,180	27,831
Jumbo prawn	214,314	80,400	30,225	Nil	–	–
Lobster	87,141	21,199	1500	2769	3162	110,639
Colia	–	–	3348	–	–	–
Dhoma	–	–	11,487	3565	–	–
Other fish	420,538	186,427	106,951	27,854	34,056	–

ASSBY Alang–Sosiya ship breaking yard

than the maximum prescribed (Mehta 1997). These highly toxic fish are not suitable for human consumption. However, they are locally caught, consumed (mainly by fishers and ship breaking workers) and go in dried or fresh form all over India and abroad. Fish can swim long distances and be caught elsewhere. High levels of butyltin, a Persistent organic pollutants (POP), have been found in fish for consumption in the entire Asian–Pacific region. Ship breaking, along with sewage disposal and antifouling paints, is considered the main source of this (Kannan et al. 1995).

There is no simple solution for fishers. They cannot easily fish elsewhere for a better quality of catch: The area of pollution is very wide, and fishers are not readily mobile. They are also constrained by legal restrictions on where they may fish. In consequence, the quality of life of all the community has worsened significantly. The most vulnerable have to work as unskilled labourers while others have emigrated in search of better opportunities. This picture is very similar to the one of Chittagong (Bangladesh), the second world largest ship breaking yard: 'As the commercially important species are replaced by low priced species and scarcity of fish, many coastal fishers are leaving their hereditary profession and moving around everyday as environmental refugees in a state of under employment and poverty to unemployment and grim poverty' (Hossain and Islam 2006).

14.4.3.4 Impacts on Villagers

The ten villages in a radius of 12 Km (Alang, Sosiya, Manar, Sathara, Kathwa, Bharapara, Mathavada, Takhatgadh, Jasapara and Madva) have experienced great economic and social changes because of ship breaking (UNESCO 2001). Previously working mainly in agriculture, after the arrival of ASSBY they could find new employment and business opportunities (often in accordance to their caste) in transportation, trade and retail (Chaudhari 1999). Some of the environmental impacts are of concern for the villagers. Those living close to the operation yards are affected by noise pollution. More generally people complain to Sarpanches (heads of villages) and local authorities about the dumping of waste from the dismantled boats as there are hundreds of dumping sites in all the surroundings. Preferred sites are waste lands, traditionally used for grazing, but also farming fields; people report that oxen and cattle have died because of eating waste. Villagers report respiratory and skin problems particularly when the waste is set on fire. Most of the villages along the coastline in this region suffer from water scarcity and salinity. The industry has worsened the problem inducing overexploitation of water reserves (through population growth and workers immigration) leading to a decrease in groundwater level. Apart from the deterioration of agriculture and animal husbandry, villagers report kidney diseases that are related to both salinity and pollutants. A number of wells are so polluted that they have been abandoned. Modern and traditional forms of agriculture coexist—for respectively large and small farms—growing mangos, chikos, coconuts and onions. Since the industry has settled, land and labour prices have increased locally. On the other hand, the quantity and the size of crops have decreased, and the flavour has changed.

14.4.4 Emergence of a Conflict: From Material Origins to Cultural Discourses

If this was the end of the story, it would simply confirm the Lawrence Summers' Principle. Instead the next section shows the emergence of a conflict where disputes about values are vocalized. The conflict has material origins that are then shaped by cultural discourses. As discussed in social movement theory, diagnosing a problem (such as ship breaking) turns out to be a very contentious process, where the different actors try to affirm and impose their interpretative frame to the detriment of representations proposed by the others (Snow et al. 1986). The construction of reality is inextricably linked to asymmetries of power (Della Porta and Diani 2006).

14.5 Looking Closer at the Ecological Distribution Conflict: The 'Blue Lady' Case at the Supreme Court (2006–2007)

14.5.1 Three Spatial Scales for the Conflict: International, National and Local

In the late 1990s, the Alang and Sosiya landscape attracted worldwide interest in terms of its aesthetics as an industrial and social inferno. Although environmental and labour groups started structuring their complaints, its socio-environmental aspects are still neglected. The conflict has developed at three different scales (international, national and local) with environmentalists playing a major role, accompanied by trade unions and human rights groups, together with industrial lobbies, the Gujarat and Indian governments, and as so often in India, the judiciary.

At the international level, environmental NGOs, including Greenpeace and Basel Action Network (BAN), carried out campaigns to raise public awareness in developed countries and lobby for the implementation of regulations (notably the Basel Convention). In 2005, the 'Platform on Shipbreaking' (www.shipbreakingplatform.org) was created as an international network of environmental, human and labour rights organizations to challenge the global shipping industry. In India, environmental NGOs (like Toxic Links, Corporate Accountability Desk and Human Rights Law Network) and independent activists (like the researcher Gopal Krishna, the activist Madhumitta Dutta, the lawyers Bushan Oza and Colin Gonsalves) engaged in judicial activism and fight to this day on the Civil Written Petition on Hazardous Waste Management first filed in 1995 to the Supreme Court by the 'Research Foundation for Science, Technology and Natural Resources policy'.

At the local level the conflict has remained latent. Seasonal workers are vulnerable because of their precarious social and economic condition and so can be easily kept under pressure and domination. A local trade union (Alang–Sosiya Ship Recycling and General Workers' Association), with limited power, exists, accepted since 2005 by ship breakers to negotiate wages. Workers report the use of violence (by the local police) against sporadic attempts of strikes over dispute about salary, safety, working and living conditions.

Villagers have expressed oral complaints to authorities, normally through the heads of villages. Some attempts of frame bridging (Snow et al. 1986) are being undertaken by national activists between villagers (environmental issues) and workers (working and living conditions). The alliance could potentially be strong, especially in case a common organizational base can be built. Ship breakers actively oppose the process with threats and a 'divide et impera' strategy.

Media coverage of the human and environmental conditions at ASSBY obliged competent Indian authorities and International Organizations (United Nations Environment Programme, UNEP; International Labour Organization, ILO; and IMO) to react. Both attempted to assess the main issues at stake, tackling them with detailed policy-making initiatives (mostly technical guidelines) and more effective implementation (Basel Convention 2002; IMO 2003; ILO 2004). The proposed practices (i.e. technology to improve labour safety and environmental protection) are similar to the ones used in developed countries. The industry left those shores to avoid the rules. Technically correct, but politically naive, none of them has been enforced. Instead, in order to understand the situation, the case of the 'Blue Lady' at the Supreme Court of India is presented hereafter. This is not the most famous one (this would be the Clemenceau in 2006) but it illustrates the issues at stake, the decision-making process and the valuation languages deployed at different scales by different actors of the conflicts over this type of waste disposal.

14.5.2 History of the 'Blue Lady' Last Voyage

SS France was built in 1960 by the French Line and was at that time the longest passenger ship ever built. It had a mass of 45,000 t, was 316 m long and 34 m wide, and had 16 floors and 1400 rooms.

In 1979, it was sold to Norwegian Cruise Line (the mother company Star Cruise Ltd—SCL), renamed SS Norway and transformed into the world's most glamorous cruise ship. Seriously damaged in 2003 by a boiler explosion in Miami, it was towed to Germany where repairs were planned. A feasibility study, in 2004, estimated that to decontaminate part of the in-built asbestos would cost 17 million—Euros. In 2005, the ship left Germany, its official destination being Singapore, for reuse. The ship owner intentions were to discard the ship, and therefore the SS Norway became 'waste' under the EU Waste Shipment Regulations. Moreover, since it contained hazardous substances, it could have been considered hazardous waste for the purpose of the Basel Convention. Under Article 9 and Basel Ban Amendment, the export of ships from Organization of Economic Co-operation and Development (OECD) countries to non-OECD countries, should be conceived as illegal traffic (Moen 2008). The ship arrived in Malaysia and was planned to be scraped in Bangladesh.

However, due to protest by BELA the sale was declared invalid. In 2006, SS Norway left Dubai the authorities being informed that it was going for repairs, as in reality it was sailing towards Alang, to be dismantled.

In June 2006, the mother company Star Cruise (Malaysia) sold it through Norwegian Cruise Line (Bermuda) to the Liberian (shell) company Bridgeed Shipping for, officially, US$ 10 (as indicated in the Bill of sale for Bahamian ships). Bridgeed sold it, after 1 month, to the Indian ship breaking company Hariyana Steel Demolition Pvt. Ltd. The ownership was then transferred again to another ship breaking company, Priya Blue Industries Pvt. Ltd. The ship, finally renamed Blue Lady, apart from the common practice of under-invoicing, had a real price of about US$ 15 million.

On May 2006, the ship was initially prevented from entering Indian waters by an application of the activist Gopal Krishna to the Supreme Court of India. On humanitarian grounds, because of the monsoon, the ship was allowed, on June 2006, to anchor at Pipavav port near Alang. It was finally beached (without permission) on the 3rd of August 2006 and allowed to be dismantled by the Final Court Order of the 11th September 2007.

14.5.3 The Case in the Supreme Court: Arguments and Languages of Valuation

The analysis of a judicial case, such as the 'Blue Lady' one, offers an insight into the framing conflict, meaning the struggle over reality construction. Different actors participated in the 'politics of signification' (Hall 1982). There were environmentalists, villagers, ship breakers and the Indian authorities. Actors involved are signifying agents engaged in the production of alternative and contentious meanings (Benford and Snow 2000). They undertook two core framing tasks: diagnostic and prognostic. The first concerns the definition of what the problem is and who is responsible; the second regards the proposed solutions. In particular, this section analyses the different attitudes expressed by these actors to the three methods of waste management, and the different valuation languages they used to frame the issue.

14.5.3.1 Environmentalists

Indian environmentalists, in alliance with international organizations, challenged ship breaking according to languages of justice, economics and legality. Using a justice discourse and invoking the Basel Convention they described it as an 'illegal export of toxic waste' from rich to poor countries highlighting impacts on the environment, and the health and livelihood of workers and local communities. 'Blue Lady' was a case of 'toxic imperialism', Gopal Krishna argued. The practice is perceived as environmental injustice or environmental racism on a global scale (Lipman 1998) and a human rights violation. Orthodox economic language (such as 'internalizing externalities' or 'polluter pays principle') was also strategically adopted by Greenpeace, BAN and the Ship Breaking Platform. While they are conscious of the pitfalls of economic values, the polluter pays principle offered a suitable language to link questions of economics with questions of justice. Finally, the last language used was one of compliance with the rule of law. Indian activists often claimed that existing legislation had been violated. Gopal Krishna, in his application, called for respect for the 2003 Supreme Court Order including prior informed consent, inventory of hazardous waste mandatory for ship owner, decontamination by the ship owner prior to export, proper removal and waste management (with special attention to asbestos) and transparent pollution monitoring by GPCB. He pointed that the Blue Lady carried 1250 t of asbestos, the import of which is banned under the Basel Convention and Indian Hazardous Waste (management and handling) Rules, 2003.

Finally, he claimed that ILO standards on occupational and environmental health hazards had not been respected.

14.5.3.2 Ship Breakers

The position of ship breaking companies has always been articulated by Mr Nagarsheth, historical president of the Iron Steel Scrap and Shipbreakers Association of India (ISCSAI). The argument was basically made on economic values, though also environmental values were employed.

In their Application on February 2006 (IA 25), they highlighted ASSBY's contributions to the economy claiming that more than 100,000 people were in direct and indirect employment, up to 2.5 million t of good quality and cheap steel (approximately 5 % of the domestic demand) had been returned to market, and that ₹20 billion (US$ 400 million) had been raised by the authorities in the form of customs duties, income and sales taxes. Mr Nagarsheth presented ship breaking as an environment-friendly activity because, by recycling the materials, it saves nonrenewable resources (such as iron ore and the energy needed to produce primary steel). In contrast to other methods of manufacture, especially steel manufacturing, it does not produce solid waste. This position is apparently supported by the IMO which has declared ship breaking a Green industry.

Mr Nagarsheth claimed that occupational hazard is the issue, and not the environmental impacts as publicized by media and activists. The recognized hazard could be met by resorting to new technologies. In fact, Mr Nagarsheth declared to be committed to proper waste management without causing any harm or damage to human life or to the environment. Finally, he refused to see ships as hazardous waste, so that national and international legislation for the transport of hazardous waste would not apply. What is interesting here is that while business interests employ the same languages (economic, environmental and distributive) as the environmental groups, they frame very different and at very different scales the issues, to come to very different evaluative conclusions.

14.5.3.3 Indian Authorities

The Ministry of Environment and Forestry (MOEF) together with three other ministries (steel, shipping and labour) are in charge of the policy making on ship breaking; GMB and GPCB deal with the local implementation of rules and regulations. Their valuation premises and positions were very close to those of ship owners and ship breakers, emphasizing public benefits in terms of economic and environmental values at the national scale. According to the authorities, ships are not waste, hazardous substances are managed in an environmentally sound manner and workers safety is under control. Since there was never any pollution, no remediation is necessary. The Menon Committee (HPC), constituted by the Supreme Court in 1997, presented a Report (background to the 2003 Court Order) that represents the sole governmental admission of severe pollution and inhuman conditions of the workers.

In the 'Blue Lady' case, the MOEF was in charge of the TEC on Management of Hazardous Wastes that the Supreme Court had established on March 2006, to investigate environmental protection, workers' safety and health. Committee ship inspections, such as the routine ones by GPCB, are visual, because in ASBBY there are no proper laboratory testing facilities. Gopal Krishna proved the conclusions of these inspections to be wrong. The committee had declared that 'presence of radioactive materials in a passenger ship like the "Blue Lady" is quite unlikely'. In fact, it contained 5500 fire detection points containing 1100 radioactive elements in the form of Americium-241.

14.5.3.4 Villagers

On March 2007, Mr Bhagavatsinh Halubha Gohil, Sarpanch of Sosiya (head of the village), filed an application on behalf of 12 sarpanches and 30,000 people who live within a distance of 1–25 Km from the ship breaking yard. The applicants were opposing the dismantling of the ship because of the damage it would do to the health of workers and villagers and the environment (the soil, sea food, water, air, flora and fauna) on which the livelihood of the people depend (the majority of the population consist of farmers and fishers).

They acknowledged that the 'scrapping of the ship was vigorously opposed by environmental groups in India, as the Indian breakers did not have the facility or technology to safely dispose off the estimated 1000 t of asbestos'. Three main reasons prompted them to take legal action. First, they presented a study that they had commissioned to a consultancy about the potential health dangers due to cancerogenic effects of the airborne dispersion of the asbestos fibres contained in the ship. Second, they mentioned how open dumping of waste into the sea had affected fishers forcing them out into the sea beyond 5 or 6 km because of the oil that spreads over the water, ruining fishing. Third, they explained how in the past 15–20 years farmers had been noticing that the yield of their crops was diminishing. Even though its cause had not been easy to pinpoint, they had come to the conclusion that this was related to air, water and soil contamination brought on by the work at ASBBY. The villagers requested their inclusion in the Civil Writ Petition No. 657 of 1995[16] (a public interest petition on the question of hazardous waste import), that the Blue Lady not be allowed to be dismantled at ASSBY and asked for social and environmental justice. Notably, in an interview for the Indian magazine Frontline, Mr Gohil, promoter of the petition and Sarpanch of Sosiya, clarified their intentions declaring:

[16] Available at http://www.elaw.org/node/1400.

'we don't want to stop ship breaking because that would mean loss of jobs for hundreds of people. All we are asking is that it should be done in a responsible manner and our lives and earnings are not affected'.[17]

Witnesses report that a Judge snubbed the petition and the validity of the knowledge of the local people commenting 'What do these people know about asbestos?'

14.6 The Final Court Order on 'Blue Lady'

The final Court Order was passed the 11th of September 2007. The villagers' petition was never taken into consideration. The Court considered whether permission should be granted for dismantling of the ship 'Blue Lady' at Alang, Gujarat. The Court mentioned that the vessel Blue Lady would give employment to 700 workmen, provide the country with 41,000 t of steel and reduce pressure on mining activity elsewhere. So that, in the framework of sustainable development (recognizing recycling as a key element), the precautionary and polluter pays principles (said to be accepted and settled in Indian Law) should be considered together with the concept of balance (between economic development and environment) under the principle of proportionality (declared to be important in an emergent economy). These considerations, together with the technical and scientific suggestions by the Technical Expert Committee (TEC), supported the conclusion that: 'It cannot be disputed that no development is possible without some adverse effect on the ecology and the environment, and the projects of public utility cannot be abandoned and it is necessary to adjust the interest of the people as well as the necessity to maintain the environment. A balance has to be struck between the two interests. Where the commercial venture or enterprise would bring in results which are far more useful for the people, difficulty of a small number of people has to be bypassed. The comparative hardships have to be balanced and the convenience and benefit to a larger section of the people has to get primacy over comparatively lesser hardship'. The intention, as declared by the Court, was to balance the priorities of development (generation of revenue, employment and public interest) on one hand and environmental protection on the other. Under a general admission that activity needed to be strictly and properly regulated, the breaking of the Blue Lady was allowed. No quantification of costs and benefits was asked for by the Supreme Court, and neither was a multicriteria evaluation carried out.

[17] Shipload of trouble, Lyla Bavadam, Frontline, 16 Nov 2007.

14.7 Conclusion

Economic development, through economic growth and globalization, has considerably increased the magnitude of the global social metabolism. The shipping industry represents the key infrastructure through which material flows travel around the world. The increase in physical trade flows leads to a proportional increase in the shipping capacity (the number and size of ships), which leads—sooner or later—to an increase in the supply of ships for scrap.

This chapter investigated ship breaking in India's greatest yard, ASSBY, contextualing it within the world's social metabolism and analyzing the social, economic and institutional logics at play. At the Supreme Court of India competing value frameworks, languages of valuation and truth claims came at clash. Environmental and civil activists as well as business interests and public authorities framed the issue as one of economic, environmental and equity values. The former emphasized the injustice of an unequal distribution of costs and benefits and the disproportionate environmental and social damages at the local scale, considering local livelihood and ecosystem losses as incommensurable with benefits at other scales. The latter instead valued monetary and environmental benefits at the national scale, assuming them commensurable with local losses, and finding a positive balance. Facts apart, different languages of valuation clashed and the Supreme Court decided in the favour of the language of the powerful, interpreting sustainable development as a positive economic benefit at the national scale. 'Development' turned out to be the dominant ideology with a substantive power to signify. The Supreme Court's decision is based on a (controversial) utilitarian reasoning rather than on (Kantian) rights and, instead of recognizing value pluralism, the so called 'principle of balance' is based on a trade-off between development and environment that does not recognize the incommensurability among the expressed values. It rests upon the idea that economic benefits can compensate for environmental degradation. It would be interesting to know how would the Court undertake a cost–benefit analysis (CBA) and how much would it count a worker's life, how much the loss of livelihood and how much the irreversible damage done to local ecosystems.

The above are not just rhetorical questions. They show the irreducible difficulty in the call made by economists to internalize externalities. Furthermore, our analysis shows that there are important reasons why externalities are not internalized. The dumping of toxic waste, rather than a market failure, can be seen as a cost shifting success (Kapp 1950) this being made possible by social asymmetries in the distribution of political and economic power, property rights and income (Martinez-Alier and O'Connor 1999). Racism should also be accounted as a driving social force for environmental inequality (Pellow 2007), both at international and national levels, where in the Indian context, caste plays an important role (losers tend to be of lower caste than winners). Ship breaking can also be seen as an ecologically unequal exchange because of the 'externalization' of environmentally damaging disposal activities to the periphery of the world system as a consequence of exchange relations with more industrialized countries (Hornborg 1998). This is based upon the usurpation of waste assimilation properties of ecological systems in a manner that enlarges the domestic carrying capacity of the industrialized countries to the detriment of peripheral societies (Rice 2009).

Last, ship breaking can also be interpreted as a case of 'accumulation by contamination'. Marxist theory of over-accumulation argues that the economic crisis is due to a lack for capital of profitable investments opportunities. Dispossession is one strategy, among others, to overcome such crisis (Harvey 2003). Accumulation by dispossession is the inherent necessity of the capital system to separate, through extra-economic means (such as a change in law or violence), the labourers from the means of production to perpetuate the capitalistic relation (e.g. bioprospecting, patent rights, privatization of public utilities, etc.). Our analysis suggests the existence of a second strategy: accumulation by contamination. This is the process by which the capital system endangers, through cost shifting, the means of existence (and subsistence) of human beings to perpetuate the capitalistic relation (e.g. marine pollution, alteration of biogeochemical cycles, etc.). In the case of dispossession, something that was pre-existing outside the capitalist system is brought inside, i.e. privatization of the public assets or commons (Harvey 2003). Normally a specific social group is dispossessed by another one to obtain profit (e.g. farmers are dispossessed of their land by land grabbers). In the case of contamination, an appropriation of de-facto property rights takes place resulting in the shifting of costs and risks, i.e. exploiting the sinks over their sustainable assimilative capacity (e.g. climate change). The consequences most likely fall upon the most vulnerable social groups (e.g. small-scale farmers or fishers in the South), but the society as a whole can be affected.

Things could be done in a different (better) way and it is technically feasible to have a proper if more expensive dismantling operation (as in Europe). There are in fact many guidelines (Basel Convention 2002; IMO 2003; ILO 2004; European Commission 2007) which are simply not implemented. Ship owners could pay a deposit (or guarantee) throughout the ship life to be spent for proper dismantling, established as a requirement for allowing entrance at any harbour. This could allow investments to take the activity off the beach because dry docks operations potentially offer better labour and environmental standards. But all this would increase their costs, which is precisely the reason why they send boats to unregulated shipyards like those of ASSBY in the first place. This is why the calls for a green ship breaking have actually failed (Neşer et al. 2008; Khan et al. 2012). Using the terminology proposed by Martinez-Alier, ASSBY is therefore an example of the 'waste disposal frontier' of the world's social metabolism, where those who maintain the power (ship owners, ship breakers and authorities) manage to perpetuate a system of 'accumulation by contamination', exercising de facto property rights. Just as in the case with climate change, the crucial question is not only who is to pay and who is to be paid, but who is the owner of the sinks?

We have attempted to highlight the complexity of the linkages among nature, economy and society, showing how ecological economics and political ecology can contribute to its understanding. Whether improvements in ASSBY or other wasteyards of the global economy will ever become true, are then a question of social and political struggle and the ability of those who currently loose to affirm their own rights. From this perspective, greater and effective opposition encountered by ship owners and ship breakers regarding their shifting of environmental costs would result in improved sustainability, potentially both locally and globally. Locally for the pollutants that would not be discharged into the environment; globally because an increment in the operations' costs for the shipping industry, might slow down the social metabolism (by increasing the costs of trade) and its multiple impacts. People who struggle for environmental justice potentially contribute to the environmental sustainability of the economy (Martínez-Alier 2002).

Acknowledgements The author thanks Joan Martinez-Alier, Gopal Krishna, Giorgios Kallis, Giacomo D'Alisa, Giuseppe Munda, Ignasi Puig Ventosa, John O'Neill, Julien F. Gerber, Christos Zografos, Leah Temper, Richard Christian, D&D and all the interviewees. This chapter contributes to the projects EJOLT (2011–15), Environmental Justice Organizations, Liabilities and Trade (www.ejolt.org) and CSO2010-21979 on Social Metabolism and Environmental Conflicts.

References

Alter H (1997) Industrial recycling and the Basel convention. Res Conserv Recycl 19(1):29–53

Antweiler W, Copeland BR, Taylor MS (2001) Is free trade good for the environment? Am Econ Rev 91:877–908

Basel Convention (2002) Technical guidelines for the environmentally sound management for full and partial dismantling of ships. UNEP, Basel Secretariat, Geneva. www.basel.int. Accessed Jan 2012

Benford RD, Snow DA (2000) Framing processes and social movements. Annu Rev Sociol 26:611–639

Bhatt PN (2004) Monitoring of marine pollution at Alang–Sosiya Seacoast with Respect to Oil and Grease. Final Progress Report, Dept. of Analytical Chemistry. Bhavnagar University, Gujarat, India

Chaudhari S (1999) A study of port and Bhavnagar city, PhD thesis. Bhavnagar University Press, Gujarat

Clapp J (1994) The toxic waste trade with less-industrialised countries: economic linkages and political alliances. Third World Q 15(3):505–518

D'Alisa G, Burgalassi D, Healy H, Walter M (2010) Campania's conflict: waste emergency or crisis of democracy. Ecol Econ 70(2): 239–249

Della Porta D, Diani M (2006) Social movements: an introduction, 2nd edn. Blackwell, Oxford

Desai A, Vyas P (1997) Preliminary studies on microbial ecology of ship-breaking yard at Alang, Gujarat. Gujarat Ecology Commission, Ecological Restoration and Planning for Alang–Sosiya Ship-Breaking Yard, Gujarat

Deshpande PC, Tilwankar AK, Asolekar AA (2012) A novel approach to estimating potential maximum heavy metal exposure to ship recycling yard workers in Alang, India. Sci Total Environ 438:304–311

Dholakia AD (1997) Studies on coastal fauna. Gujarat Ecology Commission, Ecological Restoration and Planning for Alang–Sosiya Ship-Breaking Yard, Gujarat

Dubey MK (2005) A study on the role of banks in the development of ship breaking industry at Alang & Sosiya and its resultant impact on the economy of Bhavnagar District. PhD Thesis, Bhavnagar University

European Commission, Directorate General Environment (2007) Ship Dismantling and Precleaning of Ships. Final report. . Accessed Jan 2012

Eurostat (2001) Economy-wide material flow accounts and derived indicators: a methodological guide. EU, Luxembourg
FIDH (International Federation for Human Rights) (2002) Where do the "floating dustbins" end up? Labour rights in Shipbreaking Yards in South Asia. The Case of Chittagong (Bangladesh) and Alang (India). FIDH, Report No. 348/2, Paris
Fischer-Kowalski M (1998) Society's metabolism: the intellectual history of material flow analysis. Part 1. 1860–1970. Ind Ecol 2(1):61–78
Foster JB (1999) Marx's theory of metabolic rift: classical foundations for environmental sociology. Am J Sociol 105(2):366–405
Georgescu-Roegen N (1971) The entropy law and the economic process. Harvard University Press, Cambridge
Giampietro M, Mayumi K (2009) The biofuel delusion: the fallacy of large scale agro-biofuels production. Earthscan, London
Giljum S, Eisenmenger N (2004) North–South trade and the distribution of environmental goods and burdens: a biophysical perspective. J Environ Dev 13(1):73–100
Gujarat Ecology Commission (GEC) (1997) Ecological restoration and planning for Alang–Sosiya ship-breaking yard, Gujarat, Ahmedabad
Hall S (1982) The rediscovery of ideology: return to the repressed in media studies. In: Gurevitch M, Bennett T, Curon J, Woolacott J (eds) Culture, society and media. Methuen, New York
Harvey D (2003) The New Imperialism. Oxford University Press
High Powered Committee (HPC) (2003) Report on hazardous waste management. Supreme Court of India, New Delhi
Hillyer HH (2012) The hard realisty of breaking up: the global transboundary movement of ocean vessel demolition and waste. Vt J Environ Law 13:754–795
Hornborg A (1998) Toward an ecological theory of unequal exchange. Ecol Econ 25(1):127–136
Hornborg A, McNeill J, Martinez-Alier J (eds) (2007) Rethinking environmental history: world-system history and global environmental change. Altamira, Lanham
Hossain MM, Islam MM (2006) Ship breaking activities and its impact on the Coastal Zone of Chittagong. Towards Sustainable Management. YPSA (Young Power in Social Action), Chittagong, Bangladesh
International Labour Organization (ILO) (2004) Safety and health in shipbreaking: guidelines for Asian Countries and Turkey, Geneva. http://www.ilo.org/public/english/standards/relm/gb/docs/gb289/pdf/meshs-1.pdf
International Maritime Organization (IMO) (2003) Guidelines on ship recycling. IMO, London
International Metalworkers' Federation (IMF) (2006) Status of shipbreaking workers in India—A survey. IMF, Delhi
Islam KL, Hossain MM (1986) Effect of ship scrapping activities on the soil and sea environment in the coastal area of Chittagong, Bangladesh. Mar Pollut Bull 17(10):462–463
Kannan K, Tanabe S, Iwata H, Tatsukawa R (1995) Butyltins in muscle and liver of fish collected from certain Asian and Oceanian countries. Environ Pollut 90(3):279–290
Kapp KW (1950) The social costs of private enterprise. Harvard University Press, Cambridge
Khan I, Chowdhury H, Alam F, Kumar A (2012) Sustainable design of ship breaking industry in developing countries. Asian J Water Environ Pollut 9(1):1–11
Lipman Z (1998) Trade in hazardous waste: environmental justice versus economic growth. In: Proceedings of the Conference on Environmental Justice, Melbourne. www.spartan.unimelb.edu.au/envjust/papers. Accessed Jan 2012
Majumdar J (1997) Study of recent benthic foraminifera for evaluating environmental stress in and around Alang ship-breaking yard (Ghogha-Gopnath sector) Off Saurashtra Coast, Gujarat: A Case Report. Gujarat Ecology Commission, Ecological Restoration and Planning for Alang–Sosiya Ship-Breaking Yard, Gujarat
Mandal SK (2004) Studies on the effect of ship scrapping industry waste on marine phytoplankton at Alang, Gujarat. PhD Thesis. Bhavnagar University, India
Martínez-Alier J (2002) The environmentalism of the poor: a study of ecological conflicts and valuation. Edward Elgar, Cheltenham

Martinez-Alier J (2009) Social metabolism, ecological distribution conflicts, and languages of valuation. Capital Nat Soc 20(1):58–87
Martinez-Alier J, O'Connor M (1999) Distributional issues: an overview. In: Bergh J van den (ed) Handbook of environmental and resource economics. Edward Elgar, Cheltenham
McKee DL (1996) Some reflections on the international waste trade and emerging nations. Internat J Soc Econ 23(4/5/6):235–244
Mehta SNK (1997) Studies on offshore areas. Gujarat Ecology Commission, Ecological Restoration and Planning for Alang–Sosiya Ship-Breaking Yard, Gujarat
Metallurgical and Engineering Consultants (MECON) (1997) Shipbreaking industry—present status in India and its impact on environment. Union Steel Ministry, Delhi
Moen AE (2008) Breaking Basel: the elements of the basel convention and its application to toxic ships. Mar Policy 32:1053–1062
Neşer G, Ünsalana D, Tekoğula N, Stuer-Lauridsenb F (2008) The shipbreaking industry in Turkey: environmental, safety and health issues. J Clean Prod 16(3):350–358
Neşer G, Kontas A, Ünsalan D, Altay O, Darılmaz E, Uluturhan E, Uluturhan E, Küçüksezgin F, Tekoğulc N, Yercan F (2012) Polycyclic aromatic and aliphatic hydrocarbons pollution at the coast of Aliağa (Turkey) ship recycling zone. Mar Pollut Bull 64:1055–1059
Occupational Safety and Health Administration (OSHA) (2001) Ship Breaking Fact Sheet. U.S. Department of Labor, OSHA. www.osha.gov. Accessed Jan 2012
Pathak SJ (1997) Impact of Alang ship-breaking activity on water-sediment quality of the intertidal ecosystem at Alang–Sosiya complex and surrounding areas. Gujarat Ecology Commission, Ecological Restoration and Planning for Alang–Sosiya Ship-Breaking Yard, Gujarat
Pearson CS (1987) Multinational corporations, environment and the third world. Duke University Press, Durham
Pellow DN (2007) Resisting global toxics. MIT Press, Cambridge
Reddy SM, Basha S, Sravan Kumar VG, Joshi HV, Ghosh PK (2003) Quantification and classification of ship scrapping waste at Alang–Sosiya, India. Mar Pollut Bull 46:1609–1614
Reddy MS, Joshi HV, Basha S, Sravan Kumar VG (2004a) An assessment for energy potential of solid waste generated from a ship-scrapping yard at Alang. J Solid Waste Technol Manag 30:90–99
Reddy MS, Basha S, Kumar S, Joshi HV, Ramachandraiah G (2004b) Distribution enrichment and accumulation of heavy metals in coastal sediments of Alang–Sosiya ship scrapping yard, India. Mar Pollut Bull 48:1055–1059
Reddy MS, Basha S, Joshi HV, Sravan Kumar VG, Jha B, Ghosh PK (2005a) Modeling the energy content of combustible ship-scrapping waste at Alang–Sosiya, India, using multiple regression analysis. Waste Manag 25:747–754 (Oxford)
Reddy MS, Basha S, Joshi HV, Ramachandraiah G (2005b) Seasonal distribution and contaminations levels of total PHCs, PAHs and heavy metals in coastal waters of the Alang–Sosiya ship scrapping yard, Gulf of Cambay, India. Chemosphere 61:1587–1593
Rice J (2009) The transnational organization of production and uneven environmental degradation and change in the world economy. Int J Comp Sociol 50:215–236
Robin de Bois (2011) Information bulletins on ship demolition: #19–22. www.robindesbois.org
Rousmaniere P, Raj N (2007) Shipbreaking in the developing world: problems and prospects. Int J Occup Environ Health 13:359–368
Scott Frey R (1998) The hazardous waste stream in the world system. In: Ciccantell PS, Bunker SG (eds) Space and transport in the world-system. Greenwood, Westport
Singh S (2001). Domestically prohibited goods, trade in toxic waste and technology transfer: issues and developments. Discussion paper. CUTS, Centre for International Trade, Economics & Environment, Jaipur (India)
Snow D, Rochford B, Worden S, Benford R (1986) Frame alignment processes, micromobilization, and movement participation. Am Sociol Rev 51(4):464–481
Soni A (1997) Ecology of intertidal Macrofauna and literature review for Marine Biota. Gujarat Ecology Commission, Ecological Restoration and Planning for Alang–Sosiya Ship-Breaking Yard, Gujarat

Stuer-Lauridsen F, Husum H, Jensen MP, Odgaard T, Winther KM (2004) Oil tanker phase-out and the ship scrapping industry: a study on the implications of the accelerated phase out scheme of single hull tankers proposed by the EU for the world ship scrapping and recycling industry. Final Report, Brussels

Tewari A, Joshi HV, Trivedi RH, Sravankumar VG, Ragunathan C, Khambhaty Y, Kotiwar OS, Mandal SK (2001) Studies on the effect of shipscrapping industry and its associated waste on the biomass production and biodiversity of biota "in situ" condition at Alang. Mar Pollut Bull 42:462–469

Trivedi JM (1997) Microbiological studies. Gujarat Ecology Commission, Ecological Restoration and Planning for Alang–Sosiya Ship-Breaking Yard, Gujarat

UNCTAD (United Nations Conference on Trade and Development) (2007) Review of Maritime Transport. United Nations, Geneva. http://unctad.org/en/pages/publications/Review-of-Maritime-Transport-%28Series%29.aspx, http://unctad.org/en/pages/PublicationArchive.aspx?publicationid=1676

UNCTAD (United Nations Conference on Trade and Development) (2011) Review of maritime transport. United Nations, Geneva

UNESCO (2001) impacts and challenges of a large coastal industry. Alang–Sosiya ship-breaking yard, Gujarat, India. UNESCO, Paris

Upadhyay GB (2002) The problems and prospects of ship breaking industry in India with reference to Alang ship breaking yard. PhD Thesis, Bhavnagar University

Vallejo MC (2010) Biophysical structure of the Ecuadorian economy, foreign trade, and policy implications. Ecol Econ 70(2):159–169

Wayne Hess R, Rushworth D, Hynes VM, Peters EJ (2001) Disposal options for ships. National Defense Research Institute (U.S.), Rand Corporation. http://www.isbns.co.no/isbn/9780833030146

Yin R (2003) Case study research: design and methods. Sage, Thousand Oaks

Zhijie F (1988) Pollution from Chinese shipbreaking. Mar Pollut Bull 19(10):501

Chapter 15
Environmental Cost of Shrimp Farming in the Coastal Paddy Lands of South India

Umamaheswari Leelakrishnan, Nasurudeen Pathusha, Omar Hattab Kasim and Ravirajan Karunanidhi

15.1 Introduction

The trade policies for export promotion, increasing global demand and hike in the price of marine products have enhanced shrimp exports from India since the era of economic reforms in 1991. Frozen shrimp accounts for 51.35 % of the total value of marine product exports fetching foreign exchange of ₹ 18,856 crore (US$ 3.51 billion) in 2012–2013. Shrimp is cultivated on 2 lakh ha mostly in the states of Andhra Pradesh, West Bengal, Kerala, Orissa, Maharashtra and Tamil Nadu and exported in various forms to Japan, the USA, Europe and elsewhere.

The higher earnings from shrimp have motivated the adoption of intensive shrimp culture. The use of sea water along with fresh water for shrimp culture causes salinisation of land and groundwater with adverse impact on agricultural productivity and groundwater quality. Legal restrictions to reduce the adverse impact of shrimp farming exist in India but are not effective.

Paddy as the major crop in coastal areas bears the maximum impact of environmental externalities of shrimp farming. Few studies on the same have been reported in Thailand (Flaherty et al. 1999), Vietnam (Thanh et al. 1999) and Bangladesh (Battacharya et al. 1999). In India, a comprehensive study of shrimp salinity integrating economic and soil aspects has to be undertaken to assess its environmental impact. Therefore, the present study estimates the external cost of shrimp-farm-induced salinisation of land on paddy productivity in two villages with similar agro-climatic conditions in Southern India. The paddy village near the shrimp farms and the other away from the shrimp farms formed the study villages.

U. Leelakrishnan (✉) · N. Pathusha · R. Karunanidhi
Department of Agricultural Economics, Pandit Jawaharlal Nehru College of Agriculture and Research Institute, Karaikal, Puducherry 609603, India
e-mail: luraha@rediffmail.com

O. H. Kasim
Department of Soil Science and Agricultural Chemistry, Pandit Jawaharlal Nehru College of Agriculture and Research Institute, Karaikal, Puducherry 609603, India

N. Ghosh et al. (eds.), *Nature, Economy and Society*,
DOI 10.1007/978-81-322-2404-4_15

15.2 Study Area

Nagapattinam district in the state of Tamilnadu has shrimp as the major enterprise and paddy as the main crop in the coastal tracts. Paddy is largely grown during samba season (October–February). Shrimp farming is practised during summer (February–June) and monsoon (October–January) adopting modified extensive and semi-intensive methods.

Chandrapadi in Nagapattinam district is the shrimp village. It has 14 shrimp farms with a combined pond area of 65 ha along the Nandalaru tertiary on the Tamilnadu–Puducherry boundary. Poovam is the affected adjacent paddy village, and Thiruvettakudy is the unaffected control village.

The paddy villages are located in the Kottucherry commune. Soil texture varies from sandy to sandy clay and, sub-surface texture is sandy throughout. The average annual rainfall of the commune is 1350 mm, 70 % of which is received during the Northeast monsoon which coincides with the second crop season. Both the villages are at the tail end of the Cauvery delta and the canal is the major source of irrigation with erratic supply.

Farm households that had cultivated paddy during 2005 Rabi season were surveyed using a pretested interview schedule. Poovam has a total land area of 70 ha, and paddy was cultivated on 49.36 ha. Land holding per household is 1.27 ha. Thiruvettakudy has a total land area of 228.84 ha. Paddy accounts for 88 % of gross cropped area (GCA), which is 218 ha. Land holding per household is 2.08 ha. The Farmer's Irrigation Society (FIS) with all the farmers as members has been functioning taking care of maintenance of canals and temple ponds, operating sluice gates to regulate canal water supply and settling disputes between farmers. The FIS opposed and stopped the starting of a shrimp farm on 88 ha by a private firm in 1994–1995 and the land is lying fallow.

15.3 Data

To confirm the homogeneity of paddy villages prior to the establishment of shrimp farms, secondary data on agro-climate, land use and cropping pattern of the study villages during 1990–1991 and 2003–2004 were collected from the Directorate of Economics and Statistics at Karaikal. To record the salinity during the pre-shrimp period (1994–1995) in the paddy villages, secondary data on soil salinity were collected from the Soil Testing Laboratory at Karaikal.

Primary data from paddy farmers were collected for the 2005 Rabi season farm-fragment-wise using an interview schedule. The final sample size was 165 paddy households covering 257 fragments. Of the total sample, 55 farms and 48 fragments are from Poovam, and 110 farms and 209 fragments are from Thiruvettakudy. Surface soil samples up to 30-cm depth were taken from the villages during September and October 2006. A total of 314 and 577 soil samples were collected from Poovam and Thiruvettakudy, respectively. The salinity indicator, electrical conductivity (EC) was measured after processing the soil samples.

15.3.1 Homogeneity of Paddy Villages

Poovam is adjacent to shrimp farms, whereas Thiruvettakudy is away from them. In the pre-shrimp period, Poovam had 48 % of its GCA under paddy, and it was 59 % for Thiruvettakudy. The cropping and irrigation intensity were similar and so was the ratio of net sown area to total land area.

In the post-shrimp period, area under paddy as a percentage of GCA remained stagnant in Poovam, while it increased to 83 % in Thiruvettakudy as Poovam encountered higher levels of soil salinity due to shrimp farms. There was a decline in GCA in both the villages, on account of water scarcity.

15.3.2 Soil Characteristics

An EC value less than 1 indicates that soils are highly suitable for cultivation; 1–3 is injurious to crop growth; 3–4 will cause yield reduction, and soils with an EC value of more than 4 are designated as saline soils and need reclamation for cultivation.

In the pre-shrimp period given in Table 15.1, salinity levels were below 1 in both the villages. Further, the point estimates of means are the same, which provides evidence that the selected paddy villages were similar with regard to soil salinity at this time. In Thiruvettakudy, EC values ranged from 0.01 to 0.96 as shown in Table 15.2, implying normal soils. To know the current salinity level, soil samples were taken from cultivated lands and fallow lands in Poovam in 2006 and they showed EC values ranging from 0.02 to 6.60. In areas adjacent to shrimp farm, EC levels were very high ranging from 4.95 to 15.89. Table 15.3 shows that in the cultivated lands, the EC ranged from 0.02 to 2.13.

Table 15.1 Soil salinity during pre-shrimp farming period (1994–1995)

Soil status	Soil EC (dS m^{-1})			
	Minimum	Maximum	SD	Mean[a]
Poovam	0.1	0.4	0.125	0.23
Thiruvettakudy	0.1	0.7	0.136	0.23

SD standard deviation, *EC* electrical conductivity in deci Siemens per metre
[a] *t-test*: *p*-value = 0.461
Source: Soil Testing Laboratory, Karaikal

Table 15.2 Range of EC values for Thiruvettakudy soil samples

Soil salinity indicator	Range
EC in dS m^{-1}	0.01–0.96
Total soil samples	577

EC electrical conductivity in deci Siemens per metre

Table 15.3 Range of EC values for Poovam soil samples

S.No.	Category of land	Soil samples (no)	EC range (dS m^{-1})
1.	Cultivated lands	215	0.02–2.13
2.	Current fallows	36	0.20–3.90
3.	Permanent fallows	17	1.70–6.60
4.	Near shrimp farms		
a)	Distance of 100 ft	18	5.64–15.89
b)	Distance of 200 ft	28	4.95–11.09
	Total soil samples	314	0.02–15.89

EC electrical conductivity in deci Siemens per metre

15.3.3 Economics of Paddy Cultivation

The paddy economics of the two villages are given in Table 15.4. The chief difference in the economics of paddy cultivation is evident in the per hectare net returns of ₹ 5348 for Thiruvettakudy against a loss of ₹ 5058/ha^{-1} for Poovam. The variable cost per hectare is ₹ 16,930 in Thiruvettakudy and ₹ 14,796 in Poovam. In Thiruvettakudy the average yield is 3519 kg/ha^{-1}, which is 86 % more than Poovam's 1888 kg/ha^{-1}. The productivity gap between the two villages accounts for the significant difference in net returns.

15.4 Theoretical Framework for the Study

Shrimp farming in agricultural lands causes (i) an intra-generational externality borne by the present generation due to decline in crop yields caused by increasing salinisation of land and water resources and the associated adverse socio-economic effects in the region (by adopting reclamation measures and effective water management, salinity could be reversed or minimised) and (ii) an inter-generational externality that will be borne by future generations because of the environmental damage to land and groundwater resources. This study estimates the intra-generational externality cost of salinisation of land.

The externality effect of decline in soil quality is depicted in Fig. 15.1. Given the market price P_0 and MC, the marginal cost of production (i.e. the supply curve), TR is the total revenue obtained from sale of the main produce of the paddy. TVC

Table 15.4 Descriptive statistics for paddy

S.No.	Items	Poovam ($n = 48$)		Thiruvettakudy ($n = 209$)		t-test p-value
		Mean	CV %	Mean	CV %	
1.	Seeds (kg ha^{-1})	139.32	39.48	127.27	15.02	0.140
2.	Organic manure (kg ha^{-1})	9111.28	95.43	5594.61	62.36	0.008
3.	Urea fertilizer (kg ha^{-1})	241.82	117.95	300.65	99.65	0.205
4.	DAP fertilizer (kg ha^{-1})	68.43	126.01	119.02	89.28	0.000
5.	MOP fertilizer (kg ha^{-1})	83.79	167.95	42.59	269.80	0.063
6.	Plant protection cost (₹ ha^{-1})	284.64	147.96	84.01	365.56	0.002
7.	Human labour (₹ ha^{-1})	7517.46	42.69	8770.39	33.94	0.016
8.	Machinery charges(₹ ha^{-1})	1884.20	62.35	2375.29	54.68	0.012
9.	Variable cost (₹ ha^{-1})	14796.35	40.14	16930.53	37.21	0.029
10.	Productivity of paddy (kg ha^{-1})	1888.28	49.59	3519.79	44.65	0.000
11.	Price (₹ kg^{-1})	5.42	8.15	6.29	11.28	0.000
12.	Gross returns (₹ ha^{-1})	9737.78	54.55	22278.63	49.66	0.000
13.	Net returns (₹ ha^{-1})	−5058.57	147.03	5348.09	202.03	0.000

CV % coefficient of variation in %, *DAP* diammonium phosphate, *MOP* murate of potash

is the total variable cost, which includes costs of seeds, manures, fertilizers, plant protection chemicals, human labour and machinery. The output is Q_0. OP_0EQ_0 is TR, $OAEQ_0$ is TVC, and AP_0E is the producers' surplus, which equals the sum of fixed costs and profit. With externality, MC shifts to MC', output falls and Q_1BEQ_0 is the loss in TR; ABE is both loss in profit and loss in producer surplus, because costs are fixed. The externality cost ABE caused by salinity is valued using change in the productivity method by comparing salinity affected and unaffected paddy villages.

15.5 Estimation of Externality Cost

A production function was estimated with soil salinity as an independent variable to assess the salinity externality on paddy yields.

$$Y = \alpha X_1{}^{\beta_1} X_2{}^{\beta_2} X_3{}^{\beta_3} X_4{}^{\beta_4} e^u \tag{15.1}$$

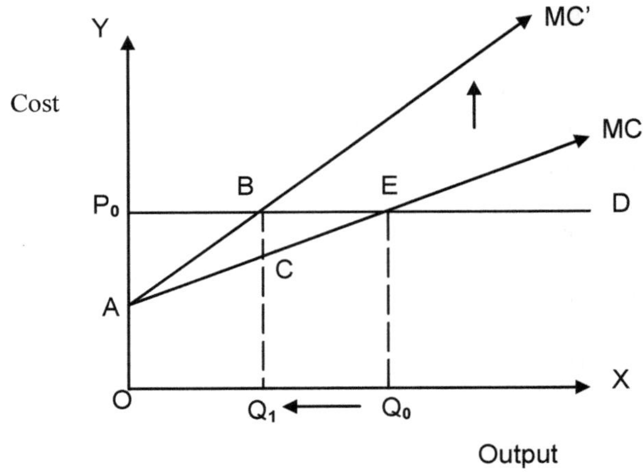

Fig. 15.1 Externality effect of decline in soil quality. P_0 market price, *MC* marginal cost of production

where

Y =	Paddy yield (kg ha^{-1})
X_1 =	Human labour cost (₹ ha^{-1})
X_2 =	Machinery cost (₹ ha^{-1})
X_3 =	Quantity of urea + diammonium phosphate (DAP in kg ha^{-1})
X_4 =	Mean EC (dS m^{-1})
D:	Village dummy variable (1 = affected village)
X_5 =	Quantity of Urea + DAP (kg ha^{-1}) × mean EC (dS m^{-1})
e^u =	Error term
β	Regression coefficient of respective variables.

In order to estimate the production function given in Table 15.5, pooled regressions of both control and affected villages were used. Yield is measured as kg of paddy per hectare. The model is a Cobb–Douglas production function and includes all the variable inputs identified in Eq. 15.1 in log form. The village dummy takes a value of 1 if the farm is a Poovam (shrimp-affected) farm and 0 if the farm is from a shrimp-unaffected village. The model showed that salinity had a significant influence on paddy yield. The estimates from various specifications indicated the non-linear relationship between salinity and land productivity as explained by logarithmic and exponential functions.

Table 15.5 Estimates of production function

Dependent variable ln Y	Model I		Model II		Model III	
Independent variable	Coefficient	t statistics	Coefficient	t statistics	Coefficient	t statistics
ln X_1	0.1372[a]	1.827	0.1340[a]	1.788	0.1343[a]	1.790
ln X_2	0.0949	1.424	0.1086[a]	1.613	0.1094[a]	1.629
ln X_3	0.1422[b]	3.986	0.1838[b]	3.894	0.1305[b]	3.517
ln X_4	– 0.0628[c]	– 2.031	– 0.1952[a]	– 1.889	–	–
D	– 0.4860[b]	– 6.297	– 0.4765[b]	– 6.158	– 0.4953[b]	– 6.600
X_4	–	–	–	–	– 0.1981[c]	– 2.084
X_5 =Ln X_3 * Ln X_4	–	–	0.0244	1.343	–	–
Constant	5.1194	7.414	4.829	6.683	5.294	7.670
R^2	0.3963		0.4007		0.3968	
Adjusted $\acute{R}^2$	0.3843		0.3863		0.3848	
F	32.957		27.853		33.029	
N	257		257		257	

[b],[c] and [a] denote significance at 1, 5 and 10 % levels, respectively

15.6 Welfare Gains from Salinity Reduction

Table 15.6 gives the estimated results of different models used to estimate welfare gains from decreases in salinity. Estimates of welfare gains are obtained by comparing predicted yields per hectare corresponding to the salinity levels of 1 and 3, given the sample mean values of the variables in the production function.

The productivity gain with the Cobb–Douglas specification of production function is 172 kg of paddy per hectare. With the production function considering the synergistic effects of fertilizers and salinity, the production gain falls to 141 kg/ha^{-1}. In the production function that considers the exponential relationship between paddy yield and salinity, a change in salinity from the maximum level of 3 to the safe level of 1 results in a gain of 836 kg/ha^{-1}.

Model IV estimates the gains if salinity decreases by comparing the productivity in the controlled farms with the affected farms. The predicted per hectare yield of unaffected and affected farms are 3252 and 1606 kg, respectively. These estimates are obtained by substituting mean values of input variables for unaffected and affected villages into the estimated Cobb–Douglas production function. By reducing the salinity levels, the average gain is 1647 kg/ha^{-1}.

The maximum average gains that can be obtained from reducing salinity ranges from ₹ 1000 to 5000/ha^{-1} depending on the specification of the production function. An analysis of the controlled means in the affected and unaffected villages reveals that the losses could be high (₹ 10,000/ha^{-1}).

Table 15.6 Estimates of losses per hectare from increased salinity obtained using different models

Estimate	Per hectare land productivity (Mean EC = 1 in dS m^{-1})	Per hectare land productivity (Mean EC = 3 in dS m^{-1})	Loss per hectare (kg)	Loss per hectare (₹)
I	2582	2410	172	1008
II	2681	2540	141	826
III	2557	1721	836	4899
IV	–	–	1647	9651

Estimates I, II and III use different production function specifications
Estimate IV is obtained by comparing farm productivity of affected and unaffected villages. Losses are calculated by assuming an average paddy price of ₹ 5.86/kg^{-1} that prevailed during 2005–2006

15.7 Conclusions

The study examines the externality effect of shrimp-induced salinity on productivity of paddy. Analysis of data showed that soil salinity was normal in both the currently affected and unaffected villages in the pre-shrimp period (1994–1995). Analysis of soil samples taken during 2006 showed that soil salinity was in the normal range in the unaffected village. In the salinity-affected village, a spatial pattern in soil salinity is observed. In the lands adjoining shrimp farms, the mean EC level is high ranging from 4.95 to 15.89 dS m^{-1}, while cultivated lands have an EC range of 0.02–3.0 dS m^{-1}.

Analysis of farm budgets indicates that the net loss in Poovam (the affected village) is ₹ 5058/ha^{-1}. In the unaffected village, net gain is ₹ 5348/ha^{-1}. This situation has forced a few farmers in Poovam to sell their lands.

Estimations of paddy production functions with different specifications implied that salinity has a negative and statistically significant influence on paddy yield. In the case of the Cobb–Douglas specification, 1 % increase in EC is associated with 0.063 % decrease in paddy yields. There is a non-linear relationship between paddy production and land salinity. The estimates of gains in paddy yield from reduced salinity increase with the non-linear specifications of production function.

What are the potential gains for Poovam from a reduction in soil salinity? Observed data from the cropped areas in Poovam show a maximum salinity level of 3, and for those farms that are growing paddy despite being near this salinity level, the average gain in yield is 172–836 kg/ha^{-1} and in monetary terms it would be ₹ 1000–5000/ha^{-1} depending on the specification of the production function. There are many farms that have not reached the maximum salinity level of 3. Their continued exposure to shrimp neighbourhood effects will increase the salinity level. The high salinity level of 6.60 in non-cropped areas of Poovam has created permanent fallows as they were no longer productive.

Thiruvettakudy, the unaffected village, has been transformed into an agricultural tract partly because of the active role of the FIS in water management.

Such an institutional mechanism with farmer's participation in salinity management is the solution for managing agriculture along the coastal tracts. Salinity can be decreased in Poovam as well. This will require maintenance and regulation of shutters, adoption of the recommended salinity control activities such as rainwater harvesting, leaching and drainage and soil-test based application of amendments like gypsum. Proper monitoring of soil and water salinity is necessary for effective implementation of the reclamation measures to prevent a further build-up of salinity.

The study suggests the need to evolve a regulatory framework on pragmatic lines for reducing the externality effects of salinisation of land and water resources for sustainable agricultural development in the study area.

Acknowledgement The authors acknowledge the financial support of SANDEE in carrying out this research work.

References

Battacharya D, Rahman M, Khatun FA (1999) Environmental impacts of trade liberalization and policies for the sustainable management of natural resources: a case study on Bangladesh's shrimp farming industry, United Nations Environment Programme

Flaherty M, Vandergeest P, Miller P (1999) Rice paddy or shrimp pond: tough decisions in rural Thailand. World Dev 27(12):2045–2060

Thanh TB, Canh LD, Brennan B (1999) Environmental costs of shrimp culture in the rice growing regions of the Mekong delta. Aquac Econ Manage 3(1):31

Chapter 16
Designing PAT as a Climate Policy in India: Issues Learnt from EU-ETS

Shyamasree Dasgupta, Frank van der Salm and Joyashree Roy

16.1 Introduction

Perform, Achieve and Trade (PAT) has emerged as an energy and climate policy mandating a decrease in specific energy consumption (SEC) (that is energy use per unit of production) in energy-intensive industries in India. Under this scheme, the mechanism of benchmarking and trading of energy efficiency certificates will allow the domestic overachievers to trade their credits with those who underachieve. Emphasising on the fact that growth in production is not obstructed by limited energy use, PAT creates an incentive for the producers to use energy with greater efficiency. From various perspectives, the mechanism design of PAT bears significant similarities with that of European Union Emission Trading Scheme (EU-ETS). EU-ETS emerged as a cap and trade policy in the EU in order to curb emission in line with Kyoto targets for developed countries. The objective of this chapter is to give an overview of the PAT scheme targeting the SEC, analyse lessons learnt during the pilot phase of EU-ETS and to identify similar issues regarding benchmarking and allocation, price volatility, rebound effect, accounting and transparency, interaction with other international mechanisms like clean development mechanism (CDM), etc. In the context of PAT. However, it should be noted that the issues discussed are not similarly applicable for all the participating member states of the EU, rather they signify some broad issues related to the scheme (Buchner et al. 2006). Section 16.2 gives an overview of the genesis of PAT and EU-ETS, Sect. 16.3 describes the mechanism design of PAT, Sect. 16.4 identifies issues related to PAT

S. Dasgupta (✉)
Department of Economics, Jadavpur University, Kolkata, India
e-mail: shyamasree.dasgupta@gmail.com

F. van der Salm
Utrecht University School of Economics, Utrecht, The Netherlands
e-mail: F.H.vanderSalm@uu.nl

J. Roy
Jadavpur University, Kolkata, India
e-mail: joyashreeju@gmail.com

N. Ghosh et al. (eds.), *Nature, Economy and Society*,
DOI 10.1007/978-81-322-2404-4_16

which need special attention on the basis of what has been learnt from the similar mechanism design of EU-ETS during its pilot phase and Sect. 16.5 concludes the chapter.

16.2 Genesis of PAT and EU-ETS

16.2.1 Perform, Achieve and Trade

India has adopted climate change mitigation and adaptation policies under the flagship programme of the 'National Action Plan on Climate Change (NAPCC)' in 2008 (National Action Plan on Climate Change 2008). Being a non-Annex I country within the United Nations Framework Convention for Climate Change (UNFCC), there is no binding commitment for India to reduce the emission of greenhouse gases (GHG) during coming years. However, as India ratifies the protocol, there is national imperative for positioning the country in the low-carbon paradigm where intervention needs to be both nationally and internationally compatible. The adoption of NAPCC is a move towards ensuring the country's energy security and to take comprehensive development policy measures with climate change mitigation and adaptation as possible co-benefits. The plan identifies eight core 'national missions' running through 2017, one of which is the National Mission for Enhanced Energy Efficiency (NMEEE). By 2014–2015, NMEEE is expected to achieve annual fuel savings in excess of 23 million t of oil equivalent (MTOE) with cumulative avoided electricity capacity addition of 19,000 Mega Watt (MW); all getting translated to a carbon dioxide (CO_2) emission mitigation of 98.55 million t per year (Bureau of Energy Efficiency 2012). Within the framework of NMEEE, PAT has emerged as a market-based mechanism to ensure energy efficiency improvement in energy-intensive large industries and facilities in a cost-effective manner through certification of energy savings that could be traded. The genesis of the PAT lies in the provisioning of the Energy Conservation Act 2001[1] (Government of India 2001). The Bureau of Energy Efficiency (BEE) of the Government of India is working as the nodal agency towards activation of the scheme. PAT has emerged as a 'baseline and credit' mechanism (Dyre et al. 2006) where the target variable is the SEC of industrial units. This is particularly relevant given the fact that the industrial sector in India consumes 40 % of commercial energy while accounting for almost 22 % of total GHG emission which is over 400 million t of CO_2 equivalent (Planning Commission, Government of India 2011). It is estimated that PAT could accelerate energy efficiency improvement for facilities accounting for 25 % of India's fossil fuel use and reduce CO_2 emissions by 25 million t per year by 2014–2015 relative to

---

[1] The Energy Conservation Act (EC Act 2001) in India is an act towards the efficient use of energy and its conservation. It was approved by the Parliament in 2001 and came into force from March, 2002.

business as usual (Ministry of Environment and Forests, Government of India 2010; Roy 2010). This is approximately 1.4 % of the country's projected total annual CO_2 emissions in 2015 (IEA 2009; Roy 2010). The Government of India has notified the energy intensity targets under PAT for 478 large industrial units, the compliance period being from April, 2012 to March, 2015 (Government of India 2012). At the end of the compliance period, overachievers will be issued energy saving certificates (ESCerts). One ESCert is planned to be issued for 1 t of oil equivalent saved by the unit in the process. These ESCerts can be sold to those who underachieve. PAT has its uniqueness in the sense that it targets the energy intensity rather than total energy use or total emission. So PAT puts emphasis on the fact that growth in production is not obstructed by limiting energy use, rather it creates an incentive mechanism for energy-intensive industrial units to use energy in an efficient manner.

16.2.2 Europe Union Emission Trading Scheme

Cap and trade has gained predominance to the policy discourse to curb GHG emission, with EU-ETS perhaps being the most prominent one. The origin of the programme dates back to the year 1997 when the UNFCCC Kyoto protocol was signed. Under Kyoto the target for the developed countries was to reduce GHG emission 5.2 % below the 1992 level on an average. The then existing 15 EU members who agreed to their 8 % reduction commitment together formed the 'EU Bubble'. But other than the UK (with a shift from coal to natural gas) and Germany, all EU nations found it difficult to meet the target. EU-ETS was designed as a way out policy measure to achieve this target (Pew Center on Global Climate Change 2009). It is basically a 'cap and trade' mechanism to restrict the overall emission level below a threshold level. However, as ten central and eastern European countries (whose emission had generally declined due to economic restructuring in the post-1991 period following the fall of Soviet Russia) joined the EU in 2004, the EU-ETS was expanded to 25 countries and the target became easier to achieve. The first phase of EU-ETS was over in 2008, the second phase ran during 2008–2012 with the same group of EU countries and the third phase began after that. Each phase started with a number of modifications of the scheme on the basis of what they had learnt during the previous phase (Betz and Sato 2006).

An analysis of the learning experience from the first phase of EU-ETS gives a scope to identify some very important aspects of the mechanism design of PAT at its early stage which need to be dealt with specific importance. Although there are inherent differences between PAT and EU-ETS, in terms of objectives and mechanism design, there are several issues in common as well. Thus, to solve the global problem of increasing concentration of GHGs in the atmosphere while acting locally, it would be important to understand what the system has already learnt from similar implementations that have taken place in other parts of the world. This kind of knowledge transfer is useful in order to design a more accurate mechanism and prevent misallocation of time and investment.

16.3 Energy Efficiency and PAT: A Brief Review of Mechanism

Energy efficiency enhances the competitiveness of a production unit by reducing its expenditure on energy at the very first place. In developing countries, capacity expansion is often considered to be more profitable than going for energy efficiency. Also, risk aversion is reflected in apathy to early adoption of new efficient technology (Mathur 2011). Under such circumstances, the competition between existing production units and between existing and entering units are hugely determined by the capacity and less by production efficiency. Therefore, the old inefficient firms could not be easily driven out of the competition by the efficient firms through market mechanism. So, in order to guarantee enhanced energy efficiency, it becomes important to design a market mechanism which gives incentive to large energy-consuming units to go for enhanced energy efficiency. PAT is designed to create such additional incentive.

Although PAT is an integral part of NAPCC, the genesis, as mentioned, lies in the EC Act 2001. This Act identified the energy-intensive units in India as designated consumers (DC) who were required to appoint or designate energy managers and to get their energy uses audited by accredited energy auditors. These DCs were also required to implement techno-economically viable recommendations and comply with energy norms. Under PAT, apart from railway 478 installations among the DCs are identified in rest of the 8 sectors: thermal power stations, fertilizer, cement, iron and steel and pulp and paper, chlor-alkali, aluminium and textile (Ministry of Power, Government of India 2012). These units are identified as they have energy consumption over and above certain threshold levels, expressed in metric tons of oil equivalent (MTOE) per year. Being high energy consuming installations they are perceived to contribute towards energy use reduction and emission mitigation to greater extent through enhanced energy efficiency. The threshold level is 30,000 MTOE per year for the thermal power stations, fertilizer, cement, iron and steel and pulp and paper industries, 12,000 MTOE per year for chlor-alkali, 7500 MTOE per year for aluminium and 3000 MTOE per year for textile industries. Although railway is one of the DCs, the benchmarking for this sector will be done in a completely different manner which is beyond the scope of this chapter. This sector will include the electric traction subsection, diesel loco shed, production units and workshops of Indian Railways having a total annual energy consumption of 30,000 MTOE or more under the Ministry of Railways (this is not particularly directly looked into by BEE; Mathur 2011).

The PAT scheme runs as follows. As per the design mechanism, there is a four-stage implementation of PAT. It starts with 'development phase' which includes energy audit and SEC target setting for the 478 DCs by BEE. Once the targets are in place the 'reduction phase' begins. The DCs are given a timeframe of 3 years to act upon their energy efficiency targets and to adopt suitable measures in order to achieve the targets set by BEE. Once the 'reduction phase' is over there will be a 'monitoring and verification phase' followed by distribution of ESCerts to the over-achievers and reporting of the magnitude of underachievement by the rest. Saving

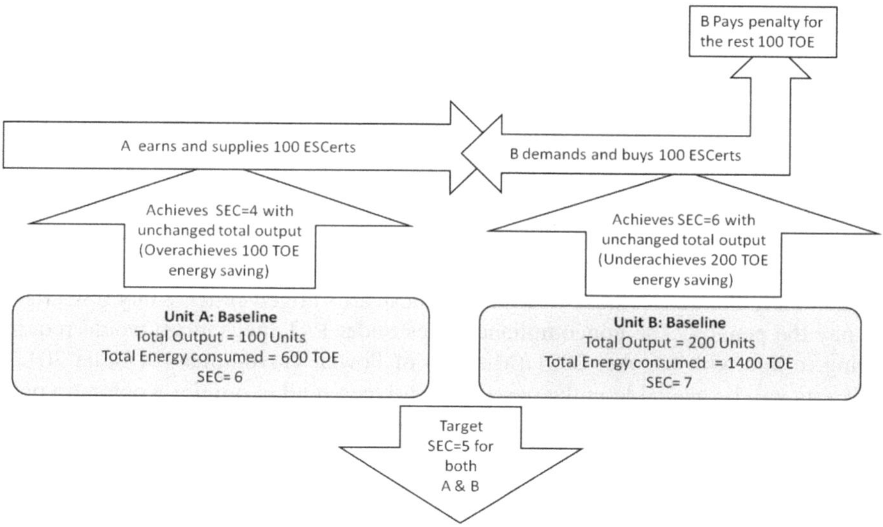

Fig. 16.1 A simple depiction of the mechanism of PAT. *PAT* perform, achieve and trade

of 1 t of oil equivalent would fetch one ESCert. Once the ESCerts are distributed, the overachievers start trading their ESCerts with the underachievers. The underachievers, however, will have two options, either to purchase ESCerts from the overachievers or to pay a penalty (Ministry of Power, Government of India 2012). Figure 16.1 gives a schematic description of the mechanism of PAT.

The target was initially to set up a single process-wise benchmark for each of the sectors. However, the studies undertaken by BEE on 'Setting up Sectoral Bandwidths for Designated Consumers' have revealed widely diversified energy-use pattern within industrial units in a particular sector (Mathur 2011). This diversity was found with respect to scale of production/ installed capacitates, use of raw material, process technology, vintage of technology, operation and management and finally the varied types of output within a sector. After reviewing the sectoral energy scenario and bandwidth, BEE concluded that baselines and reduction targets are more accurate to be unit specific (Ministry of Power, Government of India 2012). The required reduction of SEC has been formulated in such a way that it is inversely proportional to the current rate of energy efficiency, that is more efficient units will face lower proportional reduction of their SEC as compared to less energy efficient units.

Process of data collection, compilation, verification and analysis are over and the baseline format with system boundaries has been developed by BEE during the development phase. An average of SEC for the years 2007–2008, 2008–2009, 2009–2010 is considered to be the baseline SEC provided the capacity utilization is uniform and the period of compliance is from April 2012–March 2015. This achievement phase is expected to involve measures like technology transfer, installation of energy efficient technologies, research and development, etc. This will be followed by verification of the SEC of each designated consumer in the baseline year and in the target year by an accredited verification agency as mentioned before. To ensure compliance, incentive and penalty structure have been formulated by BEE. BEE accredited verification agencies would verify the SEC target achievement for each DC through detailed energy audit after the compliance period. Those who overachieve the target would be issued ESCerts in favour of them. ESCerts issued by the market registry could be traded with DCs who are unable to meet their target SEC reduction. In that sense, those who underachieve are obliged either to buy ESCerts or to pay the penalty. The noncompliance fines under PAT mechanism would require being in line with EC Act 2001 (Ministry of Power, Government of India 2012). ESCerts will be available in the open market. Open market implies a common platform like share market where the ESCerts will be traded. The exact structure of the market is yet to be decided. Power Exchange India Ltd, Energy Efficiency Ltd., etc. will be parts of the market. They will bid, fix the trade matching, discover the actual price through demand supply match, etc. One ESCert will be equivalent one ton of oil equivalent saved by the unit and the expected price of one ESCert is INR 11,000–12,000[2]. The transfer agents/depositories must hold the ESCerts under each industry in electronic form and provide client services in relation to ESCerts. Selling and buying of ESCerts would be possible either in one to one basis between the overachiever and the underachiever or through using the common platform. The exact trading mechanism will be worked out once the compliance period is over and the achievements by the plants are reported and verified[3].

16.4 PAT Relevant Issues Learnt from EU-ETS

In the context of developing nations, with expansion of output for the purpose of economic growth and development in the presence of unmet demand, it might not be possible to reduce the aggregate energy consumption and emission (as mandated in case of EU-ETS) in near future. But definitely there lies huge potential to use the energy input more efficiently so that per unit energy consumption is significantly reduced resulting in avoided energy consumption and emission. Literature shows that in Indian industries, SEC reduction achieved in the past in absence of any targeted policy was not high enough to achieve decoupling of production growth and energy use growth (Roy et al. 2013). At this juncture, as India has implemented the mechanism of PAT under NAPCC, it becomes crucial that the mechanism design is accurate enough and loopholes which have already been identified during implementation of similar mechanism are taken care of. This section thus discusses such

[2] Based on interactions with experts at BEE (7 April 2011).

[3] Based on interactions with experts at BEE (7 April 2011).

areas and how the learning from EU-ETS could be used to reduce them as much as possible. This learning could actually save a lot of time and make the investment decisions more accurate.

16.4.1 Benchmarking and Allocation

While EU-ETS is a 'cap and trade' mechanism, PAT could be more appropriately characterized as a 'baseline and credit' mechanism[4] (Dyre et al. 2006). Although there are some inherent differences between these two mechanisms, the benchmarking and the process of allocation, be it allocation of allowances or of credits, have major implication towards perceived fairness, political acceptability and economic efficiency (Sterner and Muller 2006). The political feasibility of EU-ETS is often claimed to be contingent upon the 95 % free allocation of emission allowances through 'grandfathering' in the first phase. How much each member state under the EU could emit was fixed depending on their national allocation plans (NAPs). NAPs are the instruments each member states used to decide how many allowances would be allocated and who would get them. In some of the cases, such allocation is criticized to have less clarity (Buchner et al. 2006). Then the allowances were distributed to the member states who subsequently distributed them among the relevant sectors at the domestic level. What is important is the fact that how many allowances a unit should get was decided through 'grandfathering'. The emission reduction target was directly proportional to their 1990 level of emission. This implied that the units historically producing more output or emitting more would be allocated more of permits. This accompanied by the fact that the permits were distributed free of cost led to a situation where there was an inevitable tendency towards lobbying for higher allocation through escalations of their expected emission (Egenhofer et al. 2006). This is evident in the fact that the first phase of EU-ETS was characterized by a high degree of overallocation of allowances. Several member states had allowed their covered sectors to increase emission by as much as 10–20 % of the existing level which by no means was compatible with Kyoto target (Egenhofer et al. 2006). Overallocation led to overproduction and suboptimal abatement effort. The EU's formal release of data on May 15, 2006 showed that of the

[4] In a cap and trade programme, the regulator establishes an overall limit on emissions—the 'emissions cap'. Allowances equal total emissions permitted under the cap are then distributed through free allocation or by auction. Once the allowances are distributed, they may be traded freely. On the other hand, the participants in a baseline and credit (or 'averaging') programme have to 'earn' credits before they can begin trading. An emissions baseline is first defined for each participant by the regulator. Each participant then makes reductions and monitors or calculates its actual emissions using specified procedures. At the end of the compliance period, participants whose actual emissions are lower than their baseline receive 'credits' equal to the difference. Credits can then be traded freely. A participant whose actual emissions exceed its baseline must purchase credits equal to its excess emissions to achieve (Dyre et al. 2006).

twenty one countries that reported in time, 15 countries had been 'overallocated', that is the average of their 2005–2007 allocation was greater than the actual emissions for 2005. This overallocation remained in place through to the end of 2007 as new allocations are only implemented at the commencement of the new, phase II, trading period; 2008–2012. This was enough to signal the huge excess supply of allowances in the market which actually dragged the price of allowances close to zero as discussed in the next section. In India, if we consider the mechanism of PAT, a similar problem would arise if both benchmarking and target setting are not accurate. Although one efficient point about PAT is the fact that it requires energy efficient units to reduce less as compared to their inefficient counterparts, and hence inefficient firms are not favoured at the expense of efficient ones as it happened in case of the first phase of EU-ETS. PAT also does not create any incentive for the units to over- or under-report their baseline SEC. While an inflated reported SEC would make a unit reduce its SEC in a higher proportion, a deflated SEC would also mean more stringent target as the base SEC used to calculate the target becomes lower. However, what is important to ensure is that the target SEC compels the units to reduce their energy intensity at a higher rate as compared to the historical rate of reduction. Absence of such stringency would imply overachievement by most of the units with similar implication of overallocation of allowances under EU-ETS and realised energy efficiency gain would be much less than the potential[5].

16.4.2 Price Volatility

The impact of the overallocation of free allowances under ETS inevitably resulted in instability in market and volatility in price of allowances with high fluctuations in price levels (Egenhofer et al. 2006). Free allocation does not generate any price signal beforehand. Ex-ante measurement of carbon emission turned out to be problematic and politically loaded in case of EU-ETS than it was expected to be and ex-post realization of effective pricing was found to be quite sensitive to unforeseen or exogenous conditions. In May 2006, within a period of one week, the price of EU-ETS allowances (EUAs) fell from a high of around € 30 to a low of around € 9 following the release of actual emission data of the EU member states in 2005 and the EU's formal release of emission data of sectors covered under ETS (Convery and Redmond 2007). Prior to the release of such data there was no certain signal regarding the demand for allowances. There was quite naturally the tendency to hold

[5] Windfall profit was another consequence. The burden of emission reduction by the power sector was largely transferred to the consumer which generated large windfall profit for the power sector and at the same time reduced consumer surplus hugely (14 billion € over the period 2005–2008)—this led to additional rent or supernormal profit. This happened because while fixing the price of power, the cost of carbon is already included. When excess allocations are sold then these power utilities generate some revenue out of that as well. In India, the possibility of such windfall profit is less because the market is not completely deregulated, and the carbon prices are not incorporated.

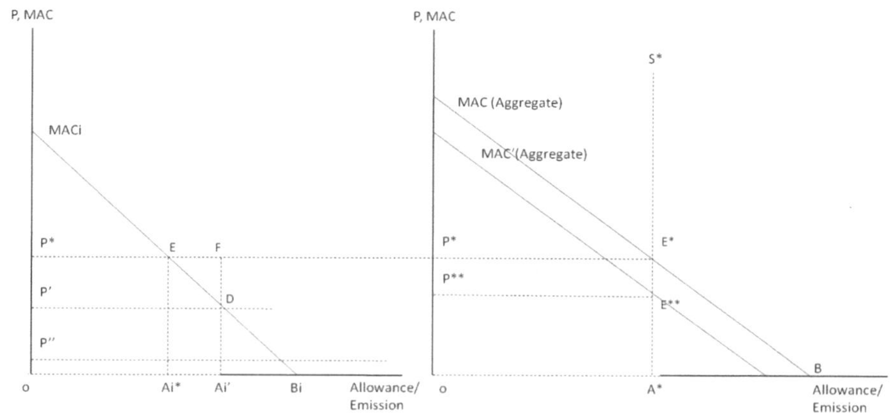

Fig. 16.2 Difference between impacts of free allocation of allowances and low MAC on allowance price. *MAC* marginal abatement cost

back the allowances by the overachievers in a forward looking market where they expected the allowance price to rise in the future. But the price collapse occurred when the market moved from what was forecast to be a ‘short position’ to the realisation that the market was ‘long’ (Dyre et al. 2006). This phenomenon is distinctly different from a sharp fall in auction value of allowances in case of the sulphur dioxide emission trading in the USA (Chan et al. 2012). The guiding difference was in the mechanism of distribution. When price of allowance falls sharply in a free allocation system like ETS, it signifies the excess supply of such allowances in the market. In case of auctioning the emitter reduces emissions to the point where the auction price of allowances is equal to the emitter's marginal abatement cost (MAC). So a low auction price in the sulphur dioxide emission trading market in the USA implied that the cost of the technology to reduce sulphur emission was actually quite low.

As depicted in Fig. 16.2, suppose the authority issues the fixed number of allowances OA* (in the right panel) so that the supply curve of allowances becomes A*S*. In an auction, the price of allowances would be determined at P* where aggregate MAC and A* S* intersect. This is because given MAC, P* is the price at which the demand for allowances matches exactly the total allocation. So, the economy as whole pollutes OA* by purchasing the permits and abates the rest A*B. Once P* is determined in the market, individual firms take their abatement decisions as a price taker. Given the MAC curve of a representative firm (MAC_i) as depicted in the left panel, and given P*, the firm optimizes by choosing to pollute OA_i* by purchasing allowances and abates the rest A_i*B_i (upto A_i* allowance prices are lower than the marginal cost to abate). This kind of a mechanism with an auction determined price of allowances a lower price of allowance (P**) is revealed when MAC of abatement is lower (e.g. MAC′). It is also guaranteed that given the price of allowances a more efficient firm with lower MAC would abate more. This was the case with respect to the sulphur dioxide emission trading scheme in the USA. Since the abatement decisions are taken on the basis of market determined optimized values of instruments, the possibility of significant over- or under-abatement activity is much less irrespective of lower or higher price of allowance. The mechanism is, however, significantly different in case of free allocation of allowances through mechanisms such as 'grandfathering'. Suppose the same representative firm gets $OA_i^* > OA^*$ allowances free of cost. This would at the first place lead to suboptimal abatement effort by the firm $A_i^{\prime} B_i < A_i^* B_i$. Not only that if the at an aggregate level allowances are distributed among less number of units with most of them abating suboptimally the aggregate emission reduction would also be lower that the efficient level. However, if the firm anticipates a high price of allowance during the trading phase, for example P*, it has every incentive to abate $A_i^* B_i$ as before and sell the excess allowance. By doing this the firm would expect to earn $A_i^{\prime} A_i^*EF$ area while the cost it has to incur would be lower than that $A_i^{\prime} A_i^*ED$. But such efficient abatement decision by a large number of firms would result in huge supply of tradable allowances in the market during the trading phase with every possibility of subsequent fall of the price of allowance. This was what happened in case of EU-ETS. If in fact the price goes below $P^{\prime}$ (e.g. $P^{\prime\prime}$), there is no possibility that the firm would be able to recover the cost of excess abatement by selling the allowances, and in the following phases the firm would lose incentive to overachieve.

Less stringent target SEC in the Indian scenario has a possibility to lead to a situation similar to what happened in case of EU-ETS. If the benchmarking has an upward bias and most of the firms overachieve the target, there would be an excess supply of ESCerts. This excess supply of ESCerts in the market would lead to a drop in the price of the same as it happened in case of EU allowances. This would actually have an extremely severe implication in the long-run price mechanism and technology innovation as the incentive of overachievement would be less if not negative. Suppose given the benchmarking under PAT, a unit invests in an energy efficient technology which helps the unit to overachieve the target and hence to earn some ESCerts. This particular unit might take the investment decision keeping in mind partial recovery of cost of the energy efficient technology through reduced energy bill and the rest through revenue generated from the selling of ESCerts. Now if the market price of such ESCerts drops to zero or close to zero, then it would generate disincentive for firms to invest in energy saving technologies and innovations. Protection against such volatility and uncertainty might be partly achieved by attaching a floor price to the ESCerts which would imply that the price would never go down below the level.

16.4.3 Rebound Effect

PAT is an intensity targeting mechanism. Unlike purely environmental target driven instruments like ETS, PAT has an inherent implication towards the energy security of the country. As the units are to reduce their SEC one possibility would be to install an energy efficient technology which requires less energy per unit of production. In case of energy-intensive industries covered under PAT energy cost account for a large proportion of total costs. So a decline in energy use per unit of production would lead to a decline in the marginal cost of production leading to an increase in total output if energy is not hugely substituted by other inputs of production. This in turn might increase in the total energy demand of the existing units hampering the prime essence of the policy to address the issue of climate change though reduced energy use and emission. This kind of 'rebound effect' is effectively found in the context of the industrial sectors of several countries such as China, India, the UK, and the USA (Chakravarty et al. 2013 and references therein). This aspect needs to be considered while designing the mechanism related to any intensity target instrument such as PAT. Important to note, this phenomenon would not arise in case of instruments that target total energy use or total emission such as EU-ETS.

16.4.4 Linking Clean Development Mechanism, Joint Implementation and Other Similar Mechanisms

There needs to be a very clear guideline regarding the interaction of any newly designed market instrument not only with those which already exists but also the forthcoming ones. The EU-ETS sets an example by clearly stating its relation with the credits earned through two other project based Kyoto mechanisms. It mentions that the linkage to other trading schemes would base on a bilateral agreement and is expected to increase liquidity in the system. However, at the first stage there were no such linking policies clearly mentioned. In terms of project mechanism, credits generated from CDM or joint implementation (JI) could constitute the part of compliance in the first phase of EU-ETS. While no limits applied to the amount of CDM credits companies could use in the first phase, in the second phase however each member state had to state beforehand how many credits are they going to use. Such linkages are yet to be specified for PAT. But this lack of specification might generate a scope for some units to earn supernormal profits. Suppose a technology is installed with CDM fund which reduces emission below the baseline through a significant decline in SEC. As a result this unit not only earns CER under CDM but at the same time might end up earning ESCerts under PAT. It generates a problem of double accounting and gives an opportunity to earn a supernormal profit. Also, it needs to be clarified, if any technology installation covered under PAT reduces SEC and emission then whether it is in conflict with the additionally criterion of CDM.

Renewable energy projects were carefully kept out of the domain of operation of ETS (Pew Center on Global Climate Change 2009). Biomass sequestration and nuclear power projects also could not be used for credit generation. Hydroelectric power projects were also restricted. Similarly, a good thought needs to be to put in order to decide what happens in India with respect to renewable energy certificates (REC). REC would be designed to drive investment in low-carbon energy projects. REC would be issued to the generators of powers using renewable energy. This REC could be sold and purchase of REC would be considered as purchase of renewable energy. Now a pertinent question could be whether any shortfall in meeting the target SEC reduction could be made up through purchase of such REC. It might turn out to be a cost effective way to achieve energy efficiency and resultant emission mitigation through establishment of some kind of a relationship between ESCerts and REC because power exchange is expected to be the major agency in both the markets and it might lead to a very low transaction cost.

16.4.5 Accounting Predictability and Transparency

MacKeinze has clearly articulated the importance of defining the nature of an 'emission permit'; whether it should be considered as an asset or not. If it is considered to be an assent whether it should be an 'intangible asset' or a 'financial instrument' (MacKeinze 2009)? The accounting frameworks would vary given the nature of the asset. Same is true in case of ESCerts. A very clear guideline needs to be laid down in order to fit it in the existing accounting structure of a firm well in advance before the trading and awarding of ESCerts. Since NAPCC mentions that banking option of ESCerts would be available this issue becomes even more important. Issues such as the length of maximum banking period and limit for banking and expiry period, etc. also need to be resolved. Creation of an efficient and transparent market for trading by the exchanges, taking measures to safeguard market integrity and enhance transparency in operations and also maintain data of traded prices, traded volumes and trends are extremely important.

16.4.6 Fraud

There are scopes for carousel fraud in ETS. Suppose a trader opens a trading account with the national carbon allowances registry in country A. This trader buys allowances in a different country and he/she is exempted from VAT. These allowances are then sold in the original country with VAT attached which is actually appropriated by the trader. This is not a likely threat to India. But corruption and fraud are generally unforeseen events and the mechanism design needs to be such that when this kind of frauds is noticed, there needs to be some ex post flexibility in the mechanism to eliminate such fraud.

16.5 Conclusion

Although there are inherent differences between PAT and ETS, in terms of objectives and hence in terms of mechanism design, there are several issues in common as well. Targeting SEC in case of PAT as compared to 'grandfathering' in the allocation of emission permits in EU-ETS has made the former free of certain inefficiencies but the target has to be stringent enough to finally deliver the efficient outcome. It thus emphasizes the role of knowledge transfer in order to design a more accurate mechanism and to prevent the wastage of time and malinvestment while designing PAT in India.

References

Betz R, Sato M (2006) Emissions trading: lessons learnt from the 1st phase of the EU ETS and prospects for the 2nd phase. Clim Policy 6:351–359

Buchner B, Carraro C, Ellerman AD (2006) The allocation of European union allowances: lessons, unifying themes and general principles. Global Science Policy Change MIT . Report no. 104. http://globalchange.mit.edu/files/document/MITJPSPGC_Rpt140.pdf. Accessed 9 May 2015

Bureau of Energy Efficiency (2012) www.beeindia.in/NMEEE/NMEEE2.ppt. Accessed 23 July 2013

Chakravarty D, Dasgupta S, Roy J (2013) Rebound effect: how much to worry? Curr Opin Environ Sustain 5:216–228

Chan G, Stavins R, Stowe R, Sweeney R (2012) The SO_2 allowance trading system and the Clean Air Act Amendments of 1990: reflections on twenty years of policy innovation. NBER Working Paper No. 17845

Convery FJ, Redmond L (2007) Market and price developments in the European Union emissions trading scheme. Rev Environ Econ Policy 1(1):88–111

Dyre M, Frazer S, Price K (2006) Confronting climate change and emissions reduction in the energy sector. Emissions trading EU ETS experience and lessons for New Zealand. Paper presented at the 8th Annual New Zealand Energy Summit, New Zealand, 2006. http://www.climatechange.co.nz/publications/Emissions%20Trading%20-%20EU%20ETS%-20Lessons%20for%20NZ.pdf 2006. Accessed 17 Jan 2014

Egenhofer C, Fujiwara N, Ahman M, Zetterberg L (2006) The EU emissions trading scheme: taking stock and looking ahead. Brussels: European Climate Platform, a joint initiative of CEPS and CLIPORE, special CEPS report

Government of India (2001) The Energy Conservation Act, 2001. India. http://manipursda.nic.in/ECact01.pdf. Accessed 9 May 2015

Government of India (2012) PAT: perform achieve and trade. Ministry of Power, Government of India. http://beenet.gov.in:90/downloadbooks.aspx?fname=BEE_PAT_Booklet_Final.pdf. Accessed 9 May 2015

IEA (2009) World Energy Outlook 2009. OECD/IEA, Paris

MacKeinze D (2009) Making things the same: gases, emission rights and the politics of carbon markets. Account Organ Soc 34:440–455

Mathur A (2011) Energy synergy with Dr. Ajay Mathur, Bureau of Energy Efficiency, Government of India. www.indiaworldenergy.org/ppt/NMEEE.pps. Accessed Jan 2014

Ministry of Environment and Forests, Government of India (2010) India: taking on climate change. Post-Copenhagen Domestic Actions. Accessed 9 May 2015

Ministry of Power, Government of India (2012) PAT: perform achieve and trade. Government of India, India

National Action Plan on Climate Change (2008) National action plan on climate change. Ministry of Environment and Forest, Government of India, New Delhi

Pew Center on Global Climate Change (2009) The European Union emissions trading scheme (EU-ETS): insights and opportunities. http://www.c2es.org/docUploads/EU-ETS%20 White%20Paper.pdf. Accessed 9 May 2015

Planning Commission, Government of India (2011) Low carbon strategies for inclusive growth: an interim report. Government of India, New Delhi

Press Information Bureau, Government of India, Ministry of Power (2012) Union power minister launches "PAT" scheme under NMEEE energy intensive industries to benefit by trading ESCerts energy saving of 6.6 million tonnes oil equivalent by 2014–15. http://www.pib.nic.in/newsite/erelease.aspx?relid=85182. Accessed 15 Jan 2014

Roy J (2010) Iron and steel sectoral approaches to the mitigation of climate change: perform achieve and trade in India. Working Paper. Climate Strategies

Roy J, Dasgupta S, Chakravarty D (2013) Energy efficiency: technology, behavior, and development. In A. Goldthau (Ed), The handbook of global energy policy. Wiley, Blackwell pp 282–302

Sterner T, Muller A (2006) Output and abatement effects of allocation readjustment in permit trade. Resources for Future, Washington DC

Chapter 17
Persistence of Jhum Cultivation: Social Capital and Labour Market

Indrani Roy Chowdhury

17.1 Introduction

This chapter examines a puzzle associated with slash and burn, or shifting cultivation (locally called Jhum cultivation), namely its persistence in several regions of the world, including the northeastern Indian states, in particular Mizoram. We argue that a simple theory of such persistence can be built around the interaction of several factors, namely Jhum's mode of production (with its emphasis on collective action), and social capital. Further, our analysis generates the testable hypothesis that such persistence is more likely if the population is not too large.

The motivating example for our study comes from Mizoram in Northeastern India, where Jhum is the predominant mode of production.[1] It is interesting that despite several initiatives by the Mizoram state government to incentivise alternative sources of livelihood (e.g. Jhum Control Projects, New Land Use Policy, etc.), the practice of Jhum has proved to be highly resilient. This is even more striking given that the wage levels of skilled and unskilled casual labour (both rural as well as urban) in the state is fairly high.[2] Coupled with high wage rate there is an indication of growing underemployment across the successive National Sample Survey (NSS) Rounds, as the rural workers seeking additional work or alternative work shows an increasing trend (See Tables 17.1, 17.2 and 17.3).

---

[1] The percentage of Jhum cultivator family out of total households in the state of Mizoram is 58 % (Statistical Abstract, 2003–2004). The share of cultivator (broadly approximated by shifting cultivator) to the main worker fell from 61 % in 1991 to 55 % in 2001 (Census data of occupational statistics). The percentage of cultivator families in all the seven districts of Mizoram (except Aizawl the capital of Mizoram where it is 32 %) ranges from 45 to 74 % shows the strong evidence of Jhum practice (Statistical Abstract of Mizoram)

[2] The wage level is substantially higher than the national average (see, e.g. NSS report No. 458, 515 on Employment and Unemployment Situation in India). In fact, the average daily wage of even casual workers (in public as well as non-public works both in rural and urban areas) is higher than most of the Indian mainland states across the NSS Rounds.

I. R. Chowdhury (✉)
Department of Economics, Jamia Millia Islamia (Central University), New Delhi 110025, India
e-mail: indranirc1@gmail.com

N. Ghosh et al. (eds.), *Nature, Economy and Society*,
DOI 10.1007/978-81-322-2404-4_17

Table 17.1 Average daily wages for rural casual workers in public works (Rupees). (Source: NSS 55, 61 Rounds, NSS Report No. 458, 515 on Employment and Unemployment Situation in India)

	1999–2000 NSS 55th Round	2004–05 NSS 61st Round
States	Person	Person
Arunachal Pradesh	80.66	71.12
Assam	59.81	70.0
Manipur	42.31	85
Meghalaya	57.8	106.98
Mizoram	77.14	92.58
Nagaland	87.67	87.67
Tripura	95.64	76.71
All India	45.55	56.33

Table 17.2 Average daily wages for casual workers in non-public works (Rupees). (Source: NSS 55, 61 Rounds, NSS Report No. 458, 515 on Employment and Unemployment Situation in India)

	2004–2005 NSS 61st Round	2004–2005 NSS 61st Round	1999–2000 NSS 55th Round	1999–2000 NSS 55th Round
States	Rural person	Urban person	Rural person	Urban person
Arunachal Pradesh	91	139	57.64	159.07
Assam	60	70	46.63	65.58
Manipur	71	134	56.88	65.86
Meghalaya	64	73	51.9	7.06
Mizoram	110	117	92.33	79.23
Nagaland	146	85	68.23	81.47
Tripura	61	72	47.83	59.1
All India	49	69	39.64	56.96

The analysis is based on the intuition that labour plays a central role in the success of Jhum. The returns to Jhum labour basically incorporates returns to all other factors of production and is therefore quite high. We next try to understand why it is so. On the one hand, Jhum involves little cash, or credit or any technological inputs. The only two inputs are labour and land. Given that Jhum, land is a non-priced factor of production, as it is held communally, and allotted for free by the community, labour picks up the returns to land as well. This is in line with Hayden (1980, 1983) who argues, in the context of the African peasant economy, that free availability of land allows the peasants to remain outside the ambit of the market economy.

The second reason that labour in Jhum has a high return, is because of social capital since the production process is based on norms of reciprocity and mutual cooperation. Given that the mode of production in Jhum is largely based on common pool resources, where collectivism plays a major role, such norms of mutual

Table 17.3 Indicators of underemployment in Mizoram. (Source: NSS 50, 55, 61 Rounds, Report No. 409, 458, 515, on Employment and Unemployment Situations in India)

	Workers seeking additional work					
	Rural			Urban		
	1993–1994	1999–2000	2004–2005	1993–1994	1999–2000	2004–2005
Male	2.1	12	12.5	1.5	3.4	5.1
Female	0.08	4.5	4.8	1.1	4.5	3.7
Persons	1.6	9.1	9.5	1.3	3.8	4.5
	Workers seeking alternative work					
	Rural			Urban		
	1993–1994	1999–2000	2004–2005	1993–1994	1999–2000	2004–2005
Male	2.6	8.8	11.9	1.8	5.1	5.1
Female	0.5	1.8	4.2	1.6	6.1	3.7
Persons	1.9	6.1	8.5	1.8	5.5	4.7

help develops quite naturally. Such norms also act as an insurance against the uncertainty intrinsic to most societies where such subsistence practice abounds. Instead of taking on these challenges, cultivators seek social insurance in the form of support from family, friends, neighbours and higher levels patrons. The uncertainties and unpredictabilities in such insular economies are dealt with via strict social norms and sanctions, which makes social capital a central element of its success. In fact, as we argue later in greater details, the presence of social capital seems to be a very natural assumption in the context of Mizoram also, our motivating example. Thus taken together, these two imply that returns to Jhum labour may actually be quite high, capturing as it does the returns to land , which is free, as well as the returns to associated social capital. This in turn causes Jhum to persist.

We next turn to formalizing these ideas. We consider a baseline model where labour is the only variable factor of production. There are two households, who simultaneously decide on how much of labour to allocate to Jhum, and how much to non-Jhum activities. As discussed earlier, social capital plays a central role in the analysis, with the social capital of household 1, say, being assumed to depend positively on the labour allocated to Jhum by both households. Further, the marginal social capital of household 1, say, is increasing in the labour employed by household 2.

We show that this game has a symmetric and stable Nash equilibrium. We find that all equilibria involve too much labour being allocated to Jhum, compared to the case where there is no social capital. The intuition relies on the fact that marginal social capital of the households are increasing in the labour being employed by the other households. This makes the amount of labour being employed by the two households strategic complements, hence the result.

Turning to the comparative statics, we find that the results depend on whether the equilibrium is a corner one or not. Suppose that the population is not too large (as for example is the case in Mizoram). We find that under the appropriate parametric restrictions there is a unique Nash equilibrium where all of the labour is employed in Jhum cultivation. In this case we find that an increase in outside wages has no impact on labour allocated to Jhum, thus providing a theory of persistence of Jhum. Interestingly, we find that in the absence of social capital, such persistence cannot arise. This is of interest given that in general social capital is seen as a positive force, e.g. Elinor Ostrom the Nobel Prize winner focusing on its role in resolving prisoner's dilemma issues. While there is recognition that social capital may have negative effects (Dasgupta 2005), it is interesting that Halpern (1999) had to resort to an example of gangland shooting to find an example. For an interior equilibrium however, an increase in outside wages leads to a lower allocation of labour to Jhum. This suggests the following testable hypothesis:

An increase in outside wages reduces labour allocated to Jhum if population density is high, but not otherwise.

We perform other comparative statics exercises, e.g. with respect to change in land productivity, etc. Further, we find that the amount of labour allocated to Jhum is very large compared to the socially optimal level, if the environmental externality is relatively large. Finally, we perform an n household extension for robustness check.

The rest of the chapter is organised as follows. The next section discusses some aspects of Jhum cultivation that provides a context to our work, and also provides a literature review. The following Sects. 17.3 and 17.4 develop the basic framework for analysis. Sect. 17.5 builds on the framework to solve for the equilibrium under social optimum. Section 17.6 derives an extension for n household case. Finally, Sect. 17.7 concludes the chapter.

17.2 Jhum Cultivation

The subsistence agricultural practice of slash and burn is widely prevelant in the tribal regions of northeastern India, characterised by hilly, difficult terrain and remote locations. It is intrinsically interwoven with the ethos of the tribal society, social relationship, cultural values and religious beliefs (Singh 1996).

Jhum has its origin in the early evolutionary stage of agriculture, preceded by hunting, gathering and followed by settled cultivation. This mode production starts with the clearing of a stretch of forest land in the dry season, which is then allowed to dry for some time, before it is set to fire. The burnt residues of the biomass replenishes the nutrient cycles. After a few production cycles the land looses the fertility and is therefore again kept as a fallow land for forest regeneration. The fallow period is critical for productivity of land, as well as for conservation of the ecosystem. There may be some local variations, but based on our observations in Mizoram Jhum is having the following characteristics.

Jhum is based on common pool resources, where the land is allocated through lotteries by the village council. Each village council has a stock of buffer forest land which is not under the regulation of the forest department. Given that remote locations and high transport costs, the system does not incentivise the production of marketable surplus, rather Jhum production is subsistence oriented, with multiple cropping of food grain. Further, unlike private settled agriculture, the success of Jhum lies in the high degree of collectivity and community control, mutual insurance and reciprocity, rather than on series of agronomic practices and individual decision-making as such. Thus, all major decisions, e.g. the length of crop cycles, the timing of the various Jhum based activities (clearing forest, setting fire, sowing of different seeds, harvesting, etc.) the pattern of crop plantation, and even ensuring that any given plot remains fallow for the required number of years, are being taken at the community level. It is a highly organic practice where virtually no cash inputs are required, with little use of chemical fertilizers and seeds being recycled from the preceding year's harvest. The extensive form of cultivation involves a total dependence on nature and therefore requires a very limited use of technology, with the use of only some basic tools like machetes, axes, hoes, makeshift shovels, etc. With its reliance on slash and burn techniques, inter-cropping and differential harvest cycles, it requires a high degree of manual labour of the unskilled type (mostly family based labour). Given the access to the common pool resources along with abundant access to land, labour seems to be the central force behind production (Rao and Roy Chowdhury 2010).[3]

The high land-to-man ratio (most of the northeastern states have forest cover above 70 percent %) and extensive form of cultivation imply that land is often not a constraint. Further, given that land allocation is random, and that production is for local consumption, there is very little incentive for investment in land. Also, the use of organic practices mean that little cash inputs are required, so even credit is not a constraint. Consequently, labour becomes the only scarce resource in the system, with the return to labour incorporating returns to all other factors of production. This pushes up the opportunity cost of labour, hindering the formation of a viable labour market.

Rao and Roy Chowdhury (2010) argue that the Mizo society is closely knit and insular with its VI-th schedule status (and inner line permits). In a near homogenous society (with four main tribes and 87 percent % of people being Christians), the Church has a large presence with almost every village having a church. Moreover, the presence of several large NGOs (with extensive membership), and village councils suggest that social capital is likely to play an important role in such societies. While these institutions promote social norms, community bonding and internal cohesion within the community, they may influence and encourage to sustain insulated communities as well.[4] People in the rural areas rely more on relatively stable

[3] Rao K. and I. Roy Chowdhury: "Jhum cultivation: Implication for the Mizo Economy and Society", Chap. 2, Mizoram State Development Report, Draft, March 2010, National Institute of Public Finance and Policy, New Delhi, sponsored by Planning Commission.

[4] Anecdotal evidence suggests that contractors in the forest Mahal system prefer to bring labour from outside the state, rather than employ locally.

reciprocal solidarities. People exchange labour at critical hours in agricultural season, clears forestlands, build common paths between their homesteads and fields and also monitor each others land at night. They also chip in to regularly assist each other at times of important family occasions, assist in time of birth, baptism of a child, youth initiation, wedding, illness and funerals. These reciprocities are a form of social insurance to cope with various uncertainties and external forces that threaten their livelihood.

17.2.1 Literature Review

In this subsection we provide a brief survey of the literature.

We first discuss the role of social capital in various forms of subsistence cultivation. Hayden (1980, 1983) studies the African economy and analyses the "peasant mode of production and the economy of affection". He argues that such a mode of production survives as a network of support, communications and interaction among structurally defined groups connected by blood, kin, community or other affiliations, for example, religion. Free availability of land allows the subsistence farmer to remain outside the market. Thus, a farmer's production decision is based on values of family, kinship, tribal obligations and interdependence, rather than market profits only.

Scott (1977, 1985) argues that social practices among peasants in Southeast Asia are based on two moral premises: (1) the norm of reciprocity, and (2) the right to subsistence. This translates into a fear of technological innovation and social change. These relations of mutual support are sometimes lateral, at other times vertical, involving a relationship with a patron. Waters (1992), draws heavily on Hayden's concepts of "peasant mode of production and the economy of affection", with his observations in remote subsistence farming villages in Tanzania from 1984 to 1987. Water's book published in 2007, "The Persistence of Subsistence Agriculture: Life beneath the level of market place", analyses the nature of modern social change in eighteenth-century Scotland, nineteenth-century USA, and twentieth-century Tanzania. He concludes that this typical inertia of persistence of subsistence agriculture in transition to market economy is transnational and transhistorical fact across the globe.

According to Putnam (1995), social capital includes "...features of social life, networks, norms, and trust—that enable participants to act together more effectively to pursue shared objectives". Although there is some uncertainty about the origin of the term social capital, many associate it with Coleman (1988) and Putnam (1993). However, the concept of such non-market interactions influencing individual behaviour within a group is getting widely acknowledged among the economists (Dasgupta (1999, 2005, 2009), Dasgupta and Serageldin (1999), Hiroe and Unai (2008), Ostrom et al. (1994), Ostrom (1990, 1999), Durlauf (1999, 2002), Glaeser et al. (2000, 2002), Portes and Landlot (1996), Sobel (2002), etc.).

A coherent study of social interactions, however, requires a clear conceptualisation of the interaction processes (see Manski 2000), and thus led to a lot of academic debate. In fact, Arrow (1999) and Solow (1999) advised in favor of the abandonment of the term social capital, arguing that it is not capital in the sense that economists define capital, even though they both agreed that social capital was a mechanism to reduce transaction costs. Hayami (2009), while acknowledging this debate, attempts to develop some operational definition of social capital based on the various concepts of capital traditionally used in economics.

Finally, we turn to discussing the literature on the economic aspects of Jhum cultivation. Unfortunately, the literature, either empirical or theoretical, is quite thin, with the existing literature being mainly focused on its ecological aspects (e.g. Ramakrishnan 1993; Tawnenga 1997; Tawnenga et al. 1996, among others). One recent paper by Saikia and Bhaduri (2012), has studied the institutional analysis of shifting cultivation in the process of transition to market in the state of Nagaland. They observed that since collectivism plays a dominant role in such society and production mechanism, the policy push by the government dictated mostly to alter the mode and system of production, though short-term incentives failed to make transition to market economy effective. Resonating our basic contention, Arnott and Stiglitz (1991) have observed that certain networks can prevent market from functioning well, or may even prevent markets from coming into existence, thereby retarding economic development. Dasgupta (2009) also argues about the flipside of kinship obligations in diluting personal incentives to invest for prosperity. However, to the best of our knowledge there is no analytical framework in the literature that addresses the formalization of the persistence of subsistence mode agriculture in a transition economy.

17.3 The Framework

There is a village consisting of two homogenous households, denoted 1 and 2. Each family has a total labour supply of $\overline{L}$, which they can allocate either between Jhum cultivation, or in wage labour. Let L_i denote the labour allocated to Jhum cultivation by family i.

We then describe the various technologies. The payoff from Jhum cultivation is $tf(L_i)$, where t is a productivity parameter which may depend both on the fertility of land, as well as the total land allocated to household i. The production function $f(L_i)$ satisfies all the standard properties.

A1. $f(L)$ is increasing, strictly concave in L and $f(0) = 0$. Further, $f(L)$ satisfies the Inada conditions that $\lim_{L\to 0} f'(L) \to \infty$ and $\lim_{L\to\infty} f'(L) = 0$.

The payoff from outside labour is $w(\overline{L} - L_i)$, where w denotes the expected wage in non-Jhum activities. This could, for example, denote the expected payoff from non-farm wages.

One feature of our framework is that we allow for social capital[5] arising from social norm, mutual insurance and mutual obligation of reciprocity which creates a positive externality to the Jhum households. For example, neighbour render their support to the incumbent household in case of sickness, hospitalization, temporary absence, during distress, marriages, child birth, funeral, repairing and fencing houses, building common meadows, provides a kind of informal insurance, which may indirectly influence the productivity. In this case, return to investment in social capital are characterised by positive externality influencing household's payoff in additive fashion[6]. Thus, return to social capital can be measured as a residual of the increased social product left unexplained by increase in the input of production factors. The social capital accruing to household i is denoted by $\gamma S^i(L_1, L_2)$, where γ captures the strength of social capital. Since a large part of the social capital is non-monetary, $\gamma S^i(L_1, L_2)$ captures the monetized equivalent of the whole of the social capital of a family.[7]

We assume that social capital accruing to household i is increasing in both L_i and L_j. An increase in L_j implies that the rest of the society has a greater ability to help household i. Whereas an increase in L_i implies that family i has a greater ability to help others, so that the rest of the society reciprocates by helping family i. We maintain the following assumptions on $S^i(L_1, L_2)$.

A2.

(i) $S^i(L_1, L_2)$ is increasing and concave in both L_1 and L_2. Further, $S^i(L_1, L_2)$ is symmetric, i.e. $S^1(x, y) = S^2(y, x)$ and $S^1_1(x, y) = S^2_2(y, x)$.

(ii) $\frac{\partial S^i(L_1,L_2)}{\partial L_i}|_{L_j=0} = 0$ and $\frac{\partial^2 S^i(L_1,L_2)}{\partial L_i L_j} > 0$. Further, $\lim_{L\to\infty} S^i_i(L, L) = 0$.

The assumption that $\frac{\partial^2 S^i(L_1,L_2)}{\partial L_i \partial L_j} > 0$ implies that an increase in L_j has a positive impact on $\frac{\partial S^i(L_1,L_2)}{\partial L_j}$, i.e. the marginal effect of L_i on household i's social capital. As we shall later find, this assumption plays a critical role in the analysis. The assumption that $\frac{\partial S^i(L_1,L_2)}{\partial L_i}|_{L_j=0} = 0$, is in the same spirit and shows that in case the rest of the society has no ability to reciprocate, the marginal social capital is zero.

Next, consider the utility function of the two households $U^i(L_1, L_2)$. Thus

$$U^i(L_1, L_2) = tf(L_i) + w(\overline{L} - L_i) + \gamma S^i(L_1, L_2), \ i = 1, 2. \tag{17.1}$$

We examine the Nash equilibria of the game where the households simultaneously decide on their labour allocation, i.e. L_i.

---

[5] The concept of social capital used here is mainly bonding type accruing from a homogenous network.

[6] see Hayami (2009)

[7] In a different but related work we analyse different formulation of the social capital, where L_i directly and indirectly enhance the productivity of household j. We find that the analysis is qualitatively unaffected by this alternative formulation.

17.4 The Analysis

We shall analyze this game by using a standard reaction function approach. From (17.1), the reaction function of the i-th household solves

$$\frac{\partial U^i(L_1, L_2)}{\partial L_i} = tf'(L_i) - w + \gamma S_i^i(L_1, L_2) = 0, \ i = 1, 2. \tag{17.2}$$

Further, observe that

$$\frac{\partial^2 U^i(L_1, L_2)}{\partial L_1 \partial L_2} = \gamma S_{12}^i(L_1, L_2) > 0, \ i = 1, 2. \tag{17.3}$$

This shows that L_1 and L_2, i.e. the allocation of Jhum labour by the households are strategic complements. Consequently, it follows that the reaction function of household i, denote it $L_i(L_j)$, $i \neq j$, is positively sloped. We then note that

$$L_i(0) = f'^{-1}\left(\frac{w}{t}\right) \ (> 0), \ i = 1, 2. \tag{17.4}$$

We can plot the reaction functions of the two households in the $L_1 - L_2$ plane. Do these reaction functions necessarily intersect?

Note that given A1, there exists some minimal $\hat{L}$ such that

$$tf'(\hat{L}) + \gamma S_1^1(\hat{L}, \hat{L}) = w. \tag{17.5}$$

This follows from the intermediate value theorem since for L small, the LHS of (17.5) exceeds the RHS (given the Inada condition).

Our next proposition demonstrate the existence of stable, symmetric Nash equilibria in pure strategies. We find that in any such equilibrium, the presence of social capital implies that the level of labour in Jhum cultivation is higher than that in the absence of social capital.

We consider two kinds of equilibria, interior and corner. Interior equilibria have the property that the labour supply constraint does not bind for either family. Whereas in a corner equilibrium the labour constraint binds for both households.

Proposition 17.1. *Let assumptions A1–A2 hold.*

1. *A stable Nash equilibrium in pure strategies exists. If $\hat{L} \leq \overline{L}$ then an interior equilibrium exists. Otherwise, a corner equilibrium exists.*
2. *Any Nash equilibrium in pure strategies must be symmetric.*
3. *In any Nash equilibrium, the labour allocated to Jhum cultivation by both households exceed that in the absence of social capital.*

Proof. (a) Suppose that $\hat{L} \leq \overline{L}$, then $L_1 = L_2 = \hat{L}$ constitutes a stable Nash equilibrium (see Fig. 17.2). In this case of course multiple Nash equilibria may exist. Whereas if $\hat{L} > \overline{L}$, then $L_1 = L_2 = \overline{L}$ constitutes a stable Nash equilibrium (see Fig. 17.1).

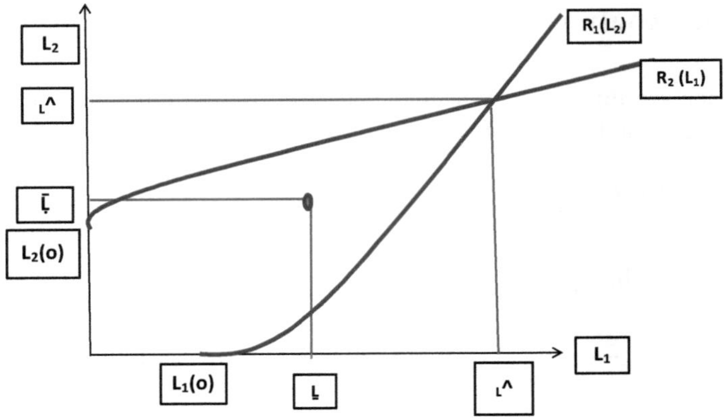

Fig. 17.1 Unique corner stable symmetric equilibrium

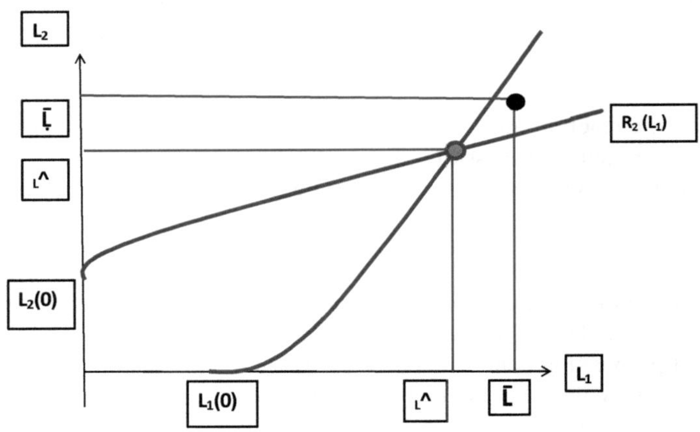

Fig. 17.2 Interior, stable, symmetric equilibrium

(b) The result follows from the symmetry of the social capital functions of the two households. Suppose to the contrary there is an asymmetric Nash equilibrium, (L_1', L_2'), with $L_1' > L_2'$. Observe that

$$\begin{aligned} 0 &= tf'\left(L_1'\right) - w + \gamma S_1^1\left(L_1', L_2'\right) \\ &< tf'\left(L_2'\right) - w + \gamma S_2^2\left(L_2', L_1'\right) \\ &< tf'\left(L_2'\right) - w + \gamma S_2^2\left(L_1', L_2'\right) = 0, \end{aligned}$$

which is a contradiction. Note the first inequality follows since (a) $f(L)$ is concave and $L_1' > L_2'$, and (ii) $S_1^1(x, y) = S_2^2(y, x)$. Whereas the second inequality follows as $S_2^2(L_1, L_2)$ is concave in L_2 and $S_{ij}^i > 0$.

(c) Note that the labour supply in case there is no social capital, i.e. $\gamma = 0$, is given by

$$f'^{-1}\left(\frac{w}{t}\right).$$

We observe that this equals $L_1(0) = L_2(0)$. Given that the reaction functions are positively sloped the result follows. □

Example. Suppose $f(L) = \sqrt{L}$ and $S^1(L_1, L_2) = \sqrt{L_1.L_2}$. Then $\hat{L} = (\frac{t}{\gamma - 2w})^2$.

17.4.1 Comparative Statics

We then turn to comparative statics properties of stable equilibria. Given Proposition 1, any equilibria must be symmetric and is denoted by $(\hat{L}, \hat{L})$.

Proposition 17.2. *Consider any stable Nash equilibrium, $(\hat{L}, \hat{L})$. Suppose there is either (a) an increase in w, or (b) a decrease in t, or (c) a decrease in γ.*

1. *In case the equilibrium is an interior one, i.e. $\hat{L} < \overline{L}$, then there is a decrease in $\hat{L}$.*
2. *In case the equilibrium is a corner one, i.e. $\hat{L} = \overline{L}$, then there is no change in $(\hat{L}, \hat{L})$.*

Proof. Consider a stable Nash equilibrium. Given that the equilibrium is stable, the reaction function of household 1 must be intersecting that of household 2 from below. The reaction function of household 1 satisfies:

$$tf'(L_1) + \gamma S_1^1(L_1, L_2) = w.$$

Suppose that there is either (a) an increase in w, or (b) a decrease in t, or (c) a decrease in γ. Given A1 and A2, we have that $f''(L) < 0$ and $S_{ii}^i(L_1, L_2) < 0$, so that the reaction function of household 1 shifts to the left, whereas that of household 2 shifts down (see Fig. 17.3). Hence the result. □

Interestingly, under a stable corner equilibria, i.e. where all the labour is employed in Jhum cultivation, there is trap in the sense that changes in w, or t does not affect Jhum labour.

Further, Proposition 2 yields an interesting testable hypothesis. Consider an economy with a large population, so that it is reasonable to assume that t would be large relative to the other parameters. In that case, Propositioin 2 suggests the following hypothesis:

An increase in outside wages reduces labour allocated to Jhum if population density is high, but not otherwise.

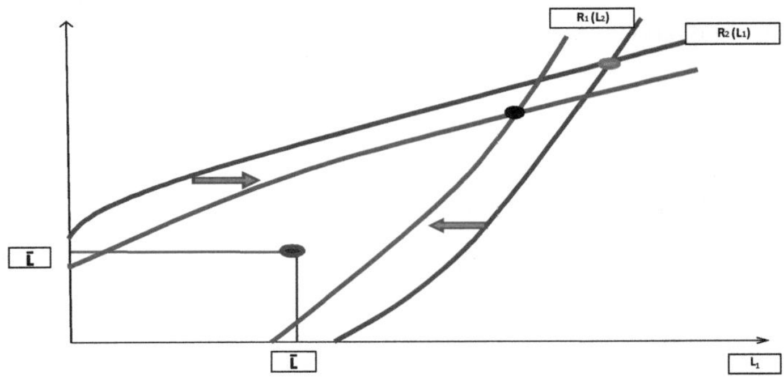

Fig. 17.3 Comparative statics of increase in w at the corner or interior equilibrium

17.5 Efficiency

We next turn to analysing the efficiency properties of the Nash equilibria. We assume, as is natural, that too intensive cultivation leads to environmental damage, e.g. a fall in land quality requiring longer periods as fallow for recovery. Further, as is realistic in many contexts, we assume that the equilibrium levels of L_i are large enough, so that the environmental damage kicks in. Let the environmental damage function be denoted by $D(L_1, L_2)$.

A3. $\frac{\partial D(L_1,L_2)}{\partial L_i} > 0$, and $\frac{\partial^2 D(L_1,L_2)}{\partial L_1 \partial L_2} = 0$.

For ease of exposition we focus on the case where there is unique and interior equilibrium.

Similarly, define L^S as the socially optimal level of L_i, so that (L^S, L^S) maximizes the aggregate utility of the two individuals *net* of the environmental damage. Thus,

$$(L^S, L^S) = argmax \;\; W(L_1, L_2) = U^1(L_1, L_2) + U^2(L_1, L_2) - D(L_1, L_2).$$

We then show that if the environmental damage is really large, then the households put in too much labour under Jhum compared to the socially optimal level.

Proposition 17.3. *Let* $D_i(L_1, L_2) > \gamma S_i(L_1, L_2)$, $\forall (L_1, L_2)$. *Then* $L^S < \hat{L}$.

Proof. Consider Fig. 17.4 and fix a point (L_1', L_2') on $L_1(L_2)$, and let us evaluate

$$\begin{aligned} \frac{\partial W(L_1, L_2)}{\partial L_1}|_{(L_1'.L_2')} &= \frac{\partial U_1(L_1', L_2')}{\partial L_1} + \frac{\partial U_2(L_1', L_2')}{\partial L_1} - D_1(L_1', L_2') \\ &= -D_1(L_1', L_2') + \gamma S_1^2(L_1', L_2') < 0, \end{aligned} \quad (17.6)$$

where the inequality follows since, $\frac{\partial U_1}{\partial L_1}|_{(L_1', L_2')} = 0$ (as (L_1', L_2') is on the reaction function $L_1(L_2)$), and the fact that $D_1(L_1, L_2) > \gamma S_1(L_1, L_2)$. Let us plot $W_1(L_2)$, where it solves $\frac{\partial S(L_1,L_2)}{\partial L_1} = 0$. Given the preceding equation $W_1(L_2)$ must lie to the left of $L_1(L_2)$, since $\frac{\partial^2 S(L_1,L_2)}{\partial L_1 \partial L_2} = \gamma S_{12}^1 + \gamma S_{12}^2 > 0$ (from A2). □

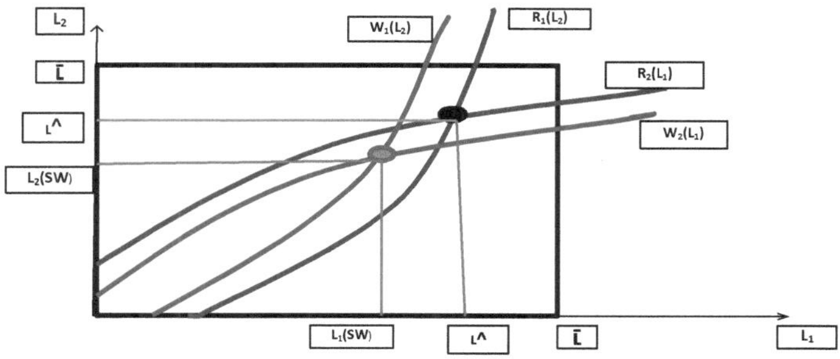

Fig. 17.4 Nash equilibrium labour allocated to Jhum is higher than the social optimum

The intuition is simple. Under a Nash equilibrium, family i does not internalise the facts that an increase in L_i has (a) beneficial effects for the social capital of household j, but (b) has harmful environmental effects. Under the given hypothesis the second effect dominates, hence the result.

17.6 Extensions: *n* Households

We then extend the analysis to the n households in a bid to establish the robustness of our analysis.

In this section, we generalize the analysis to the case of n households, where $n \geq 2$. This not only provides a robustness check, but, moreover allows us to examine the effects of an increase in population on the results.[8]

We need some notations. Let L_{-i} denote the aggregate labour supplied by households other than i. We can now write down the social capital function for household i as $\gamma S^i(L_i, L_{-i})$. Assumption 2' is the natural analogue of Assumption 2 earlier.

A2′.

(i) $S^i(L_i, L_{-i})$ is increasing and concave in both L_i, and L_{-i}. Further, $S^i(L_i, L_{-i})$ is symmetric in the sense that $S^j(x, y) = S^k(y, x)$, $j \neq k$ and $S^i_i(x, y) = S^k_k(x, y)$.

(ii) Finally, $\frac{\partial S^i(L_i, L_{-i})}{\partial L_i}|_{L_{-i}=0} = 0$, and $\frac{\partial^2 S^i(L_i, L_{-i})}{\partial L_i \partial L_{-i}} > 0$. Further, $\lim_{L\to\infty} S^i_i(L, nL) = 0$.

(iii) $S^i_{i1}(L, (n-1)L) + (n-1)S^i_{i2}(L, (n-1)L) < 0$.

[8] We are indebted to Karl Goran-Mailer for this suggestion.

It is easy to see that the reaction function of the i-th household is

$$tf'(l_i) - w + \gamma S^i_i(L_i, L_{-i}) = 0. \tag{17.7}$$

The symmetric solution in the absence of any supply constraint is given by $\hat{L}(n)$, where $\hat{L}(n)$ solves:

$$tf'(L) + \gamma S^i_i(L, (n-1)L) = w. \tag{17.8}$$

It is easy to see that an analogue of Proposition 1 holds. Thus a symmetric interior Nash equilibrium, where every household supplies $\hat{L}(n)$ exists as long as $\hat{L}(n) \leq \overline{L}$. Otherwise, the symmetric equilibrium involves all households supplying $\overline{L}$. Further, the labour allocated to Jhum by the households exceed that in the absence of social capital.

Moreover, an analogue of Proposition 2 holds here as well. Thus in case the equilibrium is an interior one, and there is either (a) an increase in w, or (b) a decrease in t or (c) an decrease in γ, then there is a decrease in labour allocated to Jhum.

We then turn to the effect of an increase in population, i.e. n on the outcome. An increase in population may of course affect the land available to an individual household and hence $t(n)$. Given that there may be constraints on the aggregate availability of land, it is natural to assume that $t(n)$ is decreasing in n.

Proposition 17.4. *Let there be an increase in n. Given an interior equilibrium, the land supplied in Jhum by an individual household may either increase or decrease.*

Proof. Given (17.8), straightforward differentiation yields that

$$\frac{d\hat{L}(n)}{dn} = -\frac{t'(n)f'(L) + \gamma L S^i_{i2}(L, (n-1)L)}{f''(L) + \gamma(s^i_{i1}(L, (n-1)L) + (n-1)S^i_{i2}(L, (n-1)L))}.$$

Given A2', the numerator is negative, however the denominator can be of either sign. □

There are two effects at play here. With an increase in population, $t(n)$ falls, so that the marginal productivity of labour in Jhum cultivation decreases. This reduces the incentive to allocate labour to Jhum. On the other hand, however with an increase in the number of households there is an increase in the social capital, which increases the marginal productivity from labour in Jhum. Because of this trade-off the net effect is unclear.

17.7 Conclusion

The study seeks to analyse the persistence of Jhum cultivation—a predominant mode of agricultural practice in many hilly regions of the world, including North-east India. We build a formal game theoretic model that tries to explain how the existence of social capital can lead to persistence in Jhum. We find that there is an intrinsic connection between social capital and the amount of labour in Jhum in that all equilibria involve too much labour being allocated to Jhum as compared to the case where there is no social capital. Turning to the issue of persistence, under the appropriate parameter values there is an equilibrium where the households devote their entire labour to Jhum cultivation. In that case, we find that an increase in non-Jhum wages has no impact on labour allocated to Jhum. Interestingly, in the absence of social capital such persistence cannot arise. Further, we find that the amount of labour allocated to Jhum is very large compared to the socially optimum level if the environmental externality is relatively large. We also examine several extensions of the basic model, showing that our results are qualitatively robust to these extensions.

We finally briefly discuss some policy implications of our analysis, in particular in the context of Mizoram, our motivating example. While the purely organic slash and burn practice has been the natural response to the agro-climatic conditions of Mizoram, with its difficult terrain, loose soil structure, remote location and poor infrastructure, over the years the system is getting stressed by population pressure and an increase in demand for alternative land uses. This is gradually leading to shorter Jhum cycles. This has serious environmental consequences with shorter cycles leading to more pressure on soil and accentuating degradation in the long run. Further, such degradation makes Jhum itself unviable in the long run, even if one is only worried about subsistence. Equally important is, however, that Jhum offers a limited scope to grow beyond subsistence. There is an inevitable threat of economic and social stagnation under the current demographic trends (decadal growth rates of population exceed national averages, Census, 1981, 1991, 2001) and severely constrained livelihood options in Mizoram. Our analysis suggests that resolving this problem involves a serious challenge. There may be cases where even reasonably large increases in outside wage may be insufficient to draw labour away from Jhum.

We finally discuss agendas for future work. One issue of interest is to see the Jhum participation by endogenising the allocation of the amount of land to each household, i.e. $t(n)$. One can think of this being allocated through some cooperative decision-making, possibly the village panchayat. Alternatively, this problem can be embedded in a dynamic framework, with the productivity of land in one period being a function of the intensity of cultivation, how much labour was used it in the earlier period, etc. Both these issues are beyond the scope of the present chapter though.

Acknowledgements: The idea of this chapter originated when I was working on the Mizoram State Development Report, 2009 (a study done by National Institute of Public Finance and Policy, jointly done by Kavita Rao, Indrani Roy Chowdhury and Gautam Naresh, sponsored by Planning Commission). A special thanks to Partha Dasgupta for his encouragements. I would like to thank in particular Karl Goran-Mailer, Gopal Kadekot and Amita Shah and other seminar participants at the INSEE VIth Biennial Conference, "Nature Economy and Society: Understanding the Linkages", 2011, held in Hyderabad, India, for their very useful comments.

References

Arnott R, Stiglitz JE (1991) Moral hazard and non market institutions: dysfunctional crowding out or peer review? Am Econ Rev 81:179–190

Arrow KJ (1999) Observations in social capital. In: Partha D, Ismail S (eds) Social capital: a multifaceted perspective. World Bank, Washington DC, pp 3–5

Coleman JS (1988) Social capital in the creation of human capital. Am J Sociol 94:95–120. The University of Chicago Press

Dasgupta P (1999) Economic progress and the idea of social capital. In: Partha D, Ismail S (eds) Social capital: a multifaceted perspective. World Bank, Washington DC, pp 325–424

Dasgupta P (2005) The economics of social capital. Econ Rec 82:2–21

Dasgupta P (2009) A matter of trust: social capital and economic development. Prepared for presentation at the Annual Bank Conference on Development Economics (ABCDE), Seoul, June 2009

Dasgupta P, Serageldin I (eds) (1999) Social capital: a multifaceted perspective. World Bank, Washington DC

Durlauf SN (1999) The case "against" social capital, Focus 20 (3). University of Wisconsin-Madison, Institute for research on poverty

Durlauf SN (2002) On the empirics of social capital. Econ J 112:F459–F479. Blackwell

Glaeser E, David L et al (2000) Measuring trust. Q J Econ 115:811–41

Glaeser E, David L et al (2002) An empirical approach to social capital. Econ J 112(483):437–458

Halpern D (1999) Social capital: the new golden goose. Faculty of Social and Political Sciences, Cambridge University, Mimeo

Hayami Y (2009) Social capital, human capital and community mechanism: towards a consensus among economists. J Dev Stud 45(1):96–123

Hayden G (1980) Beyond Ujamaa in Tanzania: underdevelopment and an uncaptured peasantry. University of California, Berkley

Hayden G (1983) No shortcuts to progress: African development management in perspective. University of California, Berkley

Hiroe I, Unai P (2008) Social capital and collective action. Environmental Economics Revisited, Environmental Economy and Policy Research, Discussion Paper Series, No. 38. Cambridge University Press, Department of Land Economy.

Manski C (2000) Economic analysis of social interactions. J Econ Perspect 14:115–136

Ostrom, E (1990) Governing the commons: evolutions of institutions for collective actions, p 280. Cambridge University Press, New York. ISBN: 9780521405997

Ostrom E, Gardner R, Walker J (eds) (1994) Rules, games and common pool resources. University of Michigan, Ann Arbor

Ostrom, E (1999) Coping with tragedies of the common. Ann Rev Polit Sci, 2:493–535

Portes A, Landolt P (1996) The downside of social capital. Am Prospect 26:18–21

Putnam R (1993) The prosperous community: social capital and public life. Am Prospect 13:35–42

Putman R (1995) Social capital, measurement and consequences. `http://www.oecd.org/innovation/research/1825848.pdf`

Ramakrishnan PS (1993) Shifting cultivation and sustainable development. Man and the Biosphere Series, 10. UNESCO/Oxford University Press, New Delhi/Paris

Rao K, Naresh G, Roy Chowdhury I (2008) Mizoram state development report. `http://www.nipfp.org.in/media/medialibrary/2013/08/Mizoram_SDR_Revised_2012.pdf`

Saikia A, Bhaduri S (2012) An institutional analysis of transition to market: the case of shifting cultivators in Mon, Nagaland. Int J Rural Manage 8:19–34

Scott JC (1977) The moral economy of the peasant: rebellion and subsistence in South East Asia. Yale University Pres. ISBN: 978-0300021905

Scott JC (1985) Weapons of the weak: everyday forms of peasant resistance. Yale University Press. http://en.wikipedia.org/wiki/Special:BookSources/0300033362. ISBN 0-300-03336-2

Sobel J (2002) can we trust Social Capital? Jr of economic literature. 40(1):139–154
Solow R (1999) Notes on social capital and economic performance. In: Dasgupta P, Serageldin I (eds) Social capital: a multifaceted perspective. World Bank, Washington DC, pp 325–424
Singh D (1996) The last frontier: people and forest in Mizoram. Tata Energy Reserach Institute, New Delhi, p 301
Tawnenga (1997) Evaluating second year cropping on jhum fallows in Mizoram, north-eastern India: energy and economic efficiencies. J Biosciences 22:605–613
Tawnenga, Shankar U, Tripathi RS (1996) Evaluating second year cropping on jhum fallows in Mizoram, north-eastern India: phytomass dynamics and primary productivity. J Biosci 21:563–575
Waters T (1992) A cultural analysis of economy of affection and the uncaptured peasantry in Tanzania. J Mod Afr Stud 30(1):163–175. Cambridge University Press
Waters T (2007) The persistence of subsistence agriculture: life beneath the level of market place. Lexington Books, Plymouth
Woolcock M (2001) The place of social capital understanding social and economic outcomes. ISUMA: Can J Policy Res 2:11–17

Chapter 18
Ecological and Socioeconomic Impacts of *Prosopis juliflora* Invasion in the Semiarid Ecosystems in Selected Villages of Ramnad District in Tamil Nadu

S. Chandrasekaran and P. S. Swamy

18.1 Introduction

Bio-invasion is the entry and colonization by certain species of new geographical locations that are not allotted by nature to them as native habitats. Such species are termed as invasive or introduced (when intentionally introduced by human beings for definite purpose) species. Generally, invasive species have profound ecological impacts on biotic communities and ecological functions of the ecosystems at invaded locations and adversely reduce the biodiversity of ecosystems. Available scientific information (Ramakrishnan 1991; Williamson 1996; Rilov and Crooks 2009) strongly suggests that these invasive species are one of the greatest and significant threats to ecosystem services generated by the native communities. Mechanical, chemical and biological control programmes are commonly employed to eradicate exotic plants from the invaded sites. However, the positive utilization of exotic organisms is one of the viable options to manage the menace of invasive plants. The term positive utilization refers to the use of huge biomass of weeds for human welfare purposes instead of destroying them either chemically or biologically. One extensively studied aquatic exotic weed is water hyacinth for its utilization in the production of biogas, vermicompost, gibberlic acid, paper and insulation board and in the treatment of sewage and industrial effluents (Singh 1989). *Chromolaena odorata* is being used as mulch for cultivating banana (Chandrasekaran and Swamy 2002) whereas *Lantana camara* is being used as an alternative to bamboo for making baskets (Singh et al. 2010). Water fern has been studied extensively for use as compost, mulch, paper pulp, livestock fodder supplement and for the treatment of sewage and effluent (Thomas and Room 1986). These activities are done by selected people of the community as accessory activity. But in the present study we report how an exotic plant *Prosopis juliflora* (popularly known as the poor man's fuel

S. Chandrasekaran (✉) · P. S. Swamy
Department of Plant Sciences, School of Biological Sciences, Madurai Kamaraj University, Madurai 625021, India
e-mail: invasionecology@gmail.com

N. Ghosh et al. (eds.), *Nature, Economy and Society*,
DOI 10.1007/978-81-322-2404-4_18

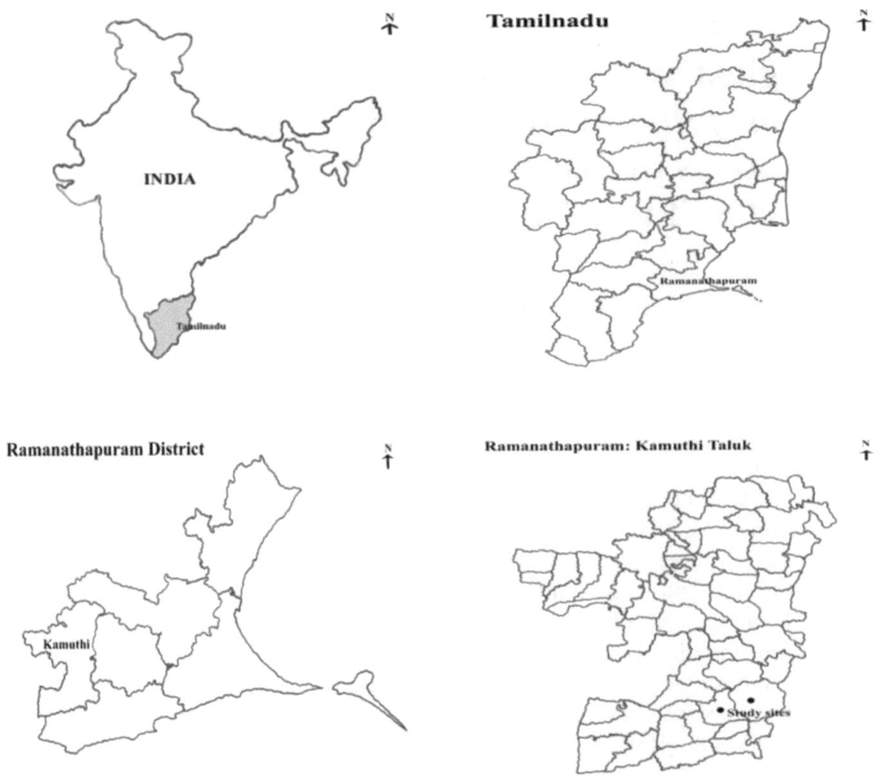

Fig. 18.1 Study area and location of study sites at Ramnad district of Tamil Nadu

wood) drives the energy and economy of selected villages in the Ramnad district of Tamil Nadu.

18.2 Materials and Methods

The study sites were located in Ramnad district of Tamil Nadu, 20 km away from Kamuthi town (9°41' N; 78°36' E), which has hot and dry weather (Fig. 18.1). The normal annual rainfall in the district is 839 mm. Two third of the rainfall in the district occurs during the period of October to December (Fig. 18.2). The soil is black loamy type. Maximum temperature is around 36 °C in summer and minimum is around 19 °C from December to January.

Vegetation analysis and herbaceous vegetation biomass were estimated in *P. juliflora* at the invaded and control sites by the quadrat method. The herbaceous plants were cut at ground level and fresh weight of biomass was obtained in the

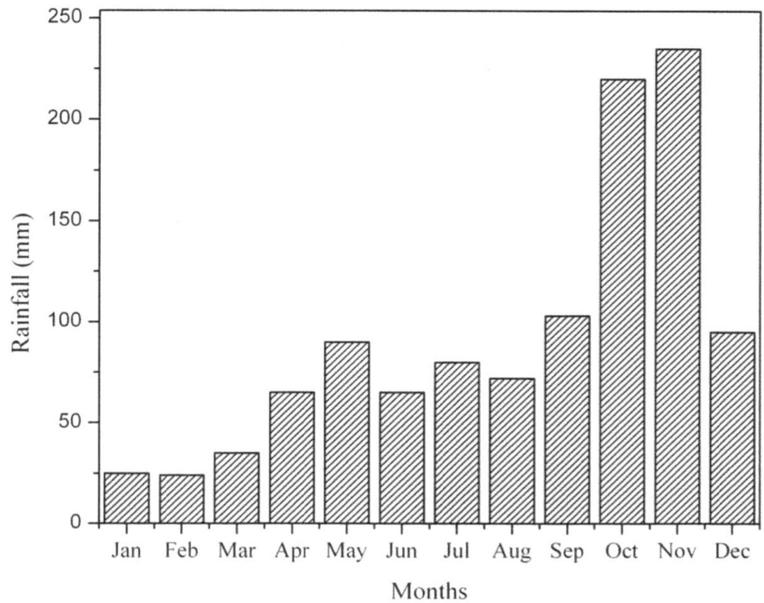

Fig. 18.2 Average rainfall pattern in the study sites at Ramnad district of Tamil Nadu

field. Dry weight was calculated on the basis of the weighed samples dried at 60 °C for 48 h and then weighed. The values were based on 20 randomly placed 50 sq. cm Quadrats and three replicate subsites were maintained for each of the study sites. The number of herbaceous species was also counted at both the study sites by using 20 randomly placed 1 sq. m. Quadrats. The area occupied by *P. juliflora* was calculated based on the information available in the Kamuthi Taluk office and our own observations in the selected villages.

Household-level data were collected on cattle population in each house of the studied villages over the study period. The cost–benefit analysis was done based on the information collected from the farmers and charcoal makers of the selected villages through regular field visits during the study period. Labour costs per man and woman were calculated on the basis of prevailing daily wages. Cost of organic manure, chemical fertilizer, seed, transportation, materials required for heap making and the monetary output were calculated based on prevailing local market rates. The reason for the decline in trend in cattle rearing and traditional agriculture practices were assessed through semi-structured interviews from local people and on the basis of our own observations. The rainfall data were collected from the GEOVANNI satellite website (http://disc.sci.gsfc.nasa.gov/giovanni).

Plate 18.1 **a** *Prosopis juliflora* invaded village pond, **b** and **c** formation of scrub jungle due to *P. juliflora* invasion, **d** loss of grazing land at the study sites at Ramnad district of Tamil Nadu

18.3 Results and Discussion

Traditionally, rice, cotton, leguminous crops and chillies have been being cultivated in the study area. However, the income generated through the yield from these crops was not guaranteed as the region is rainfed. This made the farmers to leave their lands uncultivated and promoted the growth of *P. juliflora* which provided guaranteed income, though lower than the traditional crops. This practice turns into uncontrolled proliferation of this exotic plant in almost all the ecosystems, since alkalinity, salinity and water deficit are not limiting factors for the growth of *P. juliflora* (Laxen 2007). The weed further intruded into village ponds (Plate 18.1a) and occupied vast stretches of this ecosystem leading to the reduction of the carrying capacity of the ponds. These water reservoirs are traditional rainwater harvesting structures to meet the water requirement of the villagers for both agriculture and drinking. Most of these ponds are heavily invaded by *P. juliflora* making it difficult for the cattle and people to reach the water.

This weed further reduced the cropping area (net average sown area was 58 % in the year 1993, 26 % in 2002 and only 3 % in 2010) at the study sites (Table 18.1). The rapid growth of this weed established itself as a localized thorny scrub jungle (Plate 18.1b and c). The jungles become the home for rabbits, foxes and peacocks.

Table 18.1 Changing scenario of *P. juliflora* invasion and utilization in selected villages of Ramnad district, Tamil Nadu

Village name	Village total area (ha)	Area under *P. juliflora* (ha)			% of *P. juliflora* occurrence			Total wood yield per cycle (tonnes)			Amount used as a fuel wood (tonnes)			Total Charcoal production (tonnes)		
		1993	2002	2010	1993	2002	2010	1993	2002	2010	1993	2002	2010	1993	2002	2010
Ariyamangalam	616	320	510	616	52	83	100	4000	4080	3600	182	118	30	763	775	750
Erumaikulam	438	200	280	424	46	63	96	2500	1904	1800	136	144	65	472	361	380
Kuthankulam	13	005	012	013	39	92	100	62	96	0080	009	0	0	011	018	016
Velankulam	320	100	185	299	31	57	93	1250	1665	2300	27	21	04	245	316	324
Total	1387	625	987	1352	42	74	97	7812	7745	7780	354	283	99	1491	1470	1470

The thorny scrub jungle reduced the predatory pressure on these animals and so the population of these animals increased drastically. Plants offer shelter, food and protection against predators and decrease animal stress even in marginal or suboptimal habitats (Wolff 1980; Litvaitis et al. 1985). *P. juliflora* fruits provide food for a variety of animals, including deer, foxes, rodents and rabbits (Ansley et al. 1997). As tender cotton balls are favourites of rabbits, large number of rabbits visit cultivated lands and consume large quantities of cottons balls, thus reducing the crop yield. Ojeda and Keith (1982) have reported that *Sylvilagus floridanus* (Cottontail Rabbit) population is associated with a thorny bushland in which *P. juliflora* and cactus are dominant. Similarly, increase in the population of peacock vanish the grains of paddy before harvesting. Peacocks preferred *P. juliflora* as the roost tree in Gujarat (Dodia 2011). This study also suggested that the selection of *P. juliflora* was an anti-predatory strategy against nocturnal mammalian predators such as the Jungle Cat. These reasons compelled the traditional farmers to leave the cultivation and depend on *P. juliflora*.

The intense growth of *P. juliflora* forced the village to depend on it for its economy. People make money by making charcoal from *P. juliflora* woody biomass (Plate 18.2). The wood is hard and heavy (specific gravity 0.70). It is excellent firewood (calorific value is 4800 kcal/kg) that burns slowly and evenly and holds heat well. Because of its superior quality, it is considered to be one of the best charcoals available (Vimal and Tyagi 1986). Heaping the cut *P. juliflora* wood and burning the heap will yield charcoal. Three-tier labour systems are followed in making charcoal by this method. Cutting the woody biomass, bundling the pieces and transporting them to the heap yard and heap making and maintenance are the three components of charcoal making. The heap maker is a skilled labourer who makes the heap and looks after the risky job of maintaining the heap while it burns. Other components of coal making can be carried out by both males and females based on their efficiency. People living in the study villages are one or the other way connected with this exotic plant exploitation.

The present investigation suggests that charcoal making out of *P. juliflora* is as lucrative as rainfed rice cultivation (Table 18.2) in this region. Therefore, most of the local inhabitants prefer to go for charcoal making rather than growing food crops. This leads to a change in landuse pattern in the rural villages and results in the loss of grazing land in this region (Plate 18.1d). This factor decelerates the cattle population (Fig. 18.3) in the study area. The reduction in herbaceous diversity and biomass production in the *P. juliflora* invaded sites (Figs. 18.4 and 18.5) may be due to shaded environment created by the *P. juliflora*. The greater light availability may be the reason for the higher herbaceous diversity at the control sites. Krishnankutty et al. (2006) and Chandrasekaran and Swamy (2010) recorded maximum herb density in open habitats. The allelochemicals released from *P. juliflora* also may be one of the reasons for the lower herb diversity. The higher plants synthesize allelochemicals which inhibit the growth of other plants (Whittaker 1970).

Though this non-native species overrun native diversity at a faster rate, on a long run they also suffer due to poor nutrient availability. This is because of poor renewal of nutrient resources in the land under *P. juliflora* and utilization of the existing

Plate 18.2 Process of charcoal production from *Prosopis juliflora,* **a** removal of secondary and tertiary branches, **b** cutting of wood, **c** and **d** heap making and burning, **e** and **f** charcoal produced and packed at the study sites in Ramnad district of Tamil Nadu

resources at a faster rate by it which keeps its population under stress. For the first few years, the harvesting cycle was 2 years and now the cycle period increased up to 8 years in some villages (c.f. Table 18.1). Hence, designing a suitable management strategy is essential to prevent the overrun of biomass of *P. juliflora*. The present investigation also suggests that effective management of grazing land ecosystem should be given the highest priority owing to the dangers of shrinking effective grazing land ecosystem due to invasion by this exotic plant.

Table 18.2 Cost–benefit analysis (rupees per acre) of rainfed rice cultivation and *Prosopis juliflora*-occupied lands in selected villages of Ramnad District, Tamil Nadu (1 ha = 2.5 acres)

Rice cultivation									
	Input			Output (paddy grain)			Output/input ratio		
	1993	2002	2010	1993	2002	2010	1993	2002	2010
Land preparation	120	250	650	3750	7040	19,500	2.51	2.44	3.02
Seed cost and broadcasting	70	140	340						
Weeding	425	945	2500						
Chemical fertilizer, cow dung and pesticides	425	580	1050						
Harvest and transportation	450	960	1900						
Total	1490	2875	6440	3750	7040	19,500	2.51	2.44	3.02
Charcoal making from P. juliflora									
	Input			Output (charcoal)			Output/input ratio		
	1993	2002	2010	1993	2002	2010	1993	2002	2010
Harvest	875	1340	5000	3750	6750	31,500	2.22	2.42	3.0
Transport	550	860	3000						
Charcoal making	260	580	2500						

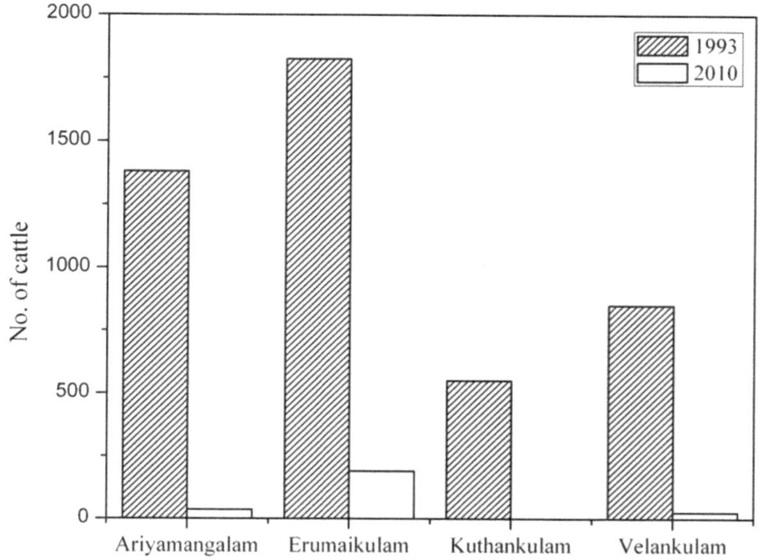

Fig. 18.3 Cattle population in selected villages in Ramnad district of Tamil Nadu

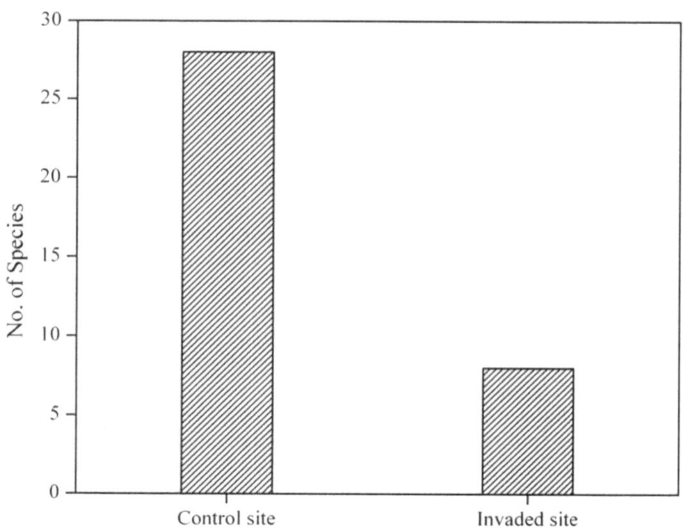

Fig. 18.4 Total number of herbaceous plant species in the control and *P. juliflora*-invaded sites in selected villages in Ramnad district of Tamil Nadu

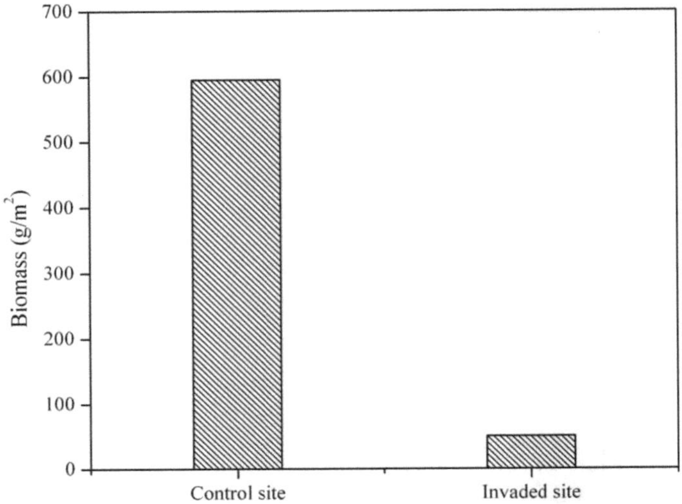

Fig. 18.5 Herbaceous plant biomass (g/sq. m) in the control and *P. juliflora*-invaded sites in selected villages in Ramnad district of Tamil Nadu

Acknowledgements We thank UGC-CAS, DST-PURSE and DBT-IPLS, Government of India, New Delhi for providing financial assistance through various research schemes.

References

Ansley RJ, Huddle JA, Kramp BA (1997) Mesquite ecology. In: Rollins D, Ueckert DN, Brown CG (eds) Proceedings of brush sculptors symposium, Texas, 1997

Chandrasekaran S, Swamy PS (2002) Biomass, litterfall and aboveground net primary productivity of herbaceous communities in varied ecosystems at Kodayar in the western ghats of Tamil Nadu. Agric Ecosyst Environ 88:61–71

Chandrasekaran S, Swamy PS (2010) Growth patterns of *Chromolaena odorata* in varied ecosystems at Kodayar in the Western Ghats, India. Acta Oecologica 36:383–392

Dodia PP (2011) Roost tree selection by the common Indian Peafowl (*Pavo cristatus*) at Bhavnagar district, Gujarat (India). Life Sci Leafl 11:346–354

Krishnankutty N, Chandrasekaran S, Jeyakumar G (2006) Evaluation of disturbance in a tropical dry deciduous forest of Alagar hill (Eastern Ghats), South India. Trop Ecol 47:47–55

Laxen J (2007) Is *Prosopis* a curse or a blessing? An—ecological-economical analysis of an invasive alien tree species in Sudan. In: Luukkanen O (ed) Tropical forestry reports VITRI. University of Helsinki, Finland, p 199

Litvaitis JA, Sherburne JS, Bissonette JA (1985) Influence of understory characteristics on snowshoe hare habitat use and density. J Wildl Manage 49:866–873

Ojeda M, Keith LB (1982) Sex and age composition and breeding biology of cottontail rabbit populations in Venezuela. Biotropica 14:99–107

Ramakrishnan PS (1991) Ecology of biological invasion in the tropics. International Scientific Publications, New Delhi

Rilov G, Crooks JA (2009) Biological invasions in marine ecosystems: ecological, management, and geographic perspectives. Springer-Verlag, Berlin

Singh SP (1989) Biological suppression of weeds. Biological Control Centre, Bangalore
Singh KP, Shukla AN, Singh JS (2010) State-level inventory of invasive alien plants, their source regions and use potential. Curr Sci 99:107–114
Thomas PA, Room PM (1986) Taxonomy and control of *Salvinia molesta*. Nature 320:581–584
Vimal OP, Tyagi PD (1986) *Prosopis juliflora* chemistry & utilization. In: Patel VJ (ed) Role of *Prosopis* in wasteland development. Jivrajbhai Patel Agroforestry Centre, Gujarat
Whittaker RH (1970) The biochemical ecology of higher plants. In: Sondheimer E, Simeone, B (eds) Chemical ecology. Academic Press, New York, pp 43–70
Williamson M (1996) Biological invasions. Chapman and Hall, London
Wolff JO (1980) The role of habitat patchiness in the population dynamics of snowshoe hare. Ecol Monogr 50:111–130

MIX
Papier aus verantwortungsvollen Quellen
Paper from responsible sources
FSC® C105338

If you have any concerns about our products,
you can contact us on
ProductSafety@springernature.com

In case Publisher is established outside the EU,
the EU authorized representative is:
Springer Nature Customer Service Center GmbH
Europaplatz 3, 69115 Heidelberg, Germany

Printed by Libri Plureos GmbH
in Hamburg, Germany